the
population
alternative

JACQUES RUFFIÉ

the
population
alternative)

a new look at competition
and the species

TRANSLATED BY
DR. LAURENCE GAREY

 PANTHEON BOOKS, NEW YORK

to
étienne wolff

First American Edition

Copyright © 1986 by Random House, Inc.

All rights reserved under International and Pan-American Copyright Conventions. Published in the United States by Pantheon Books, a division of Random House, Inc., New York, and simultaneously in Canada by Random House of Canada Limited, Toronto. Originally published in France as *Traité du Vivant* by Librairie Arthème Fayard. Copyright © 1982 by Librairie Arthème Fayard.

This edition, abridged with the author's approval, consists of Parts One, Three, and Five of *Traité du Vivant,* plus one chapter (now Chapter 7) from Part Two.

Library of Congress Cataloging-in-Publication Data

Ruffié, Jacques.
  The population alternative.

Translation of: Traité du vivant.
  Bibliography: p.
  Includes index.
  1. Population biology. 2. Sociobiology. I. Garey, Laurence. II. Title.
QH352.R8313 1986   575   85–19085
ISBN 0-394-54452-8

Book design by Jennifer Dossin.

Manufactured in the United States of America.

contents

part one: the polymorphism of populations

part three: sociobiology or biosociology?

figures and tables

FIGURE

TABLE

preface

Charles Darwin died on April 19, 1882. Few men had as great an effect on their era as he. The brilliant theoretician of biological transformism, Darwin challenged many of the assumptions that had been the basis of Western civilization for centuries. Others, like Jean-Baptiste de Lamarck in France, had foreshadowed him, but Darwin was the first to propose a reasonable model of evolution. The nineteenth century was a time of antifixism. The spirit of transformism penetrated all realms of knowledge, including geology, represented by Charles Lyell in England, and sociology, whose prophet was Karl Marx. Natural science was no exception. Darwin was not an isolated witness; he formed an integral part of the wave of evolutionary thinking of his time. In 1859 he published the *Origin of Species,* in which he presented a detailed account of his conception of evolutionary mechanisms. Independently, Alfred Russel Wallace came to similar conclusions, but Darwin's name remains the one we associate with the theory of evolution. Darwin achieved a brilliant synthesis of contemporary knowledge with the host of observations he made between 1831 and 1836 during his long voyage around the world in the *Beagle.* According to Darwin, species evolved and constantly tended to readapt to their surroundings. This fundamental concept already constituted part of Lamarck's theories, which clearly influenced the English naturalist, in spite of what has sometimes been asserted. It was in determining the initial cause of the transformations that the two diverged. Lamarck thought the use and disuse of organs caused their development or atrophy, which implied the inheritance of acquired characteristics. Without rejecting this theory entirely, Darwin attributed the origin of evolution to the spontaneous occurrence of multiple variations throughout all species, animals and plants alike. These variations had been exploited by breeders to create domestic races, and domestication by deliberate selection had struck Darwin forcibly, leading him to suggest that nature acted in the same way. But what was the basis of selection in the wild, away from man's influence?

Darwin found an answer through Thomas Malthus, an economist and minister of the church, who had lived half a century earlier. According to him, the demographic growth of a population outstripped its resources, so that there was a permanent critical danger of overpopulation and famine unless there were regulatory phenomena that would keep population size constant. This regulation was provided by individual competition, leading to the elimination of the weakest.

On the human scale, to avoid a descent toward violence, such as wars and revolutions, Malthus preached the voluntary limitation of births. Darwin saw in these theories the motive force behind selection. In natural populations competition must exercise a major role, favoring the stronger individuals at the expense of the weak. The whole of evolutionary progress was based on this struggle.

When he wrote down his findings, Darwin was totally unaware of the theories of heredity. Although Gregor Mendel was his contemporary, Mendel's laws were published only at the beginning of the next century. They revealed that hereditary characters were carried on particles, later named *genes,* that could reproduce and be transmitted from generation to generation. In most organisms, referred to as *diploid,* each of these particles exists as one of a pair — one derived from the maternal ovule, the other from the paternal sperm. The two elements are united at the time of fertilization when the egg cell is created. Often, the same gene can exist in several forms, a "normal" or *wild* form that is commonest in a given group, and rare "abnormal" varieties called *mutations.* A gene and its mutations make up an *allelic* series. Mendel's discoveries led to vast numbers of research projects, involving especially the tiny fruit fly, *Drosophila.* The "drosophilists," led by the American Thomas H. Morgan, described a host of mutations, almost all of which were expressed by morphological abnormalities. At first, mutations were considered to be rare, usually involving a profound modification of an individual, or even the creation of a monstrosity. Darwinism began to emerge in a new light, that of Mendelism. Spontaneous variations in natural populations could be explained by the appearance of mutations on which natural selection could act. There would result a competition between normals and mutants, with the "better" of the two winning and imposing its heredity on the whole group through crossbreeding. This theory has often been called *neo-Darwinism.* Being based on variation and selection, Darwinism represents essentially a trend toward uniformity. Selection must choose between a wild gene and its mutation, and preserve the one better adapted to the constraints of the environment. In the end all individuals will be identical and correspond to a *holotype* — a characteristic example of the species with all its advantageous features. Neo-Darwinian philosophy is basically *typological,* conceiving nature as made up of various groups, like species and races, each composed of similar individuals—at least for a certain time. Some geneticists, such as Hugo De Vries, went further, suggesting that a species could develop from a single mutant as long as the mutation was profound enough to confer entirely new and favorable characters on the host. This *saltationist* or *mutationist* theory was taken up by Richard Goldschmidt in 1940 and more recently, although with certain nuances, by Stephen Jay Gould and Niles Eldredge; it should not be seen as a sort of hyper-Darwinism, as some authors have tended to do, but more as anti-Darwinian. The theory involves abrupt changes and is thus radically different from Darwin's thought, which always imagined evolution as a gradual process based on successive subtle changes over long periods of time. Darwin denied the evolutionary value of single, abrupt modifications, which he called "sports," considering them rather as monstrosities with no importance for the future. As knowledge of the genetics of different species increased, it became obvious that mutations were not rare events, but existed in all populations. What was more, they were often expressed discreetly, as Darwin's "gradualist" concepts required.

Between 1920 and 1935 some mathematicians, including Ronald A. Fisher, J. B. S. Haldane, and Sewall Wright, applied the available new data to the elaboration of theoretical models of evolution. Thus population genetics was born, with the idea of studying the dynamics of living groups in relation to variations in the frequency of different known alleles, and of seeking the causes of these variations—such as selective pressure, chance, or gene flows from other populations. Although initially received with reservation by naturalists, these concepts soon attracted wide acceptance. Later, immunological and physicochemical techniques were applied to genetic analysis of individuals, which were henceforth defined not by approximate, "global" morphological criteria, but by molecular markers. These techniques revealed that mutations were even commoner than believed, and that natural populations, including those that seemed the most homogeneous, consisted of individuals differing in many alleles. This *genetic polymorphism* is encountered in all members of all species. It is undoubtedly one of the fundamental laws of life and is hardly compatible with the old idea that natural selection imposes uniformity. A new concept replaced the typological idea of species or races based on a characteristic holotype: it took into account the variations proper to a group of individuals living together and capable of reproducing. Such a group constituted a *population,* the basic unit of a species. The substitution of populational for typological theories represented the most important conceptual revolution in the natural sciences since Darwin showed that transformation was a reality. It has led to a synthetic concept of evolution, formulated just before and just after the Second World War. Four names mark this movement: the zoologists Theodosius Dobzhansky and Ernst Mayr, the paleontologist G. G. Simpson, and the botanist G. L. Stebbins.

Today we recognize that selection leading to uniformity would be incompatible with the genetic polymorphism that is found in all natural populations. Except in rare cases of profoundly deleterious mutations, selection does not seem to choose between alleles. Rather, it chooses to keep them all, even if sometimes modifying the frequency of some at the expense of others. For the geneticist the very concept of the holotype is nonsensical. Richard Lewontin's work on invertebrates, that of Maxime Lamotte on molluscs, Gérard Lucotte's on primates, and that of many others, have confirmed the permanency and extent of this "colossal multipolymorphism," as Georges Pasteur called it.

How can we explain the paradox of the genetic polymorphism of populations faced with selection? Two answers have been suggested, at first sight completely contradictory, but not really irreconcilable.

The first is the *neutral* theory proposed by the Japanese geneticist Motoo Kimura, in whose view most mutations remain neutral, all alleles in a given series having the same selective value. A neutral mutation would spread randomly, with only the rare unfavorable ones being eliminated. According to this hypothesis, selection is a sort of safety barrier. It does not construct; it protects. It does not innovate; it maintains.

The second is the *selectionist* theory, which assumes that the target of natural selection is not the individual or the gene but the whole population — that is, the ensemble of those individuals that are capable of interbreeding and live in the same place at the same time, so that they participate in the same gene pool, of which they are the outward sign.

Ecological factors, and thus constraints, that influence all groups are continually varying, either in time (with circadian and seasonal rhythms) or in space (according to relief and exposure, for instance). In a genetically monomorphic population, all individuals would have the same abilities. They would be active at the same time, would seek the same food, live in the same place, and choose the same mate. Severe competition would exist in such a peculiarly narrow habitat, with its limited resources. The disadvantages accruing from such a situation are obvious, condemning the group to merely survive at best, or disappear at worst. By contrast genetic polymorphism diversifies abilities and varies activities. It pushes back the frontiers of the ecological niche and increases the available resources. At the same time it reduces competition. The more genetically polymorphic a group, the broader are its horizons and the greater its chances of survival. That is why natural selection chose it!

The "population revolution" was not easy to accept, although it corresponds to what we see in practice and also to the latest discoveries in molecular genetics.

Earlier the gene was thought of as an independent entity, like a little isolated factory, working for itself. In a given environment, its selective value did not vary. It was a label that remained constant. The neo-Darwinian scheme was compatible with such a view — which Ernst Mayr called "beanbag genetics" — but today it is no longer valid. First of all, we now know that the selective value of a gene is not constant but varies as a function of multiple factors, particularly its frequency in a population (see the work of Claudine Petit). A given gene with a highly positive selective value, which is therefore very advantageous when rare, may become disadvantageous if its frequency increases. In addition, a gene is not an inflexible machine, obeying rigid commands, as was once thought. François Jacob and Jacques Monod showed more than twenty years ago that there exist regulatory systems in bacterial chromosomes that subject each gene to a rigid control, continually adapting its activity to the needs of the cell. Other, doubtless different, regulatory phenomena exist in cells with true nuclei *(eukaryotes)*, which may have a large number of chromosomes and an infinitely greater hereditary potential than bacteria. These regulatory systems make the action of many genes interdependent. Also, the selective value of each gene is linked to the other elements that accompany it in the same patrimony. This *genome* is no longer considered as being formed of a series of independent units lined up side by side, but as a functional whole whose different elements constitute an intimately interlinked network. In higher animals and plants, these regulatory phenomena lead to cell differentiation, as a result of which the lines descended from a single cell — which have, therefore, inherited the same information — diversify to form specialized tissues, such as muscle, bone, blood, nerves, and so on. This mechanism implies that as each cell becomes oriented in a particular direction, it retains only that fraction of its patrimony necessary to code for the construction of a muscle cell, a nerve cell, or whatever. Much progress was made in the middle of this century in this field, thanks to developments in experimental embryology. The most usual technique was to graft onto an embryo a bud from another embryo of the same species, but in the wrong place, and then to observe the graft. One of the pioneers in this particularly fertile domain was Étienne Wolff. His school has guided the transition from descriptive embryology to experimental ontogeny. This work is dedicated to him.

When one considers the many, often contradictory, forms that Darwinism has assumed over the years — whether as neo-Darwinism, mutationism (which is, strictly speaking, anti-Darwinism), or the synthetic theory of evolution — one may well wonder what remains, more than a hundred years after the death of Charles Darwin, of the model drawn up by the illustrious navigator of the *Beagle,* and wonder even more about the use it has been put to.

What strikes one most forcibly when reading Charles Darwin's works is his extraordinary power of observation. The gradual nature of variation did not escape him, nor did the tendency for a whole group, and not just a single individual, to evolve in the same way. These observations fitted rather poorly the rigid concept of selection constructed at a time when people had no idea of genetics, and Darwin was forced to introduce a little Lamarckism into his model. But he sensed intuitively the populational aspect of transformism, although contemporary knowledge did not permit him to understand its biological basis.

Nevertheless Darwinism cannot explain everything, even when its details have been modified. Certain lines seem to be privileged; some groups do not change for a very long time and then suddenly "explode" into multiple branches. Many leading zoologists, including P. P. Grassé, Albert Vandel, Hervé Harant, and, more recently, Stephen Jay Gould and Niles Eldredge, have openly criticized the concept of selection, as have paleontologists like Franck Bourdier and others. The neutral theory proposed by Motoo Kimura in 1968 has gone from strength to strength. It now seems that some species that undergo multiple mutations do not evolve, whereas others that mutate little evolve rapidly.

Certain very ancient species, the "living fossils," survive in caves or the ocean depths and have scarcely changed since the distant past, although genetically speaking they are no less polymorphic than others. Today one can hardly assume any longer that there is a relationship between the ability to mutate and the tendency to evolve.

Nevertheless, in spite of its imperfections and insufficiencies, Darwinism retains an enormous conceptual value, for Darwin's theories have repercussions beyond the realm of biology. They influenced philosophical and political thinking during the nineteenth century and continue to do so today. Darwinism was largely inspired by the ideology that reigned in England at the height of the industrial revolution. The Golden Rule was free enterprise and may the best man win! The system proved its efficiency, for in a few decades British industry led the world. Great Britain enjoyed unequaled power, building a gigantic empire and ruling the oceans. But there was another side to the coin. Within the mother country there existed a miserable proletariat, totally devoid of any form of social protection and living at the limits of survival, while the colonies were cruelly exploited to provide cheap raw materials for British industry.

In China, the Opium War had just ended, with the result that an enormous market opened up, although at the expense of a whole nation becoming drug addicts. Many people were shocked at the situation. It was at that moment that Darwin's cousin, Francis Galton, with the aid of a few others, began to apply the theories of selection to human societies. This "social Darwinism" (about which Darwin himself always manifested the utmost prudence) eased consciences. It preached the existence of a hierarchy within classes and races that was a biological necessity and resisted all moral forces. This hierarchy, the result of competition between variants, was essential for the progress of mankind, just as for plants and animals. It meant a place for everyone in the social pyramid. The proletariat was naturally destined to work much and earn little, blacks to be slaves, and yellow men to be servants, just as bats were destined to fly and dolphins to swim. Karl Marx himself, and even more so Friedrich Engels, were deeply impressed by Darwin's ideas, which supported their theory of class struggle. But they replaced biological factors with peculiarly human parameters such as work, ownership of production lines, and the redistribution of profits. According to them one class should oust another, as an old species disappears to make way for a new one. The proletariat, representing the mass of the workers, would eliminate the bourgeoisie. At the same time some people embraced Darwinism, and especially neo-Darwinism, to support racial theories that were to lead to the worst excesses of the twentieth century.

But it would be unjust if because of these abuses we did not pay homage to Charles Darwin's exceptional mind. Without him Western philosophy would not have followed the road it did and would certainly have been the poorer for it.

Today we are faced with a paradoxical situation. Social Darwinism appears to legitimize two totally opposing ideologies that have divided the world: a boundless liberalism and a hopeless Marxism. Both concepts, obviously doomed to failure, threaten humanity with conflict. We must seriously consider whether the time has come to reexamine these worn-out ideas, based on a typological vision of nature, that were conceived when nothing was known of genetic polymorphism within groups and of the true nature of selection. In other words, can the biologist remain silent, faced with the use that is being made of models that blatantly no longer correspond to reality?

This book attempts to provide the beginnings of an answer, but in all modesty.

Typological thinking, which once seemed to be a permanent fixture, has been forced to give way to populational concepts that correspond better to the discoveries of modern genetics. This revolution, whose amplitude we are just beginning to realize, will certainly not be the last. Its precarious nature should neither worry nor disappoint us. It corresponds to the image of life, which disappears just as we think we have grasped it. No discovery provides an ultimate explanation; they have all created more problems than they have resolved. A theory must not become a dogma. Science will never be a means of attaining certitude: that is the great difference between science and religion. Its fragility is its noblest feature, for it does not ensnare us and invites us to go further. The legitimacy of science is undoubted as long as it helps us understand the world and leads us toward a transformation of the condition of man.

The book is divided into three parts. The first is devoted to a study of the variations seen in living groups and their significance. It defines the concept of populations, the true units of the living world and the targets of natural selection. The second examines the mechanisms by which species evolve from polymorphic populations that, by making use of modifications in their hereditary patrimony, become able to organize new ecological niches. These mechanisms bring us to reconsider the long-established notions of species, races, and competition. The third part recalls the history of social Darwinism and the multiple, and often pernicious, uses to which it has been put throughout the last hundred years. Finally, we look at the sociological and political consequences that the replacement of typological ideas by populational concepts could have, and the possible basis of a human race that renounces, both nationally and internationally, the notion of selection, which, though it still impregnates many minds today, is far from the reality of the living world.

I wish to express my sincere thanks to all those who have kindly reviewed parts of the manuscript or who have discussed it with me, and particularly: Jean-Paul Aron, Raymond Aron, Francisco Ayala, Étienne Baulieu, Jean Bernard, François Bourlière, Jacques Caen, Yves Cambefort, Luca Cavalli-Sforza, Isac Chiva, Yves Coppens, Jean Dausset, Jean Delumeau, Jean Dorst, Pierre Drouin, Jean Ducos, René-Jean Dupuy, Jacques Durand, Bernard Dutrillaux, Jean Elleinstein, André Fontaine, Michel Foucault, Claude-Louis Gallien, Robert Görtz, François Jacob, Motoo Kimura, Georges Lambert, Jérôme Lejeune, Almerindo Lessa, Richard Lewontin, Gérard Lucotte, Ernst Mayr, Philippe Meyer, Arthur Mourant, Georges Pasteur, Jean-Claude Pecker, François Raveau, Philippe Rouger, Charles Salmon.

J.R.

PART ONE

the polymorphism of populations

types and variation

THE LIVING WORLD AND THE POWER OF RECOGNITION

Food Cycles and the Ecological Niche

All animals and plants need a constant supply of energy to survive and reproduce. Nature is continuously pervaded by an enormous energy flow derived from living organisms themselves, on which they depend but for which they pay a heavy price. This flow represents the *food cycles,* which weave inextricably in all directions throughout the animate world: all species are both consumers and suppliers at different nutritional or *trophic* levels.

In the present environmental conditions of the earth, laid down about two billion years ago, the source of the energy flow is the electromagnetic radiation of the sun, whose light is exploited by the chlorophyll cycle that enables plants to construct large organic molecules from simple mineral elements. Plants are *autotrophic;* that is, they can usually survive independently without the help of other species. They represent the first link in food cycles. Next come the animals; they cannot synthesize like plants and must have direct access to complex molecules to provide for themselves. Hence, they are *heterotrophic.* Some animals eat plants: they are herbivorous. Others, the primary carnivores, eat the herbivorous animals and are destined themselves to be consumed by secondary carnivores, and so on.

Thus any animal can find itself sometimes in the situation of the hunter or harvester, and sometimes in that of the hunted. It is part of a food cycle from which it derives its own provisions but to which it contributes. It is one link in the network. The place it occupies in this sequence is its *ecological niche.*

The concept of the ecological niche, so useful to the naturalist, was defined by Jacques Blondel and François Bourlière (1979). The niche is not only a spatial and temporal unit — that is, a part of the space and time in which a population lives and reproduces — but also represents the position of the population in the ecosystem. The niche is thus primarily a functional unit defined by a number of factors such as the environment—either physicochemical (the climate, the nature of the soil) or biological (the flora on which herbivores depend, the fauna necessary to the carnivores, competitors, predators, parasites, commensals, pathogenic complexes, and so on).[1] The nature and extent of a given population's niche depend on its ability to establish itself in a given environment, to exploit it, and to establish certain relationships with other animal or plant species.

G. E. Hutchinson stated in 1958 that the niche was only the outward objective projection of the possibilities offered by the hereditary patrimony. Any population that changes its niche changes its patrimony. We shall see in the chapters that follow how this interaction between the *genome* and the environment constitutes the fundamental drive for the development of species.

Indeed, it must be rare for any group to utilize all its genetic potential. Usually its niche is limited not only by its own capacities but also by competition from neighboring species. Thus we might distinguish the potential niche from the niche that is actually occupied. The niche represents a temporary compromise, for any given time and place, between environmental constraints and genetic capacities.

Individuals, either animals or plants, belonging to different species but related by essential biological factors form a *biocoenosis*. Within a biocoenosis a group is almost never involved in a single food cycle but in several. It is at a crossroads rather than at a point on a simple linear pathway. In addition, biocoenoses are complex groupings with, in addition to the major trophic axes, many collaterals linking the principal paths together. Any pathway may begin at a variety of places and contain numerous secondary exits.

We may recall here La Fontaine's heron that, while hunting, refused first a carp, then a pike, a tench, and a gudgeon, all of which it found not to its taste. Finally, when it was hungry, it was obliged to eat a slug! In doing this it took an "escape route," one of several that were possible. This ability to exploit several trophic pathways perhaps saved this particularly hard-to-please heron from starvation. It has an obvious selective value.

Such a system of multiple interconnected networks has great stability. A weakness in any one link, such as a reduction in numbers or

---

1. In practice the features that make up the concept of the niche can be defined fairly rigidly, even mathematically. The niche is a multidimensional model, a "hypervolume." Although it does not take into account all natural elements (since certain factors involved in the niche will be undetectable by an observer), such an analysis can be of great use, particularly when comparing phylogenetically related species.

disappearance of a species, does not threaten the whole structure, which remains functional by using other channels.

For most species nature does not offer a single menu, but a highly varied *à la carte* one. An animal can choose, and varying its dishes often involves changing food chains. Such variety in its resources provides enormous security to a living being. The more complex a biocoenosis, the less fragile it is.[2]

The Power of Recognition

To maintain itself in its ecological niche, that is to say in its place in the network, an animal must recognize and identify other species present in the same biocoenosis. This power of recognition is as indispensible at the bottom of the chain, where an individual must choose prey that will be of use to it, as at the top, where it must avoid being too easy a prey itself.

Indeed, for a food cycle to function successfully, there must be a certain equilibrium between entries and exits at all levels.[3] This implies an adequate development of the capacity to recognize and identify that permits an animal to plan its strategy. Without the power of recognition, higher organisms could not survive.[4]

In most invertebrates and lower vertebrates recognition and classification depend on innate behavior. At birth, the young know what they must eat and what they must leave, what they must pursue and what they must flee. The worker bee is capable of selecting the flowers that are useful to its harvest on its first flight from the hive. From the very

2. This explains why artificial, man-made biocoenoses involving only a few species *are* fragile. In a biocoenosis with few species the loss of a single link can stop everything, for there are no alternate or *vicariant* pathways. Since the Neolithic era, and especially since agriculture entered the industrial age, the surfaces devoted to the exploitation of single plant species have expanded considerably, which has made them extremely fragile. They have become an ideal "culture medium" for parasites and specific predators of particular domestic plants. In a natural, very varied milieu these threats would be limited by the presence of wild species to which they are not adapted and that constitute barriers to their spread. Industrial agriculture would hardly be possible without pesticides and various other defense systems. One can even consider it "antibiological."

3. This need for equilibrium, which means that at each link in the chain there is a balance between the entries and the exits, sometimes leads to astonishing consequences. Many plant species have redundant structures, such as flowers and seeds well in excess of their reproductive needs. This exuberance is a necessary response to predacity. Because of the herbivores that eat them, plants must make enough reproductive "material" to ensure the survival of the species; the surplus production guarantees perenniality. This tendency has been exploited by man in the domestication of plants. An example is the evolution of wheat, in which the number of grains has increased constantly under intentional selective pressure. Today it has no relationship with the plant's reproductive needs, only with the nutritional requirements of our society.

4. We exclude from this concept the phenomena of recognition at the molecular level—between an antibody and an antigen, for example, or between two cells—which are in no way related to conscious processes.

beginning there is no hesitation in its activity. As the nervous system becomes more elaborate, the acquired enriches this realm of the innate. In warm-blooded or *homeothermic* vertebrates such as birds and mammals, the young possess certain inherent behavioral patterns, but these are very rapidly refined by imitation of adults and by their own experience and apprenticeship. Baby wolves improve their hunting by contact with their parents, and young chimpanzees learn from adults which fruits they can consume and how to build their shelter.

We shall see in Chapter 15 the importance of the progessive replacement of innate behavior by acquired behavior at all levels of the phylogenetic tree and the selective advantage given by this substitution, so that an adequate response can be given to every new challenge.

The replacement of the innate by the acquired begins well before man's phylogenetic level (see Ruffié, 1982), but it reaches its highest expression in man, who is infinitely more capable than his predecessors of organizing the space around him and remodeling the resources of the biosphere for his own benefit.

Whatever the nature of its behavior, a living organism is most efficient when it is well informed about its environment and thus capable of adjusting its strategy according to a given situation. This adjustment depends above all on its cognitive capacity but also on the information it receives. Every increase in knowledge leads to a selective advantage. It implies a better way of life and, in practical terms, a better exploitation of the possibilities offered by the environment.

The Place of Man in Trophic Pathways

Thus one of the fundamental bases of animal behavior is the ability to recognize and classify, which is particularly well developed in predators. Just like other animals, man is part of a series of food chains. *Homo sapiens* clearly is a champion predator, for two major reasons. The first is related to his physiological makeup. As far as his nutrition is concerned, the human being is a habitual omnivore. His digestive system allows him to indulge in all sorts of diets. If we take into consideration the whole vast mammalian family, our species is perhaps the only one containing some populations that eat almost exclusively meats and fats, such as the Eskimos of Greenland, as well as other groups, such as certain Hindus and some populations in Southeast Asia, that adopt a strictly vegetarian diet. Man is extremely adaptable nutritionally and can use whatever is at hand, thus becoming involved in a multitude of food cycles. No other species occupies such a wide and varied niche. In the last chapter of this book we shall see the enormous consequences for the evolution of living beings of the spreading of the human niche to the whole world.

The second condition that makes man a champion predator concerns his intelligence. In order to survive, any species must find its right place in one or several food chains, but its adaptive possibilities are not unlimited. These depend on its physical and psychological characteristics — which means, at least in the beginning, on the nature of its innate behavior. Its inherent qualities maintain it in a given ecological niche from which its cannot escape without a fundamental modification of its hereditary patrimony and thus its development into a new species.

This was not the case with the hominids who, thanks to their intelligence, were able to develop learned behavior patterns very early, enabling them to enter a large number of food cycles and exploit them for their personal benefit. Prehistoric men harvested and hunted very efficiently in all climates. In spite of poor natural defenses — for our direct ancestors had neither claws nor fangs — and relatively modest physical strength, they dominated all other species, even the most fearsome, by making weapons that they used with great agility. Certainly, they were sometimes the unenviable victims of other predators, particularly the big carnivores alongside whom they lived in the warm regions of the world. But such a fate became more and more the exception as their culture developed. For their industry and technology permited Paleolithic men to be almost always near the final stage of any trophic chain — the stage where they gathered the fruits without having to pay for them.

Beginning in the Neolithic era, *Homo sapiens* was no longer satisfied with merely exploiting existing networks. He learned to organize his own food chains by clearing the forest and dominating nature, thanks to the domestication of plants and animals. He increased his resources considerably and had them permanently on hand by keeping herds of animals and storing his harvests. Today every farmer in the world continues to maintain similar artificial food cycles. He sows a field of alfalfa to feed the flocks that will provide his meat. He harvests oats to feed his chickens and sells their eggs. In spite of progress in agriculture, in "zootechnology," in conservation, we still live, as far as our food is concerned, according to a Neolithic cultural model set up eight or nine thousand years ago near the eastern shores of the Mediterranean and that spread from there to Europe, the Middle East, and North Africa.[5]

The ability to enter and exploit multiple chains and then to create new ones implies that primitive man was able to recognize and classify very precisely. The first hominids, probably about three or four million years ago, must have been excellent observers.

5. Other Neolithic civilizations sprang up independently at the same time in Southeast Asia and, a little later, in Central America and Africa. They were all based on the same principle: the domestication of plants and animals of use to man and the development of artificial, domestic food chains.

THE HISTORICAL CONCEPT OF SPECIES

The ability to classify living organisms precisely can be found today in primitive, isolated tribes that still lead a predatory life like our Paleolithic ancestors. Surviving as they do by hunting and harvesting, their motto could well be "Classify or perish."

Such people have an astonishing ability to recognize hundreds, if not thousands, of different species. Neolithic life was much less demanding, for it involved the rearing or cultivation of a small number of animal and plant species, often not exceeding a few dozen chosen for their particular qualities. Predatory behavior was only of secondary importance, although it did not disappear entirely.[6]

The criteria used by populations of hunter-harvesters for classifying are multiple and rigorous. They include not only the shape of an animal but also its behavior (habits, times and seasons of activity, breeding behavior, the places and tracks it frequents), as well as other characteristics particularly useful for capturing it.

This method of classification, based on patiently observed criteria, is almost always similar to that used by modern zoologists. J. M. Diamond (1966) reported that the natives of the high plateaus of New Guinea, who lead a very primitive life, distinguish 110 species of birds, whereas the best ornithologists define 120 there. Ernst Mayr (1963) goes further. One tribe he observed in New Guinea used 136 names to designate 137 species of birds that he had classified himself. What is more, this small difference concerned two groups of warblers so closely related that zoologists have only recently separated them. Similar agreement is found in the identification of amphibians and reptiles used as food. In contrast, only one name defines the many species of butterflies, which are of no use in everyday life.

During the early period of our history, or rather of our prehistory, man had a vision of living beings that was both precise and universal, defining them according to groups of characteristics (especially behavioral patterns), particularly when these beings were of direct survival value. Precise identification was a necessity.

6. Until quite recently predatory activities, involving mainly hunting but also the gathering of plants, were of considerable nutritional importance in traditional peasant societies even in Western Europe (Ruffié, 1981). Today, in industrial nations, predacity is more a sport than a necessity, as exemplified by the hunting of specially bred animals released into the wild before the real opening of the season, and the gathering of wild strawberries or mushrooms. Industrial fishing, however, remains a true predatory activity, and we are well aware of its consequences. If man wishes to continue to subsist in part from marine animals he should, at least in certain coastal waters and for some species, apply Neolithic methods of husbandry to the oceans, methods that have been used for thousands of years for birds and domestic mammals and without which civilizations could never have developed.

THE TYPOLOGICAL CONCEPT OF SPECIES

Typology and Fixism

Renaissance Europe rationalized knowledge. Naturalists approached classification more rigorously. The first comprehensive work was that by Linnaeus, who in the eighteenth century assembled contemporary knowledge into his *Systema Naturae*. Successive editions between 1735 and 1759 each contain more precise details. The Linnean approach is almost entirely based on morphological criteria. It assigns individuals that present a certain number of identical features to the same group. Higher categories of classification are obtained by putting together several groups that may have only a small number of characteristics in common, while lower categories contain few species that have many common features.

Let us look at an example of Linnean classification. Many animals are characterized by a notocord, a flexible column along the back running from the head to the tail. Above the notocord lies the nervous system, with a dilatation at its anterior end. Below the notocord is the digestive tube, whose anterior end contains the branchial clefts that allow the pharynx to communicate with the exterior. All animals with these morphological features contain common genetic information for their structural coding and make up the phylum of the chordates. Certain chordates have a cartilaginous or bony endoskeleton containing a longitudinal column made up of elements placed side by side, the vertebrae. At the anterior end of this vertebral column is the skull, enclosing the brain. These animals have four limbs for locomotion, two attached anteriorly to the shoulder girdle and two attached posteriorly to the pelvic girdle. They constitute the vertebrates and are divided into a number of classes. In the fish the limbs and tail are transformed into fins, the skin is covered with scales, and the animals lead an aquatic life. In birds the forelimbs have become wings, the body is covered with feathers, and so on. Each class can be subdivided into subclasses. Among birds there are the Ratites, which have no sternum and instead of flying run along the ground like the ostrich, and the Carinates, with a protuberant sternum on which the wing muscles are inserted. These are birds capable of sustained flight. The Carinates can be divided into about twenty orders that are in turn divided into families; these are further subdivided into genera, the genera into species, and the species sometimes divided into subspecies or races.

In summary, this classification is like a pyramid comprising upper taxonomic levels (subkingdoms, phyla, classes, orders) and the lower taxonomic levels (families, genera, and species). From its apex to its base, through the different levels, the animal kingdom is divided into more and more numerous groups, each comprising individuals of greater and greater similarity that have an increasing quantity of common genetic information. At a certain threshold the proportion of identical hereditary patrimony, or *genome,* is sufficiently large to allow individuals to interbreed: they belong to the same species. Thus, for the biologist, a species can be defined when living organisms, classified by order of increasing similarity, have a sufficiently similar genetic background to allow them to reproduce together. A species thus comprises individuals that can combine their stock of genes and give birth to new individuals; they can interbreed. All the individuals of a given species share a similar pool of genes; they are part of the same hereditary community and are its visible manifestation.

The species is the only level based on unequivocal criteria. Supraspecific levels depend on certain characteristics selected by the classifiers. Created by zoologists, they are laboratory artifacts based largely on arbitrary criteria. Families and orders do not exist as such in nature, in contrast to species, which are objective realities.

Linnaeus and his successors defined each species according to a series of constants or specific characteristics. An individual having all these characteristics, and which all the members of the same group resemble, is called the *holotype.* This ideal model is a true standard, like the standard meter at Breteuil, and is usually the specimen that was used for the first detailed description and preserved in a museum. The classifiers can always refer back to it. If an individual found in the field differs too much from it, one does not hesitate to create a new species, which itself becomes a holotype.[7]

For two centuries after Linnaeus, the essential and almost exclusive aim of the zoological and botanical sciences was classification. The whole era was marked by a fervor for classification. One captured, one drew, one preserved, one described as best one could. The most obscure Sunday-afternoon naturalist wanted to give his name to a new species.

Knowledge was confused with classification, explanation with arrangement. It was believed that an individual or a group became an entity merely by being assigned a place in the taxonomic table. The only aim of naturalists was to define keys for identification.

The idea of *typology* takes up the old Platonic idea of εἶδος, ("type") found in the essentialist concepts of Aristotle. The Greek philosophers thought that the world was made up of a limited number of universal

---

7. In fact, it was accepted for two centuries that a species could have several identical reference models, or rather models containing those features considered to be specific, in different museums or laboratories, all with the same reference value. They were called *syntypes.* In 1950 a congress was held to establish rules for nomenclature, and it was decided to have only one example represent the species—the *holotype*—while the other identical specimens could also serve as references and were called *paratypes.*

phenomena of which individuals were merely the visible manifestation, reflections of a single fundamental design.

Christianity provided a metaphysical explanation for a typological vision of the world. God created each living species, which from the beginning reached the same degree of perfection that we recognize today, for God could do nothing less than perfectly. How else could one explain the astonishing powers of adaptation that naturalists had always admired, than by divine will? The amazing ability of fish to live in water, or the equally strict adaptation of birds to flying, could not be the result of mere chance. Only God could have conceived such marvels of adaptation. The few failures that were occasionally seen were due to the intervention of the forces of evil.

The myth of the Creation is not unique to Western Christianity. It is also found in the ancient civilizations of the Far East and Precolumbian America. It is almost a cultural invariant.

For a long period the world lived with the fixist concept that species were the fruit of divine creation or appeared through spontaneous generation. They had not changed since their origin and would never change.

Typology and transformism

The transformist idea appeared among philosophers in France toward the middle of the eighteenth century and was later adopted by naturalists. Most clearly expressed by the encyclopedists, particularly by Diderot in *De l'interprétation de la nature: Rêve de d'Alembert,* it was part of the liberal, rationalist movement that shook the country and disputed the power of the church to explain by revelation. In addition, science based on observation had made much progress. Geological studies had revealed an increasing number of vanished species more or less related to modern forms, showing that the living world had not always been what it was then. From time to time groups disappeared as others replaced them. In 1793, the Convention governing the new French republic made "natural history" a part of the national program of education. The very term "history" implies a notion of evolution. In Paris the "Museum" was created from the former Royal Garden of Medicinal Plants, and two chairs of zoology were created. One, devoted to vertebrates, was occupied by Étienne Geoffroy Saint-Hilaire. The other, dedicated to invertebrates, was taken by Jean-Baptiste de Monet, chevalier de Lamarck. In 1809 Lamarck published his *Philosophie zoologique,* in which he attempted for the first time to give a positive explanation of transformism. Struck by species' remarkable power of adaptation to their environment, he assumed that the repeated use of organs caused their development, while nonuse led to atrophy, and that these modifications were preserved in an animal's descendants. But this hypothesis, which implied heredity by acquired characteristics, did not correspond to reality. Lamarck died in poverty and oblivion on December 18, 1829.[8]

Half a century passed before Charles Darwin proposed, in 1859, a rational explanation of evolution. Let us recall what happened. Between 1831 and 1836 Darwin, a young, intelligent naturalist, rich and at a loose end, undertook a long cruise around the world on the *Beagle,* a ship designed for cartographic and chronometric observations. He crossed the South Atlantic, stopping for long periods along the coast of Latin America and visiting many archipelagos; crossed the Pacific, stopping in New Zealand and Australia; and then returned via Brazil. When he arrived in London the young Darwin brought back a considerable quantity of zoological and botanical material, and a host of observations. Distributing part of his collection to various specialists and keeping the rest for himself, he began his work, and slowly developed a new explanation for evolution. But he waited a long time before revealing it. Not until 1858, twenty-two years after his return to England, did he publish a first note to the Linnean Society of London, and then the act was forced upon him, for another English naturalist, Alfred Russel Wallace, working in the Malay Archipelago, had developed an identical theory. Having almost lost his precedence, Darwin quickly finished writing a book that he had been preparing for years and that he was forced to abbreviate. The *Origin of Species* appeared on November 24, 1859, and was sold out the same day.[9]

According to Darwin, species were the product of an evolution that tended to adapt an organism more and more successfully to its environment. This idea had already been put forward by Lamarck, whose works Darwin had read over and over again, but instead of looking for the motive force of the adaptation, and thus for evolution, in the use or nonuse of organs, Darwin attributed it to two elements: *variations* appearing spontaneously in a few individuals in a population and *selection,* based on competition, that "judged" the value of the variations in such a way that subjects showing advantageous characteristics were favored over others and replaced them in subsequent generations. The idea of competition between individuals of the same group was bor-

8. In fact Lamarck was contested more on the principle of transformism (especially by Cuvier) than on the heredity of acquired characters. In 1809 nothing was known of the laws of genetics; Mendelism would only be publicized ninety years later. There was nothing shocking about inheriting the acquired. Many works have appeared recently on Lamarck's life and work, for instance those by Faure (1978) and Barthelemy-Madaule (1979). Lamarckism, in its neo-Lamarckian form, retained its supporters for a long time, and many naturalists accepted the idea of the nonheredity of acquired characters only with difficulty (see Buican, 1981).
9. The full title was *On the Origin of Species by Means of Natural Selection, or the Preservation of Favoured Races in the Struggle for Life.*

rowed from Thomas Robert Malthus, an English clergyman and economist living at the end of the eighteenth century. According to him the human population increases more rapidly than natural resources, provoking tension and merciless struggles. A great moralist, Malthus proposed voluntary birth control. Darwin's ideas about individual competition and the survival of the fittest merely extended Malthus's plan to natural history. Darwinism invoked variation, but stressed its temporary nature, while still seeing it as an essential step upon which evolutionary progress depended. It was, however, a precarious step, for variations never lasted long. Depending on its adaptive value, a variant was either successful or disappeared. In the end all the individuals of a given species would tend to have similar features, those that gave the maximum advantage in the particular environmental context. All subjects in the same group would acquire a profile that adapted them best to the constraints of their surroundings. Fundamentally, *the Darwinian concept tends toward uniformity.* It leads, by successive stages, to an optimal type, the holotype defined earlier.

The introduction of Mendel's laws at the beginning of the present century gave Darwin's theory a genetic basis that had been missing and supported the idea of typology. This was *neo-Darwinism.* Mendel demonstrated that hereditary characteristics are controlled by elementary particles, later called genes, and that a given particle could show sudden and permanent random variations or *mutations.* [10] It was easy to transpose Darwin's concepts to Mendelism. The variations that Darwin observed in all populations corresponded to Mendel's mutations and natural competition could act on them, the final result being the selection of favorable genes and the more or less rapid elimination of unfavorable ones. After a genetic mutation, selection had ultimately to make the choice between the wild gene and the mutation — that is, between *alleles.* Only the best were preserved. This law applied to all genes. So, all individuals making up the same group at the same time, and thus subject to the same selective constraints, would tend to have a similar hereditary patrimony, composed of the best adapted genes for the environmental conditions. It was this ideal genotype that produced the reference organism, the holotype, except that mutationism substituted a genetic holotype for the morphological holotype of classical Darwinism.

Thus the purest, and least threatened, species should be those that showed little variation and contained only identical subjects of a superior nature. They would show a tendency toward *monomorphism.*

10. The normal, common gene is called the *wild* gene. Together with the mutated gene it constitutes two *alleles* or *allelomorphs.* When a gene has undergone several mutations we have a *polyallelic* series or *multiple alleles.*

Apart from exceptional cases, required as a basis for progressive evolution, all departures from the basic plan were bad. Such a typological concept leads to an elitist view of the world. It does not correspond to reality. We shall come back to the work of Darwin in Chapter 12, for he was responsible for the first great conceptual revolution of the modern era in the life sciences.

Subspecific Levels

Although the nineteenth century was not, like the sixteenth, one of discovering continents, it remains, for naturalists, that of the great voyages. Darwin was not the only one to travel the world. By studying species living in widely dispersed parts of the globe,[11] even the most ardent typologists had to admit that a single species could demonstrate modifications that permitted the definition of *geographic races.* Although they have certain dissimilarities, these races can nonetheless interbreed. They are referred to as *polytypic.*

A species is called polytypic when it is made up of several geographically separated groups that differ from each other by easily recognisable characteristics.[12] Thus a polytypic species is composed of a number of geographic races called subspecies, varieties, biotypes, or ecotypes. But the term *race* is used most frequently to designate subspecific classification levels. As with the species, each race, or whatever we call it, is defined by the standard example that provided the initial description and is preserved in a laboratory or museum. The existence of races is in perfect accord with typological concepts and the principles of Darwin. If the area occupied by a species is sufficiently wide, the ecological conditions will not be the same everywhere but will vary from one place to another. Not all individuals in the species will be subject to the same selective pressure. Different groups will each tend to preserve the alleles that are of most use to them for acquiring a biological profile that will best adapt them to their particular environment.

In certain areas the wild gene will be retained, whereas elsewhere it will be eliminated in favor of a mutation better able to meet local conditions. Thus diverse environments lead to diverse groups. The species divides into races, each representing an autonomous type. Typologically speaking, races are almost always geographic because each represents a particular selective pressure. Thus the race obeys the same laws of differentiation as the species, but less rigorously. It is the first

11. Species spread over a wide area are called *eurytopes,* and those distributed over more restricted areas, such as an island or a valley, are *stenotopes.*
12. *Polytypic* must not be confused with *polymorphic,* which we shall examine later.

stage on the way to *speciation.* For the typologist, races and subspecies are synonymous.

Ecogeographical Rules

Toward the middle of the last century, even before Darwin's works, a certain number of laws had been formulated demonstrating the importance of adaptive phenomena in the constitution of species and races. These *ecogeographical rules* applied particularly to morphological features — the only ones to have been studied at the time — but can be applied equally to any characteristic having an adaptive value. These rules have been tested in very different species.

The first two to be described were Gloger's rule of 1833 and Bergmann's rule of 1847. Bergmann's rule concerned the variations in the size of an animal according to the local temperature. It can be formulated as follows: in polytypic warm-blooded species, the body size increases when the average environmental temperature decreases. The importance of this relationship is obvious, for an increase in body size will tend to lower the ratio of body surface to weight. Volume increases as the cube of a linear dimension, whereas the surface increases as its square, so that the larger the body the smaller its surface relatively. Thus heat loss is decreased.

In reality the modifications also involve an animal's overall shape: it tends to become rounded in cold climates, thus reducing the ratio of surface to volume as much as possible, and elongated in warm climates, with the opposite effect. Protuberant parts of the body such as the ears, limbs, and tail are also affected; they become smaller in cold and larger in tropical climates. In the latter case, the total surface area of the individual is increased, thus ensuring a better heat loss by convection and particularly by evaporation, in much the same way that a radiator has a large surface area. This is Allen's rule, formulated in 1877. Bergmann's rule and Allen's rule must be interpreted statistically. Their value varies with species and with regions, as Rensch (1960), Hamilton (1961), and others have shown. This is due to the fact that in order to adapt to its environment an organism can almost always "choose" between several solutions. Increasing the body surface is a good way of losing heat, but it is not the only one. The nature of the fur or feathers and the organization of the cutaneous circulation, as well as the choice of the best time for bodily activity, also play an important role.

But changes in body size do not only affect thermoregulation. They obviously have consequences in other realms of physiology and behavior. Adaptation is always an imperfect, relative phenomenon, as we have already described elsewhere (Ruffié, 1976). It is the result of a final balance sheet drawn up to include opposing selective values. Bergmann's rule is in general largely respected in warm-blooded vertebrates. It is seen in 75 to 90 percent of bird species, with the exception of some migratory varieties, and in 60 to 80 percent of mammalian species, the main exception being those living underground and thus in an almost constant temperature. In the tropics Bergmann's rule depends on the altitude. It has even been confirmed by paleontologists studying climatic variations through the centuries. Certain mammalian populations of North America, Europe, and the East Indies had larger bodies during the ice ages, only to become smaller again in the subsequent warmer periods (Mayr, 1963). Allen's rule is equally variable, and there are similar exceptions.

It is interesting to note that animal experimentation has confirmed Allen's and Bergmann's rules. As early as 1909, F. B. Sumner demonstrated the effect of cold in reducing the size of the extremities and diminishing the ratio of size to weight. Later, Harrison (1959) noted that mice whose tails had been amputated had a greater mortality in a warm environment than those with normal tails. Steegman and Platner (1968) raised young rats in low temperatures and observed craniofacial and body modifications compatible with Allen's law, such as a narrower nose, a more rounded cranium, and a shorter femur. Chernyshev (1966) kept *Macaca mulatta* and *Cercopithecus aethiops* monkeys at liberty throughout the whole year near Moscow. He was able to keep such tropical animals in the cold simply by amputating their tails.

Gloger's rule of 1833 concerns pigmentation: races living in warm, damp climates are more intensely pigmented than those living in cold, dry regions. For a long time it was believed that Gloger's rule depended on sunlight. In the tropics, a large amount of the pigment melanin in the skin would be a good protection against ultraviolet radiation, which can be carcinogenic and may provoke arteriosclerosis. On the other hand, where the light is weak, less pigmented skin would allow the much less intense ultraviolet rays to act more effectively and ensure a sufficient synthesis of vitamin D, thus diminishing the risk of rickets. In fact, the situation is more complex. Careful observation shows that black pigment is reduced in warm, dry regions, and brown pigment in cold, damp areas. Examples of Gloger's rule are found regularly, but its adaptive significance is still poorly understood. In particular, nocturnal birds, in which pigmentation could hardly have much selective value, respect it as much as diurnal species (Mayr, 1963). Aquatic animals show many variations related to changes in water temperature, and climatic racial types have been described in many insect groups.

Man did not escape ecogeographical rules, at least not until his industry became sufficiently sophisticated to allow him to free himself from natural climatic restrictions, and thus from selective pressure, by his ability to recreate his native subtropical environment wherever he goes — by clothing, housing, and special diet, for instance. For Coon *et al.* (1950) the mongoloid facies represents an application of Allen's rule. This facies is characterized by a shortening of the nose, which tends to become absorbed into the face, by slanting eyes, and by the attachment of the lower part of the earlobe, all of which tend to diminish the surface area of the extremities exposed to the air. Coon believed that the mongoloid race developed in central Asia at the time of the great Quaternary glaciations when men protected their bodies, but not their faces, with animal skins. Thus, the cold would have acted as a selective factor mainly in the face region, leading to a remodeling of its profile to adapt it better to a harsh climate. The success of this mongoloid type allowed it to invade the whole Far East. When the climate became warmer after the Ice Age, this physical type had already spread over enormous areas and remained permanently. Recent research by Roberts (1973) and others on the variations of temperature in different parts of the head exposed to cold and on the cutaneous vascular pattern support this hypothesis.

CRITICISM OF THE TYPOLOGICAL CONCEPT

The Limits of Significance of Morphological Features

Typological methodology suffers from a severe handicap in that it is almost always based on preserved specimens. As we said earlier, the essential criterion for defining a species is its ability to interbreed. To be certain that animals belong to the same species, we must be able to prove that they can breed together, and to do this we must almost always be able to rear them in the laboratory. However, many species cannot live away from their natural habitat, thus making any form of observation difficult. Typologists try to prove the capacity or inability to interbreed indirectly by describing the morphological characteristics that seem most permanent and that, by their presence in some individuals and their absence in others, they consider as signs of sexual isolation. If the number of repeated characteristics is sufficiently high (the principle of *constant correlations*), we may consider that there is sexual autonomy and thus speciation. But it remains a postulate that is impossible to prove without experimental breeding.

The concept of typology was built up by laboratory workers or museum curators working on dead specimens often initially captured by other workers. However, morphology represents the expression of only a small part of the genome, perhaps the least important part. Only to the extent that a morphological feature has a functional value — that is, only as far as it is useful — can it be a target for natural selection and thus be part of the mechanism of species formation. Such is the case, for example, of the genital appendages, the genitalia, that are used for copulation in insects, or the sound files used by Orthoptera for making their characteristic song so that pairs can recognize each other and thus make contact. Living organisms adapt to the constraints of environment chiefly by physiological and behavioral mechanisms. As was said earlier, animals initially have innate behavior that follows a genetic program. This behavior is part of the genome, just like an organ or a function, and is the result of natural selection. It is a true specific characteristic, just as much as morphological or physiological features. Its taxonomic value is great, for physiological — and above all behavioral — features are at the interface between the individual and its environment. They are at the front line of the our defenses against the environment and have a direct selective value.

Thus in systematics, we find that physiology, ethology, and psychology are certainly more significant than morphology, and yet in all three areas no data can be gathered from a preserved specimen, and the taxonomist is forced to rely on minute morphological details whose selective value may be little or nothing.

Sibling Species

The inadequacies of our traditional methods of classification were highlighted by the discovery of *sibling species.* This term, coined by Ernst Mayr in 1942, was derived from one first suggested by Lucien Cuénot in 1936 *(espèces jumelles).* It applies to related species, either completely or nearly morphologically indistinguishable, that are totally incapable of interbreeding.[13] Very early, workers studying the fruit fly noted that certain of them, all classified as belonging to the species *Drosophila melanogaster,* could not breed with other *melanogaster.* Either the breeding failed entirely or the resulting products were sterile. In 1919, A. H. Sturtevant suggested that, in spite of their almost identical phenotype,[14] these flies belonged to different species. The concept of sibling species was born. Later, minute morphological differences were discovered between the two groups (see Tsacas and Bocquet, 1976). A similar situation is seen with the American fruit flies *Drosophila pseudoobscura* and *persimilis,* both carefully studied by Theo-

13. Some authors used the term *cryptic species,* for both need not necessarily appear at the same time. However, the expression should not be employed, for it was introduced around the turn of the century to describe species whose color allowed them to blend with their milieu and thus escape from predators.
14. The *phenotype* represents the tangible, external appearance of an individual, and the *genotype* its actual hereditary patrimony. We shall see in Chapter 4 that in some circumstances so-called *recessive* genes in the patrimony are not expressed in the phenotype when the homologous allele is *dominant.* Only the dominant feature is visible.

dosius Dobzhansky. These two tiny flies belong to two independent species, totally isolated as far as reproduction is concerned, but almost identical morphologically.

The opposite phenomenon is also well known. Two morphologically very different individuals can belong to the same species and be perfectly capable of interbreeding. As an example one might again cite *Drosophila melanogaster,* in which there exists a white mutation with colorless eyes instead of the normal red ones. Another mutation, called vestigial, has very small wings that are reduced to mere stumps. Flies that are both white and vestigial (doubly homozygous) differ greatly from the wild or "standard" individual because they have white eyes and cannot fly but move by jumping. Many nonspecialists would classify these two types as two distinct species and yet, although very different, they can interbreed without difficulty and their offspring are normally fertile. And there is more. The fruit fly belongs to the order Diptera — insects with only one pair of wings and not two as in most other groups, whose second pair is present as atrophied vestigial organs, the balancers. But there is a mutation of the fruit fly in which the second pair of wings is present, and nothing prevents the interbreeding of these mutants with the wild forms. In this case, if we consider only their morphological characteristics, these two types of flies would seem to belong not only to a different species but to a different order.

The dispersion of former species into sibling species on the grounds of ethological, ecological, psychological, genetic, immunological, physiological, and other criteria has been most spectacular in those groups that have been studied in most detail in the field on account of their interest for medicine or agriculture.

The importance of the Culicidae, the mosquitoes, in human pathology is well known. They are responsible for the transmission of malaria, one of the most widespread scourges of mankind, of many forms of crippling filariasis, and of various viral diseases, among the most important of which is yellow fever. Until just before the Second World War mosquitoes were mainly studied in museums. The requirements of the war effort in the Near East, Africa, Burma, and the Pacific provoked many field studies that have continued ever since. The result was a rapid multiplication of groups until then considered homogeneous by morphologists, but now known to be made up of a large number of independent species. The Culicidae comprise 34 genera and 120 subgenera, with perhaps 4,000 species, of which almost 3,000 have been positively identified.

Let us take the case of *Anopheles maculipennis,* responsible for the transmission of malaria in Europe and the Mediterranean area. It was long considered a single species. However, these mosquitoes have been observed to exist in regions where malaria has never been encountered.

In reality, *A. maculipennis* is not a single species but a series of sibling species, difficult to distinguish by morphological examination of the adults but identifiable by differences in their eggs as well as by ecological, behavioral, and chromosomal features. Each type has a specific geographic distribution. The real *A. maculipennis* lives in the mountains of Europe and does not normally transmit malaria, for it is not anthropophilic. It only bites animals: it is zoophilic. On the other hand, *A. labranchiae,* almost identical and found in southern Europe and the Mediterranean countries, is very anthropophilic and an excellent vector for malaria. *A. melanoon, subalpinus, messeae, atroparvus,* and *sachorowi (elutus)* all represent other sibling species whose adults are impossible to distinguish from *labranchiae.* All, however, have their own biological characteristics (the larval habitat, for example), thanks to which the former species *maculipennis* can be divided into six or seven sibling species. The *maculipennis* complex also exists in North America (the nearctic group), where it is represented by five species more easily distinguishable than those of the Old World, for they became separated earlier.

One could cite many other examples of this systematic fragmentation. They are found in all zoological groups. Careful study beyond the limits of pure morphology often allows one to break up a hitherto homogeneous species into a number of sibling species (Pasteur, 1977b).[15]

Recently techniques have become available for the direct study of chromosomes and their structure — identifiable as a series of characteristic light and dark bands — thus facilitating identification. Indeed it is rare that the chromosome pattern, or karyotype, of two sister species presents exactly the same appearance. Nevertheless, it seems that in bats, for example, whole genera may have the same chromosomal topography when they are compared band by band (Baker and Bicham, 1980). In the present state of our knowledge, and in spite of rapid progress in cytogenetics, it is still too early for a final analysis of *homosequential* species (those with identical chromosome patterns) and *heterosequential* species (those with different patterns). Whatever the outcome, all goes to show that it is of little use to study fixed material in the laboratory or in the museum when the individual was created to live naturally and its primary characteristics reflect the way it has adapted to the environment. The collector may assemble a host of specimens but will always lose the most important part of the genetic information, that which controls everyday activity. Faced with dead

15. According to Pasteur, many species are considered siblings because their morphological analysis was not pursued far enough. They are really almost always slightly different to the specialist's eye.

specimens from sibling species, a taxonomist can *easily* make a mistake. In the presence of the same individuals living, a sexual partner or a predator *never* makes a mistake. In order to understand the biology of an animal—that is, the full expression of its genetic makeup—one must see it living, eating, attacking, defending itself, reproducing, making its nest, or feeding its young. In other words, one must evaluate the way it fulfills its basic functions of nutrition, protection, reproduction, and organization of the space around it. Morphological features are only one element among others that are useful for identification. Clearly, this handicap is a major one when considering fossil species.

In a few cases, the traditional method of identification has led to some horrifying results. One has only to recall the story of the fish *Sinodontis xiphias* (Gunther, 1864, cited by Daget and Bauchot, 1976), whose holotype was deposited at the Museum of Natural History in London as number 1863.11.9.1. toward the end of the nineteenth century. It was characterized by a pointed snout. After the primary description no other specimens were captured. This was not surprising, for a later X-ray of the example in the museum showed that a thick wire used in mounting the fish had pushed its head forward and stretched its snout totally artificially!

It is clear that some anomalous individuals, sorts of monsters, have led their discoverers to describe a new species that never existed. In human obstetrics, one sometimes encounters neonates with gross bony malformations, especially of the skull, related to chromosomal anomalies that can easily be detected in the karyotype. One can be sure that if such skeletons had been discovered by certain paleontologists in the last century they would have considered each of these pathological cases as belonging to a species different from *Homo sapiens*. Once again, here is a danger of classifying merely by relying on the morphological description of a standard specimen.

These reservations, however, do not diminish the importance of paleontology, without which the reality of evolution would probably never have been demonstrated. But they stress the difficulties and underline the precautions that have to be taken when extinct species are being studied.

Certain large zoological groups, such as the spiders (of which there are nearly 30,000 species) or the gastropod molluscs (which now form about 40,000 species), and even more the insects (more than a million), contain many specimens identified from a single holotype with no ecological, physiological, or behavioral data. For bird species described over a period of twenty-five years, François Vuilleumier (1976) estimated that 35 percent were based on the study of a single specimen, 50 percent on one or two specimens, 35 percent on five individuals, and scarcely 12 percent on twelve or more individuals. Of the species named at the present time, 62 percent are defined purely typologically. Taxonomic tables drawn up on such fragmentary evidence should be considered very cautiously.

Only an animal living in its natural surroundings can reveal to the observer the multiplicity of physiological and behavioral characteristics that play a primary role in its adaptation to its environment. Genetic determination is an obvious factor in many cases. Structural reorganization with morphological modification or the development of new organs comes later, when the group has exhausted its possibilities of behavioral and physiological adaptation. Before changing its shell or its skin, the animal will change its timetable, its habitat, or its food.[16]

The appearance of sibling species must be due to the divergence of functional characteristics under the pressure of natural selection. Let us look again at the example of the American fruit flies mentioned earlier. Although *Drosophila persimilis* and *pseudoobscura* are morphologically almost identical, the former is found more to the north. It lives at a higher altitude and in climates where the mean temperature is lower (see Mayr, 1963). Thus, the two species have to adapt physiologically in different ways, and this can hardly be expressed in terms of shape or color.

Before real divergence becomes evident, new species usually pass through the stages of subspecies and sibling species, which represent the first stages of differentiation.

If one day we can study all animal and plant populations using all available biological methods, there is no doubt that some species will be combined, whereas others will be split into sibling species. This new form of systematics will give a different and more objective vision of the living world, but it will need much interdisciplinary research and require a great deal of time. Between now and then many species will have disappeared before we are able to recognize them, and others will have been born.

The foregoing shows how arbitrary is the present-day classification of groups into species and races. While we must admit that morphological features are not totally devoid of value for classification purposes, they are often insufficient to define species. The appearance of morphological differences does not always necessarily coincide with a reproductive barrier. In some cases morphological differentiation comes later than sexual isolation, and apparently identical individuals cannot interbreed. They belong to sibling species. In other cases, morphologically

---

16. Except in cases of genetic preadaptation, where a preexisting but environmentally neutral morphological, physiological, or psychological feature can become highly favorable when the environment changes (Ruffié, 1976).

different subjects can reproduce together. They are part of the same polytypic species.

Intraspecific Variations

As we have just seen, one of the weaknesses of morphological methods of defining a species is that they ignore the most important part of the genetic information — that responsible for defining physiological or ethological characteristics.

But another, equally serious error is failing to take into account the sometimes very important morphological variation found in individuals belonging to the same population.[17] A classifier using a typological methodology to study a group of interfertile individuals seeks out the characteristics common to all and defines them as *specific* characteristics. He eliminates the others, considering them to be without value for his classification. However, careful observation demonstrates that many characteristics have variations, and that their genetic determinism is just as strict as that related to invariant features. In other words, there is no reason not to use them for defining a species, because they too express the activity of the genetic patrimony. There is no valid biological argument for using only monomorphic features and ignoring *polymorphic* features. In reality, a species does not correspond to a collection of ideal genotypes, all identical to each other and to the holotype in the museum. Rather, it represents a pool of varied polymorphic genes capable of endowing the species with a rich variety of features. Thanks to this polymorphism, many animals living together in groups can recognize individual members of the group. It is why we can identify our friends even in a nudist camp!

Contrary to the polytypism that allows us to distinguish individuals belonging to geographically isolated populations, the variations that we are discussing now exist within a single population, one that is exposed to the same environment. They represent true polymorphism. It is true that even this polymorphism can vary within given spatial limits: a particular variety may be more frequent in one place than in another. But this is simply a statistical problem, for all populations are truly polymorphic even at a local level.[18]

17. Such variations, genetic in origin, also involve physiological and psychological features — in fact, all features governed by the hereditary patrimony.
18. In fact one finds all shades of transition between monotypic and polytypic species. The former have the same genetic variability over their entire area of distribution, while the latter have different degrees of variability according to a particular zone. All depends on how efficiently the gene pool of the species has been mixed. The existence of a polytypic species implies a degree of isolation between different zones within its area of distribution. Then the species will split into geographic subspecies which, as we shall see later, may be new emerging species if the isolation

Classically we have adopted the habit of studying polymorphism in two ways: quantitatively, looking for variations in size and color, for instance; or qualitatively, recognizing the presence of specific characteristics or mutations, like blood groups and enzymes. Such distinctions have been confirmed by time, and we can usefully accept them as long as we bear in mind their arbitrary nature. For the geneticist, only characteristics that can be described by a precise genetic model — and are thus qualitative in nature — are really of use in the analysis of populations.

QUANTITATIVE POLYMORPHISM

If we make precise measurements of a morphological characteristic in a large number of individuals of the same species, we will never obtain the same figure in all cases, but rather a spectrum of measurements between extreme values that are usually, if not always, distributed according to a characteristic bell-shaped curve, the *Gaussian* or *normal distribution.* Its peak represents a mean value, valid for the largest number of individuals. As one moves away from the mean the number of subjects in each class diminishes regularly, until at the extreme values only a very small number of subjects are represented, and occasionally merely a single one.

As a simple example let us consider the variation in the size of an appendage in a population P1 of a species of insects A. After examining a number of animals, we plot on the abscissa the number of individuals whose appendages were of the size expressed as the value Q on the ordinate. We shall obtain a bell-shaped curve, as in Figure 1.

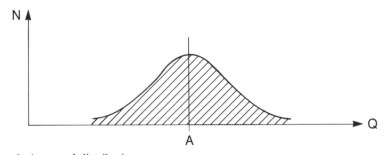

Figure 1. A normal distribution curve.

is prolonged and pronounced. On the other hand, very mobile groups that migrate freely, such as certain birds and mammals, are monotypic, sometimes with broad genetic variability that is, however, spread uniformly without distinct geographic limits. Man occupies an intermediate position, at least in his present form.

This type of curve is found when a feature is determined by a large number of factors that act in an additive way but still independently. What factors might play a role in the growth of an appendage? There are environmental conditions, such as climate, habitat, and nutrition, but we may accept that, for a population living in a relatively homogeneous area, ambient factors are about equal for all. In this case we may consider that the only valid factor to explain the variability of the feature under consideration is a genetic one involved in the expression of the feature.

To oversimplify, let us suppose that the growth of the appendage we have measured is controlled by a series of independent genes, *A1, A2, A3,* and *A4,* each responsible for an increase in size of 2 percent, and that they have been subjected to mutations *a1, a2, a3, a4,* each having no effect on growth. Since they are independent, all these factors can combine in all possible ways, as shown in Table 1.

We can see that six combinations will lead to an increase in growth of +4. They represent the mean of the curve. Four combinations lead to an increase of +2, and four others of +6. Only one combination causes a growth of +8 and one other a growth of 0.

Table 1. Combinations of Genetic Factors

Possible combinations	Frequency of combinations	Resulting change in growth
A1 A2 A3 A4	1	+ 8
A1 A2 A3 a4 *A1 A2 a3 A4* *A1 a2 A3 A4* *a1 A2 A3 A4*	4	+ 6
A1 A2 a3 a4 *A1 a2 a3 A4* *a1 a2 A3 A4* *A1 a2 A3 a4* *a1 A2 a3 A4* *a1 A2 A3 a4*	6	+ 4
a1 a2 a3 A4 *a1 a2 A3 a4* *a1 A2 a3 a4* *A1 a2 a3 a4*	4	+ 2
a1 a2 a3 a4	1	0

Thus the probability of the chance meeting of different genes responsible for a particular characteristic results in a Gaussian distribution
that represents, at least partially, the genetic polymorphism of the
characteristic. Nevertheless our example is hopelessly simple and, in
order to reflect reality, must be corrected in several ways:

1. In practice most morphological features depend on a much higher
number of genes. The possible number of combinations increases very
rapidly, thus tending to give an ideal bell-shaped curve (see Chapter 4).

2. For each gene we have only considered one allele, but most species
are *diploid* (see Chapter 4), and all individuals will therefore have two
sets of chromosomes, one paternal and the other maternal. Thus all the
hereditary characteristics depend not on one single gene but on two, one
from the father and the other from the mother, thus increasing by a
power of two the number of possible combinations.

3. Our theoretical example implies that all genes have the same
probability of interacting with others and that all combinations have
the same chance of taking place, which supposes that all genes are
found equally frequently. However, this is almost never the case. Most
often two alleles at the same locus have different frequencies of occurrence. Such variations in frequency change not the bell shape of the
curve but its position on the coordinates. Now let us study the variations of the same feature not in a single population P1 but in two, P1
and P2, widely separate geographically and with little possibility for
interbreeding. We might obtain the curves seen in Figure 2.

The two curves are similar in shape but cannot be superimposed.
There may be a common area of overlap of the two distributions, Z, in
which it is not possible to say whether an individual belongs to population 1 or population 2.

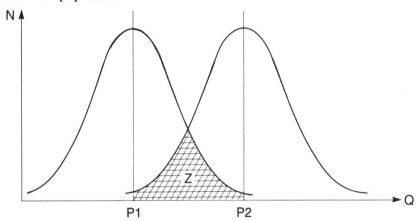

Figure 2. Normal distributions of two populations, P1 and P2.

What is the origin of this noncorrespondence? Doubtless, the fact that, since its selective constraints are different because of its geographic isolation, each genetically polymorphic group responds by a different distribution of its genes, the distribution that produces the best-adapted phenotypes for its particular environment. In other words a longer appendage may be an advantage in zone 1, so the corresponding population would retain mainly, although not exclusively, the alleles *A1, A2, A3, A4*. If the short form is more favorable in zone 2, the indigenous populations would preserve mainly the alleles *a1, a2, a3, a4*. Nevertheless, the unfavorable alleles are never totally eliminated, which may seem to conflict with the Darwinian view of selection. We shall see the reason for this in Chapter 4 when discussing the mechanism for the maintenance of unfavorable genes in the genotype.

For the sake of simplicity we have accepted that environmental conditions, and therefore selective forces, are the same for all individuals in a given population. In fact this is a theoretical situation that is never encountered in nature, for the different individuals making up a reproductive group are never at the same place at the same time, never have exactly the same habitat or the same food, and are never victims of the same form of aggression. This "ecological mosaic" will add another diversifying character to genetic polymorphism, thus helping to explain the quantitative variation seen so frequently.

Quantitative morphological features are relatively remote from the basic genetic information, for they depend on models that imply a large number of genes and in addition they are the final result of both hereditary and environmental factors. For the observer it is difficult to determine the contribution of the innate — the only contribution of use to the taxonomist — and that of the acquired, which has no value for classification. Most macroscopic features formed during embryogenesis, in terms both of organs and functions, are remote in this sense. The data that an observer can obtain from them are always approximate.

QUALITATIVE POLYMORPHISM AND ELECTROPHORESIS

Everything changes if we get closer to the actual source of the information by studying the virtually direct products of genetic activity: the proteins, such as enzymes, that are easy to identify. Generally, protein synthesis involves only one gene and is virtually independent of environmental influences, for they have hardly had the time to act at this early stage. Any mutations are immediately perceptible. Research into protein synthesis produces qualitative results concerning the action of genes. It gives an immediate key to the frequency and type of mutations and thus the extent of polymorphism of the individual or the group.

Genetic polymorphism of living species was for long unrecognized by zoologists and anthropologists, who had a typological vision of the world. Nevertheless, beginning in 1937, the Soviet geneticist N. P. Dubinin demonstrated that in apparently stable, wild populations of fruit flies there were at least 2 percent of phenotypically identifiable mutants, which implies a much higher number of mutations of individual genes when one takes into account that certain *recessive* combinations will not express the mutation. But the frequency of each *individual* mutation described by Dubinin was so low that one could scarcely speak of true polymorphism.

Ten years later, in 1947, W. P. Spencer dealt a fatal blow to the traditional concept of a mutation, which had been considered until then as an exceptional event with the mutant destined to invade all or to disappear. Thanks to painstaking interbreeding, he demonstrated that every fly had a detectable mutation somewhere in its patrimony. Shortly thereafter, Theodosius Dobzhansky (1951) demonstrated that every animal had a considerable number of *lethal* genes permanently preserved in the heterozygous state.[19] But it was the application of *electrophoretic* techniques to population genetics, techniques that have been widely used in the last fifteen years, that permitted the discovery of genetic polymorphism in all groups.

Electrophoretic techniques were described in 1966 by Harris, Lewontin and Hubby, and Johnson *et al.,* and later by many other workers. They are based on the following principle: all individuals possess proteins, often with enzymatic activity, in their serum and most of their body fluids, such as hemolymph. The proteins are composed of chains of peptides made of amino acids linked end to end like the coaches of a train. Twenty different amino acids can be used for making proteins and are always the same in all living species. Their presence and their order in the chains are subject to simple genetic control. Each peptide chain is determined by a single gene. When exposed to an electrical field, these proteins migrate toward the positive anode if they are negatively charged, or toward the negative cathode if they are positively charged. The direction of their migration depends on the nature of the overall electrical charge of a protein in the particular experimental

19. As we have seen, hereditary characters are controlled by two genes, one from the father, the other from the mother. When both are identical *(A / A),* the subject is a *homozygote.* When they are different *(A / a),* it is a *heterozygote.* A lethal gene can cause the death of an individual in the homozygous state. A sublethal gene in the same circumstances only diminishes the subject's vitality. The regular, and probably inevitable, presence of lethal and sublethal genes in natural populations is one of the most astonishing findings of population genetics. If such genes have not been selectively eliminated in spite of the considerable disadvantage they confer to the homozygote, they must be of some advantage to the heterozygote. We shall take up this point in Chapter 6.

conditions. Its mobility varies with the size of the charge and with the volume of the molecule and even its shape.

A mutation affecting a gene concerned in the synthesis of a peptide chain is generally expressed by the substitution of one amino acid for another. This often modifies the electrical charge of the molecule. Thus the mutant protein will not have the same mobility as the initial wild, or standard, protein. In practice, one allows a number of proteins present in a series samples from individuals belonging to a single species to migrate in an electrical field on a particular substrate and for a given time. Nonmutant proteins have the same mobility in all the subjects: after the required time, they are always in the same position on the substrate used for the electrophoresis (usually starch, acrylamide gel, or paper soaked in solutions that allow the passage of current). A mutation may show up as a differential migration of the protein.[20] Thanks to electrophoresis it is easy to detect if a gene has undergone one or several mutations, in which case it is a polyallelic series.

These techniques proved that all species contained many gene mutations, and that very often all the possible alleles for a given locus could be found spread throughout the population. Thus, far from being exceptional, the presence of various mutations is a common phenomenon in all living groups. Polymorphism is found everywhere, even in populations subject to particularly severe selective constraints. This concept is in opposition to traditional neo-Darwinism, according to which the process of selection should tend toward uniformity. Faced with two alleles at the same locus, selective pressure makes the choice, choosing in favor of the better adapted. Nevertheless, Darwin and the neo-Darwinists have always admitted that evolution is the result of a long series of small, successive, cumulated variations. But this was not the opinion of Hugo De Vries (1901), originator of the mutationist school, for whom a single major mutation was sufficient to create a new species. In fact, at least in the animal kingdom, this single mutant, the "hopeful monster" of Richard W. Goldschmidt (1940), carrying the whole future of the group at the price of the destruction of its ancestors and cousins, has never been found. This "catastrophist" theory is based as much on romanticism as on science and represents a caricature of Darwin's concept. Today almost all evolutionists accept the idea of numerous, repeated, small variations.[21] Certainly, we can find a number of genetic monstrosities in all groups, but nobody has been able to show that they are at the origin of a new line, with the possible exception of

20. Statistically, one time in four, which means that, in spite of its precision, there are quite a few mutations that electrophoresis cannot detect, as we shall see later.
21. See Simpson (1944), Dobzhansky (1951), Stebbins (1950), Rensch (1960).

certain chromosomal reorganizations that, appearing in a single individual, have played a role in its sexual isolation, thus helping the process of speciation; we shall discuss these later.[22]

Since the turn of the century, beginning with the school of Thomas Hunt Morgan, geneticists have used the fruit fly to study Mendelian genetics. In their breeding colonies they tried especially to isolate particularly gross mutations that were easy to identify, for this sort of material was particularly suitable for their studies. They neglected or ignored the others that were much more frequent but much less spectacular. These "discrete" mutations do not have the same exceptional and "sensational" character as the grosser ones. They are found in all individuals and involve only small modifications of the phenotype. Such changes are progressive, continuous, and in general well tolerated. They are never widely diffused or suddenly eliminated. As we shall see later, *they* form the foundations on which evolution is built, not the enormous phenotypic revolutions studied by the drosophilists.

Today we must assume that *permanent polymorphism constitutes one of the fundamental attributes of all living organisms.*

THE SIZE AND EXTENT OF GENETIC POLYMORPHISM

The technique of electrophoresis opened a new chapter in genetics unsuspected by the purely Mendelian approach: analysis of the polymorphism of populations. The Mendelian technique is to cross subjects that differ in one or several features and observe the way in which these features are transmitted in their descendants. It can only detect mutations causing qualitative changes that are easy to observe macroscopically. Other mutations, which as we have seen are by far the most numerous and the most commonly present, may be missed entirely by the experimenter. What is more, Mendelian genetics ignores non-mutated genes, which probably make up the largest part of the genome. Electrophoresis, on the other hand, bears direct witness to the structure of peptide chains and thus obtains the genetic information close to its source. Whatever the final effect on the phenotype, nothing is missed, for the products of the mutant gene are read directly, as are also the nonmutant, monomorphic factors. Thus electrophoresis of the serum or the hemolymph of an animal allows immediate characterization of a certain number of its proteins, each of which corresponds to the activity of a given gene. A whole population studied in this way will

---

22. Goldschmidt's model is not compatible with observations made in the field, and is subject as well to many theoretical objections. For instance, the exceptional mutant that appears suddenly and creates a new species immediately must, by definition, remain sexually isolated and die without progeny. Only self-fertilizing plants could comply with this model.

provide evidence of the frequency and nature of the mutations present.

Thanks to such methods, populations can be defined not on the basis of a uniform morphological profile but by a description of the *frequency* of characteristic genes. These frequencies allow us to compare neighboring populations and to detect the direction and volume of sexual exchanges between them — the *interpopulational gene flows,* to be considered in detail later.[23]

Results published so far already involve a large number of species. They are all in agreement: the degree of polymorphism detected by electrophoresis is much greater than could have been predicted by studying morphological variation — a more distant, more complex witness to the activity of a part of the genome.

We can assume that the genes responsible for the synthesis of serum or hemolymph proteins, whose properties are revealed by electrophoresis, are representative and based on random sampling of the total number of genes in an individual genotype. Thus the relative frequency of polymorphic loci[24] studied by electrophoresis should indicate the overall incidence of polymorphism in the whole genotype, rather as a public opinion poll uses the voting intentions of a few thousand electors to speculate, often in quite a rigorous manner, on how the whole electoral body, including perhaps tens of millions of people, intends to vote.

By definition, a locus is considered to be polymorphic when the frequency of the commonest allele is equal or inferior to 0.99, that is, 99 percent (Lucotte, 1977). In other words, if a mutation is very rare — less than 1 percent — its presence is considered to be accidental and the corresponding locus is taken to be genetically monomorphic.

The data obtained by electrophoresis permit us to define a certain number of coefficients to measure the extent of polymorphism in a population, in an individual, or at a locus.

The three most often used are:

P: the level of polymorphism. It indicates the proportion of polymorphic loci in a given species — that is, roughly the ratio of the number of mutated loci to the number of total loci studied, which is equivalent to the frequency of mutated loci for the whole of the genome.

H: the mean level of heterozygosity, or coefficient of heterogeneity. It corresponds to the mean frequency of heterozygous loci compared with the total number of loci in a randomly chosen individual.

23. What is more, the results of electrophoresis are virtually instantaneous, whereas Mendelian techniques involve experimental breeding, which requires much time and is only possible for species that can reproduce in captivity.
24. A *locus* is the site of a gene on a chromosome. It is occupied either by the original, wild gene or by one of its mutations. All genes that can occupy a given locus belong to the same series of alleles.

A: the mean number of alleles per locus, that is to say the mean number of different mutations able to occupy mutated loci.

Table 2 shows the results obtained for some groups of vertebrates and fruit flies (from Lucotte, 1977) and for primates (from Lucotte and Guillon, 1979).

Table 2. Electrophoretic Variability in Different Animal Species

Species	number of genes	level of poly- morphism	mean level of hetero- zygosity	mean number of alleles per locus
Vertebrates				
Primates				
Man *(Homo sapiens)*	71	0.28	0.06	—
Chimpanzee *(Pan trog-* *lodytes)*	—	0.181	0.022	1.24
Macaques				
Macaca fuscata	—	0.10	0.01	—
M. cyclopis	—	0.24	0.04	—
M. mulatta	—	0.37	0.09	—
M. fascicularis	—	0.41	0.10	—
Baboons				
Papio papio	—	0.40	0.031	1.56
P. anubis	—	0.17	0.04 ⌉ 0.018 ⌋	1.05
P. cynocephalus	—	—	0.028	—
P. hamadryas	—	0.22	0.05 ⌉ 0.055 ⌋	1.05
Rodents: Mice				
Mus musculus	41	0.29	0.09	—
Peromyscus polionotus	23	0.23	0.05	—
Birds: Pheasant				
Phasianus colchicus	31	0.32	0.09	—
Amphibians: Toad				
Bufo viridis	26	0.42	0.13	—
Invertebrates				
Insects: Fruit flies				
Drosophila obscura	30	0.53	0.10	—
D. pseudoobscura	24	0.43	0.12	—
D. persimilis	24	0.25	0.10	—
D. willistoni	20	0.81	0.17	—
D. melanogaster	19	0.11	0.11	—

(Dashes represent data not gathered in the various studies from which the table was compiled. Different figures were reported for heterozygosity in two species of baboon by Lucotte and by Shotake et al.)

An exhaustive study of research on heterozygosity in natural populations was published by Eviator Nevo in 1978. For all groups P is around 33 percent and H about 10 percent. However, there are differences between the major populations. If we consider the mean values of H and P for the principal subdivisions of living organisms, we find the highest figures in the plant kingdom and the lowest among vertebrates, with invertebrates taking an intermediate position. In plants, for instance, H = 17; in invertebrates, H = 13.4; and in vertebrates, H = 6.6 for 100 loci.

This means, basically, that vertebrates are genetically less polymorphic than invertebrates, which are themselves less polymorphic than plants. These differences are perhaps due to the adaptive necessities of each of these groups, for a high degree of polymorphism ensures a wide range of responses and therefore a large number of adaptive possibilities. This large genetic "fortune" is indispensable for plants, which cannot move around, have no homeostatic system, and must suffer the full onslaught of selective constraints. A seed has no choice other than to germinate where chance has put it, or to disappear. For it, to be genetically polymorphic and able to face a range of situations, is of prime importance. On the other hand an animal, being capable of an active search for a site that suits it, is less exposed to the constraints of selection than the plant. And, within the framework of this greater choice, vertebrates, with their greater mobility and their more developed nervous system, are better provided for than invertebrates.[25] We shall discuss this problem in Chapter 4 when we deal with introgression.

But there is more. Higher vertebrates, such as mammals and birds, have regulatory systems, such as homeothermic mechanisms, that free them to some extent from the influences of their environment. They provide for the nourishment of their young, thus greatly reducing the nutritional problems faced by the offspring of other groups, and thus have less need for a wide range of responses. Also, higher vertebrates have better-developed mental capacities, related to the growth of their brain. They can adapt their behavior to different situations, refining it by learning and transmitting it to others by examples. They can organize the space around them according to their requirements, making the most of the local conditions, and they sometimes group themselves

---

25. These results, and the following discussion, are valid insofar as future work confirms the means for H and P for plants, invertebrates, and vertebrates. Many species have yet to be studied electrophoretically, and it is conceivable that the figures published so far will be modified. For the moment, mainly invertebrates have been studied, in terms of both the number of species and the number of animals from each species. The high values of H and P for invertebrates could be influenced by this larger sample. Nevertheless, at the moment, the "selective" explanation for the preservation of polymorphism remains the most satisfactory. See Chapter 6.

into complex societies. This type of learned behavior, an offshoot of a sophisticated nervous system, is extremely efficient in practice and allows an individual to avoid a certain number of environmental constraints.

All this might be the explanation for the reduction of genetic polymorphism as we climb the phylogenetic scale. At the same time the role of learning becomes more important—a relationship that is certainly not coincidental. It is possible that in the face of the forces of natural selection, behavioral adaptation replaces genetic preadaptation.

Adaptation and Acclimatization

The response of an individual or a group to the constraints of the environment can be located at three levels, corresponding to three evolutionary stages.

1. The genetic level, or the "programs" the subject receives with its hereditary patrimony. It will have more possible responses, and thus be more adaptable, if it is genetically more polymorphic. At this level individual "freedom" is virtually nonexistent. The subject obeys its genes, no more, no less. This is *adaptation* in the strictest sense, the irreversible result of natural selection.[26]

2. The physiological level. With the development of homeostatic mechanisms, such as thermoregulation, the range of possible reactions widens. The animal can modify its physiological processes to face new constraints. This is a contingency reaction and is not permanent. It will regress when the conditions that provoked it regress. It has been called *acclimatization.* For example, someone going up high mountains from the plains will develop more blood corpuscles with altitude, and someone entering a warm atmosphere will begin to sweat.

3. The cultural level—that is, the voluntary, conscious use of new behavioral patterns (involving, for example, food, habitat, or times and forms of activity) adapted to the circumstances. This type of adaptation, essentially modifiable, easily acquired and easily lost, leaves a large margin of freedom. It presupposes a sufficiently developed brain and reaches its highest expressions in human cultures.

These three processes are not mutually exclusive. They can exist in a more or less related way in the same animal. Adaptation is a balance. In the final analysis it is enough that the individual can survive and procreate in its environment. The nature of its responses to the con-

---

26. The fact that a subject may carry genes in its hereditary patrimony that are of no immediate adaptive value but that might become indispensable in the face of new selective pressures, constitutes genetic *preadaptation.* It explains the appearance of resistant strains in most bacterial colonies subjected to an antibiotic. See Boesiger and Ruffié (1971).

straints of selection are of little importance. For example, penguins and Eskimos do not use the same methods to combat the cold polar climate. Penguins live mainly in the Antarctic and are subject to conditions that are just as severe, if not more so, than those of the Eskimos of Greenland. They have used genetic combinations that ensure a minimum of heat loss by providing them with a rounded shape and skin with a fat layer designed to protect them from the cold, for instance. What is more, they can, when necessary, bring into play thermogenetic mechanisms. Even when the outside temperature is $-45°$ C, penguins can maintain an internal body temperature of around $40°$ C, which means a difference of $85°$ C or more between the outside and the inside. While hatching their eggs, which they carry between their feet, protecting them from the cold under their feathers, penguins huddle together in large groups, turning slowly in a precise manner such that each animal has its back to the blizzard in turn, thus protecting the rest of the group from the wind. The young, during their growth period, will also benefit from this same protection. All these attributes — morphological, physiological, and behavioral — are genetically determined. There is no hope of survival in the Sahara for a penguin from Adélie Land.

At the North Pole Eskimos live in just as severe a climate, but their adaptive behavior is very different. They protect themselves mainly with their fur clothing, their igloo, the use of fire, and food that provides a large number of calories. They may not entirely abandon organic acclimatization, but it is of secondary importance. It would be quite possible for them to live in the tropics if they learned to change their life-style.

Physiological acclimatization and cultural adaptation provide economical ways for an animal to acquire a wide polymorphism. Nevertheless, important differences in the values of the coefficients H, P, and A can be observed among related species. In particular, small groups that are highly inbred, such as certain cave-dwelling fish, insular rodents, or sea elephants, are weakly polymorphic. But even these populations always preserve a certain degree of polymorphism, below which they never descend. A too-rigid monomorphism seems incompatible with life. Many laboratory experiments prove this. A single strain of domestic crickets, reared in captivity for more than twenty-five years and always inbred, preserved an astonishing degree of polymorphism during all these endogamic generations that brought no addition of foreign genes (Grassé, 1978). Similar observations are possible in colonies of nondomesticated animals. Very ancient, widely distributed species, such as the horseshoe crab *(Limulus)* and club moss *(Lycopodium)*, have just as much genetic polymorphism as much more recent species. In general, the more a species is distributed over a wide and diversified habitat, the more varied it is morphologically and physiologically, and the greater is its polymorphism as revealed by electrophoresis. This is not surprising, for all hereditary features, even those most complex and furthest removed from the basic genetic information, depend on protein synthesis in the early stages of their expression.

Nevertheless, there are some exceptions to this rule. For example, the gecko, *Hemidactylus brooki,* is very widely distributed, but is feebly polymorphic (Pasteur *et al.,* 1978). We do not know why.

THE INFINITE VARIABILITY OF LIVING BEINGS

In the end, generally speaking, most living groups exhibit a "colossal multipolymorphism," to use the splendid expression of Georges Pasteur (1974). A simple calculation can give an idea of the extent of the phenomenon. In man, H = 6.7, which means that at least 6.7 percent of his loci have mutated. Thus he has 1 heterozygous locus in 15. If man has 100,000 loci in his genotype, about 6,700 of them are heterozygous. Such a man can theoretically produce gametes capable of unbelievably large numbers of combinations — in fact, $2^{6,700}$ or $10^{2,000}$ different types of germ cells! Even if we reduce the number of loci to 10,000 — probably a considerable underestimate — an individual can still produce 10^{200} types of gametes, which is much more than the number of atoms present in the universe as we know it, (about 10^{80}) (Ayala, 1978). We are now very far removed from the typological theory and the idea that Darwinian selection tends toward uniformity!

In fact it has now been demonstrated that even electrophoretic analysis leads to an underestimation.

1. Certain modifications in genetic macromolecules, the DNA chains, are not expressed phenotypically, for there exist a certain number of DNA sequences that are *synonyms* and control the synthesis of the same peptide. Any mutation that changes a gene for a synonymous one will have no effect on the phenotype.

2. Electrophoresis can differentiate only mutations that have provoked a change in the electrical charge of the molecules. It misses all the others — perhaps two-thirds or three-quarters of the total (Lucotte, 1978). In fact, the values of H and P may represent only a quarter of the mutations that really exist. A fairly direct proof has been given by a study of the globin chains of hemoglobin. Their structure is now well known, and the exact order of the constituent amino acids can be determined. Studies of a large number of cases demonstrate that less than one-third of the substitutions that can be detected in these chains have an electrophoretic effect. Thus different mutations can leave the molecular mobility unaltered, presenting the same "electromorphic" picture. Samuel Boyer (1972) and his collaborators, in their studies of primate hemoglobin, estimate that there are at least five times more mutations in the chains than electrophoresis can demonstrate.

George Johnson (1977) studied polymorphism in certain groups of butterflies and showed that 70 percent of mutations are undetectable by their electromorphs. These observations prove that, in spite of their great interest, electrophoretic techniques miss a large number of mutations. The "stupefying extent of biochemical polymorphism" (Lucotte, 1978) is much greater than even the figures astronomers are accustomed to use and largely exceeds what the human mind can imagine.

With the exception of identical twins derived from a single egg, no species that reproduces sexually can include two genetically identical individuals, has never included any, and never will include any. This is true for man just as for any other group. No human being is exactly the same as any other. This is why a teacher can recognize different pupils in his class and an officer different soldiers in his company. This is true throughout the world. Variations are found in all aspects of hereditary characteristics, including the immunological patrimony, the enzyme pool, fingerprints (which are one of the most precise morphological characteristics), the electroencephalogram, and sleep patterns. We all have our own very personal pool of genes. Biologically speaking, each of us represents a unique adventure that has no precedent and will never happen again. Every birth is an exclusivity and every death an irreparable loss.

As we have seen, polymorphism is one of the fundamental laws of living beings. Its consistency, its extent, the multitude of ways in which it is maintained (which we shall see in Chapter 4), all lead to the idea that it must be of considerable selective importance and even a biological necessity. It is responsible for our wide powers of adaptation and provides all groups with many possibilities for change. Genetic polymorphism has permitted evolution. Without it, the living world would have been condemned to immobility. Typological philosophy is incompatible with the best-established biological data.

COMPARISON OF POLYMORPHISM IN RELATED SPECIES

It is interesting to compare genetic polymorphism in phylogenetically related species and to investigate the extent to which polymorphic differences could be responsible for the sexual isolation of each group and thus the formation of species. Thanks to data obtained by electrophoresis, this problem has been taken up by a number of authors, for example Nevo and Cleve (1978) and Pasteur and Pasteur (1980).

Let us consider a series of loci common to a few related species. Several situations can occur.

1. Sometimes the same loci consistently carry the same gene, for example gene A. In this case all the species will have an absolute monomorphism for the given locus and will be indistinguishable. All specimens belonging to this group will have the characteristic A. This monomorphism could affect a large number of species, involving for instance a whole higher taxonomic level — a genus, an order, a subkingdom, even the whole animal kingdom. It is due to DNA sequences inherited from a common ancestor in the distant past that have not varied for a very long time. In particular, they have not undergone any mutations in the course of evolutionary diversification that split the phylum into a number of branches. The resulting proteins must be assumed to be molecules that are necessary for life and cannot be modified in any way. Their mutation would be lethal. Thus we can say that A is indispensable to a given cell and no mutation is possible.

2. The species have polymorphic loci but the incidence of alleles is about the same in all species; that is, the interspecific variations in frequency are no larger than the variations observed in different populations within a given species. Let us suppose that A has undergone a mutation a and that the two alleles are seen with the following frequencies: species S1: $A = 0.8$, $a = 0.2$; species S2: $A = 0.79$, $a = 0.21$. These differences are minimal and may be explained by variations due to sampling. In such a case only electrophoresis can decide to which species a given population should be attributed.

3. A third possibility is that the interspecific differences in the frequency of alleles are considerable. Each species can then be defined according to its characteristic frequency. The results of electrophoresis will allow a population selected randomly to be attributed to its species with a high degree of probability. For example, in S1, $A = 0.8$, $a = 0.2$; in S2, $A = 0.15$, $a = 0.85$.

4. In other, rarer cases, different alleles are found in each species and in each race. The presence of a marker allele allows us to attribute a subject to its species unequivocally. In this case, in S1, $A = 1$, $a = 0$; while in S2, $A = 0$, $a = 1$. Any individual carrying A belongs to S1, and any with a to S2.

THE DISTRIBUTION OF POLYMORPHISM WITHIN A SPECIES

We have already seen that all species are genetically polymorphic throughout the whole of their geographic distribution. Let us now consider the case of polytypic species made up of several geographically localized races. If we compare these different races by electrophoresis, we obtain precisely the same results as those above. Two races can be monomorphic for a given locus. When this locus has mutated the races can have the same frequency of each allele, but in other cases each race can be typified by a dominant distribution of one of the alleles. When this locus has mutated the races can have the same frequency of each allele, but in other cases each race can be typified by a dominant distribution of one of the alleles. No allele is exclusive for one particular geographic zone, but each has a preferential distribution in a zone where it will be present at a higher frequency. This is the commonest situation.

More rarely one race may be characterized by the presence of a gene, while the other race always has the allele. Such genes are then racial markers; their locus is a *diagnostic* locus. In this case, the race can be considered a subspecies.

In fact, electrophoresis demonstrates that there is no absolute genetic frontier between species and races. Two related species may differ by a very small number of genes related, perhaps, to reproductive behavior and sufficient to cause the sexual isolation of the two groups. Other factors will all have their own characteristic frequencies. Populations of *Drosophila pseudoobscura* in Bogota, Colombia, and others in North America are virtually incapable of interbreeding, for they represent emerging new species. Both possess the same enzymatic alleles, but their respective frequency is different.

GENETIC DISTANCE

Using the results of electrophoresis, we can define the *genetic distance* that separates different groups. The more loci that fall into categories 1 and 2 listed above, the more genetically close the groups will be. The more loci there are in categories 3 and especially 4, the more distant the groups will be. This reasoning is just as valid for comparing species and races as for higher taxonomic levels.

Some recent studies involving different groups of mice may be cited here (Chaline and Thaler, 1977; Britton and Thaler, 1978; Britton-Davidian *et al.*, 1978). The common mouse *(Mus musculus)* originated in the steppes of central Asia. In the wild state it is encountered in dry areas and eats grass and seeds. Its wide geographic distribution has provoked its splitting into a series of subspecies that are easily recognizable by their characteristic regional localization. But they have remained capable of interbreeding, at least in part. Mice have adapted very successfully to the human environment, so that wherever man is present, wild forms have given rise to commensal varieties. In some places some members of commensal subspecies have returned to the wild, giving secondary feral varieties. This commensalization has widened, their niche and accentuated the splitting of groups into a large number of secondary subspecies both geographic and ecoethological.

In western Europe there are two commensal subspecies, *Mus musculus domesticus,* found through most of the area from Holland to the Basque region, and *M.m. brevirostris,* in the Mediterranean area. Both are derived from the wild species *M.m. wagneri,* found in central Asia. In eastern Europe there is another commensal form, *M.m. musculus,* derived from a different wild species, *M.m. spicilegus.* (Figure 3.) Finally, around the Mediterranean one finds also *M.m. spretus,* a wild form living in the same areas as *M.m. brevirostris,* but which has never given birth to a commensal form. Certain human populations coming

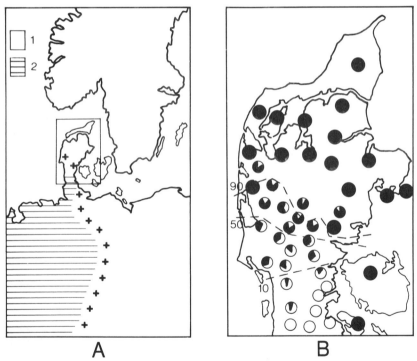

A   B

Figure 3. Zones of hybridization of *Mus musculus.* **A.** Distribution of two subspecies of *M. musculus* in northern Europe. 1 = *M.m. musculus,* 2 = *M.m. domesticus.* The zones of hybridization are indicated by crosses. **B.** Introgression of alleles of two gene loci in the hybridization of subspecies of *M. musculus* in Denmark. The black segments in the circles are proportional to the frequency of the alleles. Isofrequency lines for 10 percent, 50 percent, and 90 percent are indicated. (From Chaline and Thaler, 1977; isofrequencies from Hunt and Selander, 1973.)

from the east and establishing themselves around the western Mediterranean during prehistory must have brought with them their own commensal forms of *brevirostris*, which meant that there was no room for *spretus* to develop a commensal subspecies.[27]

Electrophoresis of these different groups gives the following results:

1. *M.m. domesticus* of northern Europe and *brevirostris* of the Mediterranean area are geographically distant but genetically closely related, for 83 percent of the loci studied were in categories 1 and 2; both are derived from the same wild species, *M.m. wagneri*.

2. On the other hand, *M.m. domesticus* and *musculus*, which share a long common frontier in Denmark and Germany with a small zone of hybridization, are genetically much more remote, only 68 percent of their loci belonging to categories 1 and 2. This is not surprising, for each is derived from a different wild species, *domesticus* from *wagneri* and *musculus* from *spicilegus*, as we saw earlier.

3. If we compare the two groups *M.m. brevirostris* (a commensal) and *M.m. spretus* (a wild form) living in harmony in the Mediterranean area, we find that their genetic distance is even greater. They only have 55 percent of their genes in categories 1 and 2.

This last point is worth stressing. Two related species or two subspecies living in close contact *(sympatry)* tend to increase their genetic distance. Any modification that tends to make them biologically more different is an advantage and favored by natural selection. We shall see later the precise significance of this phenomenon — the law of *reinforcement,* which seems quite general — as well as how we must now rethink the idea of competition (Chapter 13).

Other equally remarkable examples could be cited. The baboons, living in Africa south of the Sahara, were for long considered by taxonomists as forming five distinct species, each with its own geographic distribution, but with very limited zones of hybridization along the borders. They are *Papio papio, anubis, cynocephalus, ursinus,* and *hamadryas* (Figure 4). Electrophoretic studies made by Lucotte and Guillon (1979) and Lucotte and Lefebvre (1980) have shown that in reality there is a single "superspecies" in the process of being split. *Papio papio* and *anubis* are genetically the most closely related, while *hamadryas* seems the most remote from the others.

---

27. *M.m. spretus* now seems to be a real species. Its particular physiological resistance to dehydration allows it to inhabit localities forbidden to *brevirostris* and *musculus,* according to current work being carried out by Crozet and Thaler.

THE TRUE NATURE OF POLYMORPHISM

E. B. Ford (1945) defined polymorphism as the simultaneous presence in one place of two or more discontinuous forms within a single species, in such proportions that the rarest form cannot be maintained just by the pressure of mutations. This definition thus excludes polymorphism related to variations in time (seasonal forms) or space (geographic forms). On the other hand polytypism, which implies a particular geographic distribution for each form, is certainly compatible with Darwinian concepts, as we have seen. However the polymorphism found in all wild populations is of a different nature. It presupposes the presence at the same time and at the same place of several alleles for a given locus. It shows that genetic heterogeneity is possible within a group subjected to the same selective forces. This situation, constantly

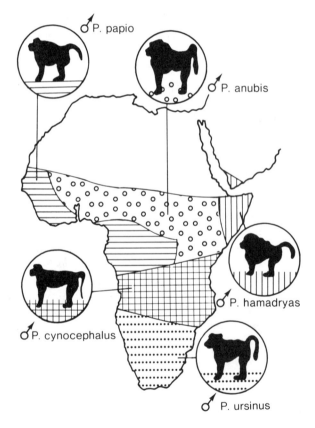

Figure 4. Approximate geographic distribution of the five species of baboon. (From Jolly and Dandelot.)

found in the wild and affecting all animal and plant species, cannot be explained by the traditional Darwinian arguments that postulate the elimination of the less favorable alleles by the more favorable. Thus, studies of natural population genetics produce data incompatible with the idea that selection tends toward uniformity. In Chapter 5 we shall see how we can now explain this paradox.

the species: biological or dimensional

THE SPECIES'S THREE LEVELS OF DEFINITION

A species can be defined on three levels.

The *genetic* level involves the hereditary patrimony, which confers a number of abilities and specific racial and individual characteristics. This patrimony is varied. Individuals within a species are capable of interbreeding but are not identical. They certainly cannot be standardized to the level of a prototype model. Rather, a species shows more or less wide variability around a central theme. This variability is visible in all domains — morphological, physiological, and psychological — and, as we have seen, it is analyzable at the molecular level, where the genes are expressed directly.

The *ecological* level is the niche, the functional unit that ensures that a species has the means to survive, to reproduce, and sometimes to transform itself. The nature and extent of the niche depend upon the possibilities imparted by the hereditary patrimony. The niche is a material reflection of this patrimony.

The *ethological* level unites the other two factors. It is represented by specific behavioral characteristics that allow a group with certain hereditary potentialities to become part of a particular niche, to exploit its resources, to grow, and to reproduce.

It is useful to analyze these three levels briefly.

1. For the geneticist a species is defined principally by a pool of more or less varied genes, capable of organizing themselves into coherent combinations. These combinations create the individuals that make up the species and are adapted to the environment that they exploit for their personal need. The more diversified the genetic pool, the wider are the capacities of the group.

These genetic combinations, the fruits of a long period of natural selection, are differentiated from the ones that characterize other species by a barrier of mutual sterility. Normally, there can be no exchange between two integrated genetic pools making up two different species. The species is a ghetto, a data bank that cannot be robbed, at least in theory. Nevertheless, we shall see in Chapter 4 that certain species can accept foreign genes by the process of hybridization — the phenomenon of *introgression*. In fact, introgression is always limited by mechanisms that slow the intermixing of two heterospecific patrimonies, such as reduced life expectancy or fertility of the hybrids.[1] Often the hybridization zone goes no further than a narrow strip following the border between the two groups. It is fortunate that this should be so. Without this ability for sexual isolation, the beautifully equilibrated ensembles created by species would be continuously threatened. No evolution would be possible, for *the species is the unit of evolution.* Evolution has jumped from species to species like a sailor from island to island.

In the absence of barriers maintaining the integrity of the species, the living world would be reduced to a mass of solitary bits of information coming together and separating again haphazardly, with no chance of building a permanent edifice. It would be an "antizoological" world, fostering "antievolution." A science-fiction writer might imagine a planet where this type of life has developed.

2. A species also implies an *ecological* definition. A species is always characterized by its ecological niche, that is to say by the position it occupies in the different food chains in which it is involved. We saw in Chapter 1 that its limits are fixed by its hereditary patrimony and by competition from other species. The different niches allowed an initial distinction to be made between sibling species making up two related, but not identical, adaptive complexes. The nature and geographic location of the habitat only represent one parameter of the niche: the spatial niche.

If we consider the distribution of several species, different situations may be encountered. Two species may live in different localities. We say that they are *allopatric.* Sometimes their habitats are adjacent. The species are then *parapatric,* and we are here most likely to encounter introgression. Two species that share the same geographic location are *sympatric,* but even then they occupy different niches. For example, one may be nocturnal and the other diurnal, or one may eat seeds and the other insects, so that each adapts to its own food chain. The barrier between the two niches is always sufficient to avoid any ambiguity and

---

1. Not to be confused with *luxuriance* or *crossbreed heterosis,* an opposite phenomenon that we shall discuss in detail later. The term *hybrid* should be used exclusively for subjects whose two parents belong to different *species,* such as the mule produced by crossing a donkey with a mare. Crossbreeds have parents of different *races,* but of the same species. Unfortunately the two terms are often confused in practice. See Chapter 4.

ensure the isolation of each group. We shall see the evolutionary signifi-cance of allopatric and sympatric species in Chapter 11, dealing with the development of species and transspecific evolution.

The concepts of allopatry, parapatry, and sympatry are only applica-ble to contemporary, or *synchronic,* species, for these are the only ones that have the physical possibility of meeting. Species that have lived at different times are called *allochronic.*

3. The third set of characters that define the species are *ethological.* A species is characterized by particular behavioral patterns. Whether they are innate or acquired, these patterns are adapted to its niche and allow it to exploit the niche. In consequence they contribute to its isolation from related species.

We have seen that behavior constitutes an ensemble of essential characteristics, because it is the primary way that living organisms react to their environment. Behavior is doubtless more important for natural selection than morphological features. An individual taps its resources by using suitable behavioral patterns.

We have already defined sibling species as species that are morpho-logically approximately identical, but barred from interbreeding. In addition, sibling species almost always differ in their niche and behav-ior, the two phenomena being, after all, related. Sibling species were defined by classifiers basing their argument on morphological features, but they do not exist for the ethologist or the geneticist, to whom they represent in fact quite clearly separated species.

THE GENETIC STRUCTURE OF THE SPECIES

Thus from the genetic point of view, a species is the result of a series of varying alleles. Far from having a single gene type at each locus, many have mutations that differ from one individual to another. What can we learn from a study of the spatial distribution of alleles?

Distribution of Genetic Polymorphism over Long Distances: Clines and Gradients

In Chapter 1 we examined the variation of certain species by geographic distribution, a concept that led to polytypic theories and the definition of races. Typological methodology would suggest that such variations are usually abrupt, with racial frontiers as clearly delimited as frontiers between species.

In reality, the modifications brought about by ecogeographical rules can affect all features, whether morphological, physiological, or behav-ioral. They are always progressive in nature. Their adaptive value is clear, at least in the majority of cases.

Frequently, a spatial ecological variation, or ecological *gradient,* is reflected by a gradual transformation of phenotypic features, or *morphological* gradient, itself related to a variation in frequency of certain alleles, or genetic *cline*—a term created by J. S. Huxley in 1939.

Some species show a color gradient that corresponds to the gradient of their background, each population tending to change so as to be least conspicuous to its predators. There are many examples, such as the observations of Cain and Sheppard (1950) and Lamotte (1966) of *Cepaea nemoralis,* the banded snail. Its shell varies from dark brown to a pale greenish-yellow, and it may have up to five or six dark bands. The populations making up this species are usually territorially very limited, perhaps to even one bush, and the shell pattern and color are adapted to give the maximum camouflage in the animal's local environment. Similarly, the great tit *(Parus major)* varies in color throughout its whole enormous Palearctic geographic distribution.

Other gradients have been described in relation to competitors or predators, certain parasitic, bacterial, or viral cycles in which the animals are involved, and the type and quantity of food that they have at their disposal. All these characteristics play a fundamental role in natural selection.

Nevertheless there are morphological clines in urodeles, for example, whose adaptive value, if there is one, is not clear to us. Such is the case of the Japanese salamander, *Cynops pyrrhogaster,* whose colored patches diminish regularly as one progresses from the north to the south of the Japanese archipelago—with, however, a new increase in coloration in the extreme south of Honshu and in the islands of Shikoku and Kyushu. This does not seem to be related to climatic factors.

The existence of these morphological clines implies a corresponding variation in the frequencies of certain genes, variations that have been demonstrated in some cases. The polytypic species *Ensatina eschscholtzi* is found throughout the whole length of California. Electrophoresis reveals a considerable degree of polymorphism in its serum albumin, the mobility of which decreases regularly in samples from the north to the south of the state (Gasser, 1977).

In fact, this is a very generalized phenomenon. The study of the distribution of frequency of certain alleles over long distances often reveals a progressive and continuous evolution. One can represent it schematically as follows: let us take two regions P1 and Pn at the two extremes of an area over which a species is distributed, separated by a series of intermediate zones P2, P3, P4, etc. If we study the evolution of the frequency of two alleles A and a in the populations from P1 to Pn we might find in P1, $A = 0.9$, $a = 0.1$, and in Pn, $A = 0.1$, $a = 0.9$. For the intermediate areas, in P2, $A = 0.8$, $a = 0.2$; in P3, $A = 0.7$, $a = 0.3$, etc.

This distribution of alleles in clines can be seen in all animal or plant species. It can be found, for instance, in the geographic distribution of blood factors in man. The fact that frequently these genetic clines correspond quite closely to ecological gradients betrays their adaptive value and bears witness to the progressive advance of the equilibrium between genes and the environment.

If we come back to the example we have just described, in P1 the gene *A* that had a high frequency, 0.9, must represent a favorable feature giving a certain advantage to the phenotypes that carry it. Its allele *a* has a low frequency, 0.1. In P*n* the opposite is true. The ecological conditions have changed, and now *a* is advantageous. The ratios between the two are reversed. The exceptional individual in P1 is mediocre in P*n*, and *vice versa*. Populations in intermediate regions have intermediate frequencies, related to the progressive modification of environmental conditions.

This shift of the equilibrium between genes and environment is a factor every time we move into a new ecological zone—that is, each time the selective pressure changes notably. It seems abrupt if we compare distant populations and then we speak of polytypism, but it is seen to be more continuous when we study more closely related neighboring groups.

Today we must interpret the ecogeographical rules described above in the light of these data. In fact, polytypism, like the description of races thought to be truly independent, is often due to the fact that the classifier is not aware of the intermediate stages. As an example, if we compare a group of people from Brittany and another group from the Congo, the incidence of various blood factors will be different in the two populations. This finding suggests a polytypic model. But for this to be really the case, there must be no other populations between Brittany and the Congo. However, the whole region between the two has been inhabited by man for a very long time. If we examine the gene frequencies of these populations over such large distances, we see that there is a regular and continuous variation, such that it is impossible to be sure, using the excellent genetic criteria provided by blood groups, where the white populations stop and the black ones begin. In practice, clinal variations can involve all features, morphological, physiological, and ecological.

Variation and Intersterility

When a species is distributed over a very wide geographic area, groups at the two extremes of the cline may become so different that they are intersterile, although retaining mutual fertility with their immediate neighbors. They may thus be regarded as newly emerging species, incapable of direct interbreeding but capable of exchanging their genes through neighboring populations.

One of the most remarkable examples is that of the gulls of the genus *Larus,* which are distributed throughout a circular area around the North Pole, in both the New World and the Old (Figure 5). The primitive species originated in central Asia and seems to have invaded the coast of northern Asia and then northern Europe at the end of the Tertiary or the beginning of the Quaternary, progressively forming a series of geographic races, all belonging to the species *Larus fuscus,* the lesser black-backed gull. Later they spread to America via Alaska and occupied a large part of the west coast of Canada. In this new biotope their divergence was amplified, and finally a new species was born, *Larus argentatus,* the herring gull, which for a while remained in America. In a later phase the herring gull crossed the Atlantic and arrived in Europe — that is, Great Britain, northern France, and Scandinavia — where it now lives in sympatry with the black-backed

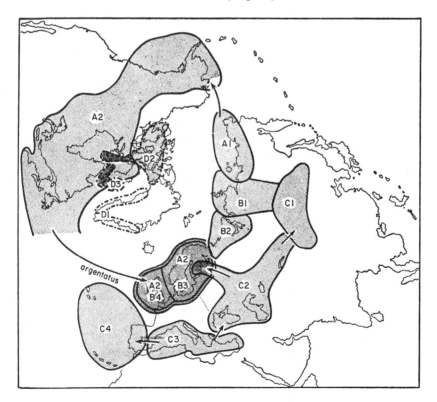

Figure 5. Circular distribution of gulls, whose origin was probably central Asia. Although cohabiting, the forms A2 *(Larus argentatus)* and B3-B4 *(L. fuscus),* representing the two extremities of the chain, cannot interbreed. (From Mayr, 1963.)

gull, but without interbreeding. Nevertheless genetic exchange is possible by the intermediary of the links in the circumpolar chain, which reflects the prehistoric migration of the group. In the light of this migration, a zoologist studying the gulls from Asia and Europe to America might consider *L. fuscus* and *L. argentatus* as two geographic races; but the ornithologist who limits his study to animals found in the north of France would define them as two distinct species.

Another example is that of the great tit, of which there are two species, *Parus major* in the west and *Parus minor* in the east. They represent the two extremes of the same cline and are linked across their very wide geographic distribution by a series of intermediate races and subspecies. However, the two typical species meet in the region of the Amur River in eastern Asia, where they live together but never interbreed (Grassé, 1978).

In addition to these spatial variations there are often temporal variations that may be seasonal in nature. One type of animal may predominate in the warm season and another when it is colder. This is true for the ladybird *Adalia bipunctata,* of which the red variety predominates in winter and the black in summer (Timofeeff-Ressovsky, 1940). We shall discuss the evolutionary significance of these temporal variations in Chapter 6. Other variations occupy a much wider time span. They have been called *secular,* but it may be better to describe them as *millennial.* They can only be detected by the paleontologist and are probably due to climatic variation. Their origin is adaptive and is fundamentally no different from that of spatial variations.

The Distribution of Genetic Polymorphism over Small Areas: Populations

If instead of considering what happens over large distances, we study the distribution of genes over smaller areas, we find local variations that do not have the same continuous nature. Frequently they correspond to natural geographic units (such as, for insects, a clearing or a valley, or for snails, even a bush), within which individuals tend to interbreed preferentially. In the human context they are often related to cultural divisions, such as tribes or ethnic groups. For man, the obstacles presented by culture, language, and religion represent a much more formidable barrier than those imposed by geography.

Such reproductive groups are referred to as *populations.* In practice a species never forms a biological continuum, as was believed for a long time, but a series of populations. *The population is the true basic unit of the living world.* This concept, formulated by the geneticist J. B. S. Haldane in 1932, met with hostility from naturalists clinging to typological definitions and was not accepted for half a century.

CONDITIONS FOR THE EXISTENCE OF
A POPULATIONAL STRUCTURE

The existence of populations implies the action of two opposite forces between which a balance must be established: first, a certain degree of sexual isolation, without which the population would merge with the rest of the species and lose its biological individuality; and second, exchanges with neighboring populations, without which the populations might diverge sufficiently to become incapable of interbreeding, leading to the breaking up of the species into several new species.

Let us examine these two elements.

1. Partial sexual isolation. If this partial isolation did not exist, the species would simply form a single population within which individuals would have equal chances of interbreeding. Gene flow would be unlimited and widespread within the species's area of distribution, and one would find the same genetic frequencies in specimens from all over the area. The species would be in a state of *panmixia* and genetically homogeneous. Certainly there would be some polymorphism, for there would be different alleles for various loci, but the degree of polymorphism would be the same over the whole area of distribution; that is, one would find similar frequencies for all genes studied. In practice, this state of affairs is never attained, and panmixia is restricted to groups that are geographically very limited. As soon as a species occupies a wider and more varied area, more and more obstacles to free interbreeding are encountered. The probability of a given individual encountering another particular individual varies with distance. Two animals of different sexes living side by side have in general more chance of mating than two others living far apart. This probability also varies with geographic features such as highlands, watercourses, and other natural obstacles that are more or less difficult to cross. In higher species the probability can vary with behavior. Individuals are more likely to meet when their maximum activity occurs at the same times, in the same season, and in the same places. For men, as we have seen, the probability of meeting also depends on cultural factors.

As an example take a boy and a girl who would be attracted to each other and make love if they met; this would be improbable if the boy lived in Brittany and the girl in Rio de Janeiro, but the chances would be greater if they both worked together in the same laboratory. The probability would be even greater if, four times a day, they took the same elevator to leave their office on the seventy-eighth floor of the Empire State Building, and if every Saturday from nine to eleven they took a swim in the same pool. The result would be almost inevitable if they took a holiday together!

Thus the general ability to interbreed that is characteristic of individuals of the same species is more a theoretical possibility than a real situation.

2. But in order to maintain the structure of a population, there must be a minimum of gene exchange between the populations making up the species — an *interpopulational gene flow*. Otherwise, the more isolated populations would diverge to the point of intersterility. The biological unity of the species would be disturbed, and it would split into daughter species. We shall see in Chapter 3 that the isolation of peripheral populations and their progressive divergence is probably the most common pathway to speciation.

THE "DIMENSIONAL" SPECIES AND ITS LIMITS

All we have said so far shows to what extent the concept of a species must be modified today. Until recently, because of the dominance of the typological way of thinking, many zoologists subdivided a species into subspecies, races, or biotypes that were purely descriptive categories, just like those defined at other levels of classification such as genera, families, or orders. They considered species as static entities in equilibrium with their environment, well protected in their sexual isolation, and virtually incapable of evolution. This represented the *nondimensional* species. But such a rigid structure cannot explain the process of evolution. The recent introduction of biochemical genetics into zoology brings with it the concept of the *dimensional* species made up of populations — that is, reproductive units that are variable in space and in time. (We prefer the term "dimensional" species to "biological" species, which has been adopted by some authors, for the latter term is a pleonasm; there cannot be "nonbiological" species any more than there can be "nonbiological" agriculture!) This dimensional species has both wide polymorphism and a dynamism of its own — a dynamism made possible by polymorphism, for a monomorphic group has no possibility of evolving. These ideas will be developed in the next chapter.

Faced with the reality and the extent of the variability of individuals making up natural populations, we can adopt two attitudes. The first, which was for long that of the typologists, amounts to a denial that fluctuating characteristics have any value in the definition of the species; the classifier accepts only the small number of fixed, or apparently fixed, characteristics present in all members of the group. As we saw in Chapter 1, such an approach is very arbitrary. We now know that many characteristics that vary from one individual to another are under just as strict a hereditary control as invariable ones. There is no reason for not considering them in the definition of species in the same way that we consider nonfluctuating features. If not, we make artificial species, which is not difficult to do. We simply take into account not *all* the characteristics that exist in reality but only those that fit the definition we have chosen to adhere to, the one that corresponds to the holotype. Such a purely subjective approach has no scientific value. Geneticists try to define reality objectively by considering *all* information, whether variable from one individual to another or not. In practice the definition of a holotype can be of some use to the naturalist in the field as a guideline, but we must not give it the importance that it had previously, and make of it the ideal model of all the attributes of a species or a race. Some zoologists who are also population geneticists, such as Simpson (1961) and Mayr (1969), have proposed certain other terms, such as hypodigma and onomatophore, to emphasize the limits of definitions. In reality the groups studied by the systematist are not permanently fixed in an unchangable genetic context. They are a series of *taxons* — that is, lines that have not all attained the same stage of evolution. Some that have reached the level of a true species live in total sexual isolation and cannot evolve any more at the moment. But others are still more or less capable of interbreeding with related groups. Depending on the degree of isolation, we can speak of subspecies, semispecies, or sibling species, all of which can be grouped into a larger ensemble, the *superspecies*. Until recently, the error of systematics was to put all known groups on precisely the same footing, whereas in fact they are *not* all equally significant, for they are not all at the same evolutionary level. The species is simply the end product of a dynamic process that many groups have not yet attained and that some will perhaps never attain, for they will have disappeared long before achieving complete sexual isolation. If we take into account the speed of extinction of certain lines — due in particular to the activity of man — it seems that many groups that are splitting up at the present time will never reach the final stage of speciation. At the present, to try to fit all wild groups into the narrow typological framework of a species, subspecies, or race is a fanciful enterprise of no interest for the geneticist. But whatever degree of sexual isolation the groups have achieved, they are all made up of populations. The populational structure is the only one that approaches real objectivity.

populations

THE BIOLOGICAL ORIGINS OF POPULATIONS

A species, then, is not a homogeneous ensemble but is composed of more or less independent basic elements, the populations. Field research has shown that a species is nearly *always* made up of population units. We have already defined the population as the reproductive unit —that is, the group of individuals that have more chance of breeding together than with other individuals. The members of the same population are thus linked by a greater probability of interfertility. As J. Gomila wrote in 1976, the population represents "the minimum unit in which one can consider panmixia to prevail and exist," a panmictic unit being one in which all individuals have a more or less equal chance of interbreeding.

For Ernst Mayr (1963) the population represents "the community of potentially interbreeding individuals at a given locality. All members of a local population share in a single gene pool."

Nevertheless, classifiers retain infraspecific taxonomic levels (subspecies, biotypes, ecotypes, etc., for they are useful in systematic studies despite their essentially arbitrary nature. But in the final analysis, all are made up of populations. As we saw in the previous chapter, the existence of populations depends on two opposite but complementary conditions, the first being a certain degree of sexual isolation that allows all populations to become independent, and the second, sufficient genetic exchange to prevent them becoming autonomous species. We can now analyze these two factors in detail.

The Sexual Isolation of Populations

A certain degree of sexual isolation is a basic condition for the formation of populations within a group. It guarantees that each gene pool remains closed so that it can evolve in its own way, differently from neighboring pools. Any form of exchange on a massive level would tend toward uniformity. This rule, which we have already mentioned, is so obvious that we need not emphasize it. Isolation is nearly always due to geographic factors, such as distance and natural obstacles. Isolating factors may vary with time, as when glaciers appear or disappear following climatic modifications. Eras of great ecological change are also those when species are split and new populations formed. Many zoologists think, like Ernst Mayr, that geographic barriers are essential for the breaking up of species into populations that could subsequently become new species (see Chapter 8).

The emergence of autonomous species from a common ancestral line implies that gene pools belonging to isolated populations evolve in different directions. This divergence is due to two factors: *chance* and *selective pressure.*

First of all, *chance.* In small, fairly closed populations, chance can play a capital role in the evolution of the original genetic patrimony. It is sufficient that certain genes are lost or become rare quite randomly, such as through fatal accidents to their bearers, while their alleles are widely diffused. The most extreme effect of chance can be seen in the *founder effect,* whose principle is as follows. Take a population that can have a gene A or its allele a at a given locus. Let us suppose that A has a much higher frequency than a; for example, $A = 0.95$, $a = 0.05$. If a fertilized female bearing only homozygous eggs a / a leaves the group and founds another population, the initial gene frequency in the daughter population will be $A = 0$, $a = 1$. For this particular locus there will have been a complete reversal of frequency in a single generation. It is possible for the gene A to be reintroduced later from neighboring populations, but unless it has a considerable selective advantage the chances are that it will only reappear at a low frequency or even be completely lost again.

This founder effect is an extreme, almost caricatural example of what is called *genetic drift.* This drift is most obvious in small groups, reaching a maximum when the founder is reduced to a single individual who obviously cannot transmit a complete gene sample representative of the whole population. This problem will be discussed in Chapter 6, when we study the role of chance in the preservation of polymorphism.

Selection is another cause of divergence of populations belonging to the same species. As we saw in Chapter 2, when a species is present throughout a large and varied geographic area, environmental conditions, and therefore selective pressures, are not necessarily the same throughout the whole area.

As each group is subjected to different constraints, each will tend to "choose" the genetic baggage that is likely to allow it to face local conditions in the most efficient way. Thus varied environments give rise to varied populations. Over large areas there may be as many populations as there are sexual enclaves and ecological variations. To preserve the population's structure it is sufficient that gene exchange between populations should not exceed a certain threshold, beyond which the tendency would be toward homogeneity. Once homogeneity was reestablished, the whole species would become a single population.

How far can selective pressure push the diversification of the population? There must be limits. Let us take a simple feature, such as the size of the animal, and suppose that in a series of neighboring populations it increases regularly according to a particular gradient. Such a change is usually linked to some ecological modification that is also progressive. The increase in size constitutes an adaptive response but cannot continue indefinitely. It cannot exceed certain limits compatible with the anatomical and physiological organization of the individual. When these limits are reached, the species has reached the limits of the area in which it can live. They represent a maximum for the adaptive variations that the organism can exhibit in response to ecological constraints.

On the continental scale, the limits for the distribution of a species are almost always its limits of adaptation. If these limits did not exist the species would tend to invade everywhere. In a world in which ecological conditions were uniform, living beings would be distributed equally in all areas.[1] A situation approaching this has been achieved by man, who has artificially removed ecological barriers to his expansion and established everywhere the microclimate that he needs. By organizing food chains that are useful to him, man has enlarged his niche on a worldwide scale. He has colonized almost the whole of the landmasses, carring with him a host of domestic and commensal animals such as the rats, mice, flies, and sparrows that he introduced to the New World, New Zealand, Australia, and South Africa, all of which have taken advantage of the constancy, richness, and protection of the human environment.

Gene Exchanges Between Populations

Gene exchanges between populations guarantee the unity of a species and are very variable. They are poor in closed populations that become distributed according to the "archipelago" model: in oceanic conditions, such as on islands, or for ecological reasons, such as in lakes and clearings. Exchanges can be much richer in continental populations not separated by major obstacles. In natural conditions a population is

1. This state of affairs is rather theoretical, for even in an ecologically uniform world some degree of geographic isolation would persist, related to the limited powers of locomotion of various species. It is unlikely that all members of a given group could explore the whole landmass during their lifetimes, and there would never be absolute panmixia.

never absolutely isolated. All are more or less open, or at least ajar! Totally closed populations exist only in breeding colonies. Wild populations are never subjected to more than relative isolation. They are linked to neighboring populations by a series of gene flows. These progressive flows form a sort of "escalator" for interpopulational genes, which are thereby transmitted throughout the species and maintain its cohesion and biological homogeneity. The flows differ in their volume — they are small for almost closed populations, massive for open populations — and in their direction — they either permit more or less equal exchanges with all peripheral populations or have a privileged direction, along a natural route of entry, for example. When interpopulational gene flows are intensive and multiple (go in all directions), the limits of the populations tend to disappear and the species seems remarkably homogeneous. When the opposite is the case, the populations are better demarcated and may even slowly diverge.

Some populationists tend to study living organisms exclusively through populations, which they consider as closed units. If this concept corresponded to reality, all species would have split into autonomous *microspecies* long ago, with each corresponding to a well-isolated local population. In fact, although a population can be defined by its structure and extent, it is also characterized by its permanent exchanges with neighboring populations. These exchanges ensure the cohesion of the species and, as we shall see later, are an essential element of its dynamic nature.

VARIATIONS IN POLYMORPHISM IN RELATION TO POPULATION LOCATION

The extent of interpopulational gene flows depends above all on the location of a population within the species's area of distribution. In general, populations near the center of the area are relatively open and are subject to numerous massive flows from all sides. In addition, the center is the place where environmental conditions are most favorable, where the species is best adapted. Selective pressure is weak or nonexistent. In this "ideal" living area many variants are tolerated. Polymorphism among groups is at its maximum, but the differences between neighboring groups are always discrete.[2]

Peripheral populations, on the other hand, are more or less closed.

---

2. This rule applies not only to populations but to different species. Indeed, by far the largest number of species is found in the tropics, where environmental conditions such as the absence of extreme variations in temperature and humidity favor the development of life. In the taiga near the Arctic Circle, by contrast, the climate is much harsher and demands a far greater degree of adaptation, so there are few species. The same is true in hot, dry deserts like the Sahara.

They receive weak gene flows in one direction only and may receive almost none at all. They are subjected to more severe selective pressure, for they are in zones that require extreme degrees of adaptation and they cannot cross the "ecological frontier," beyond which conditions become intolerable. Being obliged to adapt their genomes to extreme situations, peripheral populations are much less monomorphic than central populations. But this monomorphism is not the same in all such populations.

Let us take a species occupying a large geographic area. Its widely separated frontiers are situated at the four points of the compass, but it is also restricted by considerations of altitude. The factors limiting its expansion are not necessarily the same in all sectors. They might include, for example, too warm a climate in the south, the absence of a certain type of food in the west, the presence of competitors in the east, and cold in the north. Thus peripheral groups will be subjected to different selective pressures and powerful diversifying forces, with the extreme peripheral populations tending to be most specialized in terms of adaptation to local limiting factors. This is the price of their survival, and they will finally diverge considerably from each other.

Population Size and Stability

The potential for change in a population depends on its size. When it is small, foreign influences can rapidly cause a profound modification of the genetic structure. We have already seen that such a population is very susceptible to drift. In such a case, phenomena of a random and apparently quite secondary nature, such as the accidental loss of a few individuals of reproductive age, may have serious consequences. On the other hand, large populations have a relatively great biological inertia; they are remarkably stable, with respect to both external influences and chance fluctuations. In an extreme case, a species occupying an enormous area and made up of a large number of individuals forming a single panmictic population would be extremely stable. The human race, composed of an infinite number of populations traveling and interbreeding at will, enjoys at one and the same time wide genetic polymorphism and great stability.

THE DESTINY OF PERIPHERAL POPULATIONS

Peripheral populations are more or less closed and are subjected to powerful selective constraints. They have three possible destinies:

1. Some remain where they are, for the adaptive possibilities offered by their genetic pool do not permit them to go beyond their geographic area. They cannot face the conditions beyond their frontiers. If they invade this inhospitable territory they disappear, perhaps because they can no longer reproduce. This is their commonest fate and explains the relative stability of the zones of distribution of many species. Nevertheless, temporary excursions beyond these ecological barriers can happen. They are most often seen during periods when the group is able to exploit a particular region temporarily, perhaps during a favorable season, only to retreat later. These backward and forward movements may take the form of *migrations,* especially in birds; they increase a population's resources and diminish its competition.

2. But it may happen that other marginal populations have the potentiality in their patrimony to implant themselves successfully beyond their "traditional" frontiers. In this case, under the influence of divergent selective forces, their genetic pool tends to become specialized as a response to their new constraints. If these populations that have successfully penetrated into new territory acquire a certain sexual isolation, they diverge genetically more and more from the mother species and become first of all subspecies, with relative intersterility, and then true species, with total intersterility. This seems to be the commonest way that species are formed. It involves not a single mutant, but the whole group making up the population.

These "pioneer" movements explain the *adaptive radiations* seen periodically in all lines, during which a mother species splits into daughter species specialized for different environments or ways of life. In fact the peripheral populations of a species tend to change not only their area of distribution but also the limits of their ecological niche — which are, as we have seen, just as much functional frontiers as geographic ones. For instance, insect-eaters may become grain-eaters, and nocturnal or dusk activity may replace diurnal activity.

3. As a last eventuality, frontier populations that were beginning to become autonomous may "retreat" and become reabsorbed into their original species, such as for example when a massive climatic change, like the end of a period of glaciation, modifies the selective constraints so that barriers that had imposed a certain isolation are removed. Then if genetic exchange is resumed sufficiently intensively, the homogeneity of the group will be reestablished. This reinsertion into the "melting pot" is only possible if the preceding separation and divergence between groups was not so great that interbreeding has become permanently impossible.

We shall discuss the mechanisms of speciation in more detail in Chapter 8. In practice one encounters all degrees of differentiation within populations belonging to the same species. Populations living in the center of an area of distribution and subject to similar selective pressures will be almost identical, indulging in considerable genetic exchange, whereas those on the periphery, subjected to strong diversifying forces and having few sexual contacts, will be more differentiated.

THE TRUE NATURE OF THE POPULATION

We are now in a position to better understand the precise nature of the population, the basic unit of the living world.

1. The population is always a transient structure. Its characteristics and its limits are fixed by ecological factors that are never constant over long periods or large distances. Populations are formed and dismantled according to circumstances. Thanks to their genetic polymorphism, which allows them to adjust their gene frequency at all times in order to respond most effectively to any new environmental pressure, they have a remarkable adaptive ability. They are in constant flux, always seeking to adapt themselves to changing ambient conditions. They constantly tend to outgrow their niche to exploit new resources. This pioneering spirit is one of the characteristics of living organisms and is reflected in the dynamism that is an integral part of population structures.

If, as François Jacob (1977) wrote, "the dream of any cell is to give birth to another cell," the dream of any population is to give birth to new populations.

2. Variations between populations are progressive and continuous. In nature one almost never sees the abrupt breaks that were for a long time the *credo* of the mutationists and correspond to the "catastrophist" theory we discussed earlier. Many zoologists, struck by the discrete and continuous variation of nature, rejected the selective scheme of evolution and, like P. P. Grassé (1973), moved toward a neo-Lamarckism that to them was the only way of explaining the progressive phenomena, something the idea of a single mutation could not do. Others, like Motoo Kimura, adopted stochastic or so-called *neutral* models, in which selection is relegated to a secondary level. But if we consider that, as we said in Chapter 1, most of the features under study are controlled by a large number of genes (*polymeric* or *polygenic* features) and that in a given population all permutations are possible between the different alleles, one can clearly see the origin of variation and the role it can play in gradual selection. We also saw in Chapter 1 the importance accorded by the mutationists to the single massive mutation to explain the evolutionary process, influenced as they were by the work of the drosophilists. This view is erroneous. Evolution is not saltatory; it is achieved not by "hopeful monsters" but by successive small steps affecting discrete features throughout a whole population. Darwin and his contemporary Wallace were aware of this phenomenon and described it. Mutationism, sometimes presented as a sort of "hyper-Darwinism," is in fact only a caricature of what the famous navigator of the *Beagle* suspected.

In nature — apart from pathological cases that are irrelevant here — variation is never brutal, and selection operates discretely and quantitatively rather than crudely and qualitatively. Any crude changes that may appear are almost always eliminated irrevocably. Their impact on the future of the group is negligible. *The concept of the population has reconciled genetics and evolution.*

THE POPULATION AS A BIOLOGICAL CONCEPT

Populations, with their internal variability and their exchanges, thus represent the fundamental units of the living world. Unfortunately the typological point of view, the exact opposite of this concept, still has a considerable influence on Western science. Most morphologists, physiologists, embryologists, and even psychologists and clinicians reason from a typological point of view.[3] Yet their disciplines concern polymorphic populations, and the rigorous phenomena that such sciences are called upon to study can only be described in practice in terms of variation and probability. In spite of the foregoing data and the numerous discoveries in the domains of immunology and genetics, many biologists and clinicians persist in thinking of the nonexistent ideal prototype. At the present time only population genetics allows an objective analysis of living groups.

3. Clinicians strongly resist the idea of polymorphism. Diseases are still described like species according to a number of constant, indispensable diagnostic signs. Diseases are divided into clinical "forms," rather as species are divided into races, always with reference to a theoretical "holotype" that possesses all the signs. In fact, every patient is his own unique clinical "form."

the polymorphism of genes

THE SOURCES OF THE POLYMORPHISM OF POPULATIONS

The populations composing a given species are always characterized by an extensive genetic polymorphism that selection can "sink its teeth into" in order to adapt the group constantly to environmental constraints. The origin of this polymorphism must be sought at two levels, corresponding to two complementary mechanisms:

1. At the level of *the gene itself,* since all populations have several possible alleles for many loci. If there were no process of renewal, some mutations would be lost, either by chance or by selection, and the genetic patrimony would gradually be impoverished. In fact, a population is constantly receiving new genes through either endogenous mechanisms, such as mutations, or exogenous ones, such as interpopulational gene flows from neighboring populations or even, more rarely, from neighboring species in the form of interspecific gene flow *(introgression).*

2. At the level of *genetic combinations* in each individual's hereditary patrimony, its genotype. Selection does not act on the gene itself but on the features for which it is responsible. Only tangible, expressed features, constituting the phenotype, can be "judged" by the process of selection. In Chapter 1 we mentioned that most features — morphological, physiological, and probably behavioral — depend not on a single gene but on a series of genes working in synergy and following a certain order. This situation is certainly the most frequent one and allows selection to act on a combination of particular genes or genotypes rather than on isolated genes.

As we shall see in Chapter 5, sexual reproduction breaks up old combinations at each generation and creates new ones. This process ensures a permanent renewal of genotypes and thus the presence of extremely varied phenotypes on which selection can act. Sexuality is a source of infinite innovation.

We may now examine these two mechanisms separately. This chapter will deal with the origin of the polymorphism of the genes themselves and the following chapter with the origin of the polymorphism of combinations of genes.

VARIABILITY THROUGH THE ADDITION OF NEW GENES

All populations, whatever their size and location, are in permanent receipt of new alleles that maintain the variability of their patrimony by replacing those that have been lost or eliminated. As we have already seen, these genes come either from inside (mutations) or from outside (intraspecific or interspecific gene flow). Let us consider these three possibilities.

Endogenous Additions: Mutations

A gene can undergo structural modification as a result of accidents during cellular division at the moment when the chromosome is being reproduced, or *replicated,* in order to produce two daughter chromosomes. If such modifications are compatible with life, these copying errors are often accompanied by a change in genetic information, leading to a variation of the subject's phenotype. If this variation is an advantage, it may be preserved and even spread through the population; but it can just as easily be lost. If it represents a disadvantage or is neutral it will usually be lost, although the chances of its propagation never fall to zero. The probability that a new mutation will be retained depends on its selective value and also on the size of the population in which it appears. The probability is greater in a small population. In Chapter 7 we shall look at the biochemical structure of genes and the molecular basis of mutation.

The frequency of mutations varies with the genes. Certain ones seem to mutate more easily than others. There are thus remarkably stable loci and others that are much less stable. This explains the variability of the figures quoted by geneticists — between 1 in 50,000 and 1 in 200,000 or more per gene, per individual, per generation (Mayr, 1963). According to Kimura (1979), they could be even higher. If one takes into account the number of genes in a given subject, a fairly large population would produce a considerable number of mutants in each generation.

In Chapter 1 we saw that the fruit fly geneticists, usually working in the laboratory with selected individuals, for a long time considered mutations as rare events leading to important modifications. In fact this type of mutation is exceptional and usually produces pathological monsters that are rapidly eliminated. This phenomenon was described by H. J. Muller (1950). We shall consider it again when we study the "neutral theory" in Chapter 6. The main usefulness of such large anomalies is to provide the experimenter with suitable material that he can easily identify and trace. But they are without importance for the development of species.

In reality, if we consider the pool of genes present in the genotype of all individuals of the same population, the appearance of new mutations with little or no immediate consequence for the phenotype is a banal and unremarkable event. But it represents an endogenous source of new genetic information and thus plays a fundamental role in the maintenance or widening of polymorphism.

Exogenous Intraspecific Additions: The Interpopulational Gene Flow

Except in the case of rigidly closed populations, such as those found in laboratory colonies but only exceptionally in nature, interpopulational gene flows are a relatively constant phenomenon. In Chapter 3 we said that they were necessary to maintain the homogeneity and the coherence of a species. As soon as they disappear or fall below a certain threshold, there is a great risk of the species being fragmented into races or subspecies that, if they later become incapable of interbreeding, will give rise to new species. The gene flow ensures a permanent addition of varied genes because the distribution of alleles is not exactly the same in all populations of a given species. We have seen that this difference is due to variations in selective pressure on different widely distributed and ecologically heterogeneous groups. We shall reconsider this problem in the chapter concerned with the mechanisms of speciation. Any gene *A* found frequently in a given area because it is favorable there may be unfavorable elsewhere and be replaced by its allele *a*. But *A will never disappear completely* if it is regularly reintroduced by the gene flow from neighboring populations.

We should emphasize here that the distinction often made between wild genes and mutant genes is arbitrary. Influenced by typological concepts, workers tended for a long time to distinguish in a series of alleles a widespread wild gene possessing the best selective value, as well as one or several rarer mutants of less selective value. These labels are very relative. A gene considered commonplace in one population will be exceptional in another, and *vice versa.*

Many interpopulational gene flows responsible for the reintroduction of genes of little or no selective value have been described in natural populations. As an example, one can quote the classic case of the water snake *Natrix sipedon,* closely studied by Camin and Ehrlich (1958). It can either bear a number of colored stripes or be all one color. In Ontario the striped form is commonest, merging easily with its background and thus providing an effective camouflage against predatory birds. On the contrary, on the islands of Lake Erie one encounters mainly the unstriped form, almost invisible on the uniformly colored rocks. Nevertheless, in these same rocky areas one can find a few striped examples. This is due to the fact that the striped snakes sometimes migrate from the mainland to the islands, where they are responsible for the distribution of the genes of this particular morph, although these may be unfavorable and thus are preferentially eliminated in the rocky biotope.

The size of interpopulational gene flows depends on the size and nature of the area of distribution of a species, the presence of natural obstacles, and other factors. But it will also depend on the ability of the species to migrate. Migration can be *active,* like voluntary journeys over long distances accomplished by flying, running, or swimming. It can also be *passive,* and just as efficient as active migration.

Small size, low weight, a protective coat, and more or less prolonged dormant periods are all factors facilitating dispersal (Simpson, 1952). The most migratory species may be present on a worldwide scale. For obvious reasons — such as the greater stability of their environment — this situation is more often encountered in aquatic, and particularly marine, animals than in terrestrial varieties. Many marine invertebrates pass through a pelagic larval stage that favors their dispersion, particularly when this stage is prolonged.

The reef-building corals *Galaxia aspersa* have larvae *(planulae)* that can float for two months. In an ocean current moving at 3 kilometers per hour, they can travel 4,500 kilometers before becoming fixed. There is therefore nothing surprising in the fact that the populations making up this species are very homogeneous; it is found in almost identical form in such widely separated areas as Ascension Island, Saint Helena, and Hawaii (Mayr, 1963).

There are very widespread spiders that, in spite of the diversity of the biotopes they inhabit, show no tendency to differentiate into geographic races. The maintenance of this uniformity throughout their whole area implies the existence of powerful gene flows, something that might appear surprising, since at least in the adult state these animals do not travel very far. In fact, they have a very efficient passive means of dispersion, particularly a sort of "flight" mechanism in which young spiders are attached to a long thread of silk and can thus be carried by the wind over long distances (Blandin, 1977).

The Egyptian cricket *Anacridium aegyptium* does not migrate massively like some related species, yet occupies an immense territory that extends from Turkestan in the east to the Cape Verde Islands and Madeira in the west, taking in the Caucasus, southern Europe, Somalia, northern Sudan, Egypt, and North Africa. Although this zone is characterized by a marked ecological (and therefore selective) heterogeneity, *A. aegyptium* constitutes a monotypic species with the exception of one geographic race, *A.a. rubrispinum,* found in Pakistan and Afghanistan — a fact that implies considerable interpopulational exchanges.

When the circumstances are particularly favorable it is possible to evaluate the interpopulational gene flow, at least approximately, by studying the number of migrants. On the campus of Stanford University in California, there is a grassy area no larger than a square kilometer in which live three populations of checkerspot butterflies, *Euphydryas editha.* The two main colonies at the two extremities of the area are about 500 meters apart. The third, smaller colony is between them. These groups have been studied since 1959. Marking experiments have shown that in fifteen years these colonies exchanged only 1.7 percent of their males and 4.8 percent of their females, which represents a rate of exchange of 1 percent per generation! No additions from outside the area were detected (Ehrlich *et al.,* 1975).

Interpopulational gene flow must represent an important phenomenon for the maintenance of polymorphism. Mayr (1963) estimated that it was ten to a hundred times more efficient than the appearance of new mutations. According to him, within a local population mutation would only cause genetic change at a rate of 10^{-5} per locus per generation, whereas the change due to gene flow would be of the order of 10^{-3} to 10^{-2}. Thus *gene flow seems more diversifying than mutation.* The more capable of dispersion a species is, the more important is the role of interpopulational gene flow. In any species divided into populations there is a constant conflict between the tendency to diversify and form new races (and ultimately species) that results from the existence of different selective pressures in the different locations making up the area of distribution, and the tendency to uniformity due to interpopulational flows.

In man, a particularly dispersed species, interpopulational flows seem to play a considerable role, as can be demonstrated by studying blood groups; this justifies comparison of populations over large distances by the study of hemotype frequencies.

Exogenous Interspecific Additions: Introgression

We said in Chapter 2 that a species is a real genetic ghetto thanks to its sexual isolation. This isolation is a condition *sine qua non* for the maintenance of specific individuality, but it is less absolute than was once thought. By interspecific interbreeding, genes from one species can be transferred to another species and persist. This phenomenon is known as *introgression*.

Interspecific breeding is always quite rare; it represents true *hybridization*. Breeding between subjects within the same species that differ by only a few characteristics is more properly termed *crossbreeding*. [1]

Crossbreeding is frequent and almost always results in a more vigorous product. This vigor, called *heterosis,* is doubtless related to the increase in the number of heterozygous loci in a subject born of two parents of the same species but with different genotypes, such as individuals from distantly related populations or from different races. This heterozygosity endows crossbred individuals with a very varied genetic background, thus enabling them to respond to very varied constraints. It is a form of ecological polyvalency. When we discuss polymorphism in the face of selection in Chapter 6 we shall see the selective advantages of such polyvalency.

The hybrid, on the other hand, receives two sets of chromosomes from two different species. We shall see in Chapter 7 that a given species almost always has a particular karyotype, that is to say a singular chromosomal pattern. This particularity plays an essential role in its sexual isolation, ensuring that it will not outbreed. In the hybrid the noncorrespondence between the two sets of chromosomes inherited from parents of different species presents grave difficulties for the formation of germ cells.

In the somatic cells of most living organisms, chromosomes are found in pairs, each pair consisting of two homologous chromosomes, one from the father, the other from the mother. These cells are called *diploid*; we say that they have $2n$ chromosomes. When somatic cells divide by *mitosis,* or *equational division,* each chromosome provides a copy of itself, so that during mitosis each chromosomal pair is for a short time formed of four elements that finally separate into two pairs, each of which enters a daughter cell. Once the division is over, these daughter cells thus have exactly the same genetic composition as the mother cell. The formation of the germ cells is rather different; here,

---

1. Since there are no genetically identical individuals, apart from identical twins, all breeding is really crossbreeding. We are all crossbreeds.

when the chromosomal pairs divide, they give only a single element to each daughter cell, which thus, instead of having the $2n$ chromosomes of the diploid cell, has n chromosomes and is a *haploid* cell. This is *reductional division,* or *meiosis.* During meiosis, the two chromosomes that will separate, first of all align themselves precisely so that their corresponding loci are face to face and there can be an exchange of homologous segments. This is known as *crossing-over* and will be discussed in Chapter 5. The essential preparations for these exchanges imply that the loci are arranged in the same order on the two chromosomes making up a pair (Figure 6A).[2] This matching is easy in crossbred animals, whose parents always belong to the same species and have similar karyotypes, differing only in a few alleles. It is much more difficult in a hybrid, whose two parents are from different species and usually have incompatible parent chromosomes. During meiosis, corresponding loci cannot be matched by simply aligning the chromosomes side by side; there must be complicated solutions, such as the formation of *inversion loops* between a chromosome from one species and one from the other (Figure 6B). Such figures make the karyotype more fragile, and there can be accidents such as breaks. In some cases the disparity between the two karyotypes of the hybrid is such that matching is impossible, so the hybrids will remain intersterile or less fertile than usual, although their own vitality may be increased. This is the case of the mule, produced intentionally by man by crossing a donkey and a mare.

There are examples of spontaneous hybrids in all groups. For instance, we all know the common cricket, *Gryllus campestris,* found throughout the temperate regions of Europe, around the Mediterranean, and as far as the Sahara. It has only one cycle per year, hibernates as a nymph, and can survive quite cold climates. Another species, easy to distinguish morphologically, is *G. bimaculatus.* It is a native of Asia and tropical Africa but can be found in southern Europe and in Mediterranean regions, living in sympatry with the common species. Although the zone of contact is large, hybrids have never been described; but when reared in laboratory conditions, the two species can interbreed and produce fertile offspring. The sexual isolation observed in the wild is perhaps due to the lack of correspondence of their cycle, for unlike *G. campestris* with its one annual cycle, *G. bimaculatus* has several. However, hybrid forms were recently discovered in the south of Algeria in two locations situated at high altitude about thirty kilometers apart (Dreux, 1977).

2. In fact, things are a little more complicated. As the two homologous chromosomes separate, each splits into two daughter chromosomes. Thus the closely aligned "pairs" are really four elements, the *chromatids,* lined up lengthwise to form a *tetrad.* Normally, crossing-over occurs between the two inner chromatids of each tetrad.

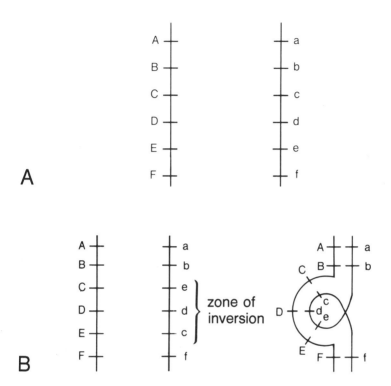

Figure 6. The different behavior of homologous chromosomes in crossbred and hybrid individuals during meiosis.

A. In crossbreeds the loci are in the same order.

Crossbred individuals have parents from the same species but differing in a number of features, so many loci will be heterozygous. At a given locus one chromosome will bear a gene and the homologous chromosome will bear its allele, but the order of the loci will remain the same on each homologous chromosome. Since the sequences are identical, there will be no difficulty when the homologous chromosomes are matched during meiosis.

B. Hybrids.

Hybrids come from parents of two different species. Therefore not only are many loci heterozygous, but their order may not be the same in homologous chromosomes. In the example here, there is a zone of inversion in the middle of the chromosome. The second homologous chromosome has the same loci as the first but in the opposite order: *e, d, c* instead of *C, D, E.* During meiosis the matching of the two chromosomes can only occur with the formation of a loop, called an *inversion loop,* that will make the karyotype more fragile and perhaps increase the frequency of accidents (such as unequal crossing-over) that would lead to duplication and deficiency.

European green frogs are just as familiar as crickets. They were for a long time grouped together into a single species *Rana esculenta,* but with quite marked geographic variations. We now know that there are three autonomous species (see Figure 7):

R. lessonae, originating in central Europe and spreading to the northeast part of France and the whole of Italy.

R. ridibunda, with a much wider distribution toward the east, even as far as the Urals, and to the south around much of the Mediterranean, including North Africa and a large part of the Sahara. In central Europe it lives in sympatry with *R. lessonae.*

Finally, *R. esculenta* is found in western Europe and has morphological and ecological features intermediate between the other two species.

These three forms can be interbred in the laboratory, where it can be shown that *R. esculenta* is none other than a hybrid of *ridibunda* and *lessonae.* This hybridization is found in the wild over a very wide area, but there is a distinction. In the sympatric areas *R. esculenta* is often accompanied by a few *lessonae* and more rarely by *ridibunda.* Elsewhere, over wide areas, *esculenta* is found alone with no sign of the reappearance of the ancestral form, just as Mendel's laws would predict. It is possible that we are dealing here with a fixed hybrid form that in its zone of distribution in western Europe benefits from positive selective pressure and has retained a mechanism of stabilization whose details are still far from understood.[3]

Electrophoretic study confirms this hybridization. There are alleles

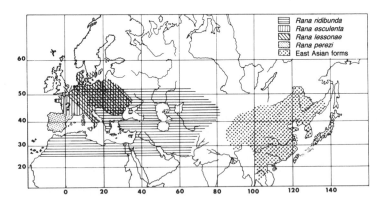

Figure 7. Distribution of species from the *Rana esculenta* complex. (From Dubois, 1977; modified from Boulenger, 1898; Berger, 1969; and Hotz, 1974a,b.)

3. Thanks to this stabilization, the hybrids produce a fixed population. The two sets of different chromosomes they bear show no tendency to segregate independently during subsequent generations, which could mean a resurgence of certain parental characters as required by Mendel's second law. A similar phenomenon of stabilized hybridization is seen in some groups of fish. Although it seems that this mechanism is now rare, it could possibly have played an important role at the beginning of metazoan evolution.

peculiar to *R. ridibunda* and others peculiar to *R. lessonae* that can be found regularly in the heterozygous state in *esculenta*. Paleontological arguments suggest that *lessonae* and *ridibunda* are derived from a common form that split into two groups that were unable to make any contact during the Pleistocene glaciations. Thanks to their prolonged isolation, and doubtless to different selective pressures, these two groups finally became intersterile and gave rise to two autonomous species. At the end of the ice ages, when geographic barriers were reduced, the two forms met again and gave rise to *R. esculenta* by hybridization, for their specific isolation was not yet so advanced that their intersterility was rigid — although it was sufficiently severe that the two species did not reunite again completely. For a general review of the problem see Dubois (1977).

Recent research has revealed an even more complicated situation. According to a personal communication from Alain Dubois, in the Italian peninsula south of the plain of the river Po one finds neither *R. esculenta* nor *lessonae,* but another, as yet unnamed, species and another hybrid. The unamed species seems to inhabit only Corsica and Sardinia, where no others are found. In North Africa and perhaps even further east one does not encounter the true *R. ridibunda* but one or two other distinct forms. What is more, it seems that two varieties of *R. lessonae* exist, perhaps representing two autonomous species, one in northern Europe and one in the south. In France one of these species is found as far south as the Massif Central. Further south and in Camargue there is a new hybrid, *R. perezi,* which is found also along the southwestern coast of France as far as Bordeaux.

The urodele *Triturus cristatus,* the crested newt, is spread over a wide area from the Urals to the British Isles (Figure 8). It is found in the northeastern part of France, while the southwestern part is inhabited by the related marbled newt, *T. marmoratus,* also found in the Iberian Peninsula. There is an overlap between the two areas stretching from Brittany to Auvergne, where one can find a third species, *T. blasii.* Electrophoresis has shown that this is really a hybrid form of the other two (Gasser, 1977), but — contrary to what happened with the green frog — these hybrids seem to have a reduced fertility, which may explain the relatively limited zone of introgression.

Many other cases of hybridization along borders are found among mammals and birds. European crows belong to two allopatric species: the carrion crow, *Corvus corone,* is completely black and inhabits western Europe; the hooded crow, *C. cornix,* is a gray variety with black head, wings, and tail that inhabits eastern Europe and most of the Mediterranean area (Figure 9). There is a fairly narrow zone of hybridization stretching approximately north to south along the whole zone of contact of these two species. Beginning in Scotland, it crosses Denmark, central Germany, and Austria, skirts the southern Alps, and finishes in the Gulf of Genoa. Except within this belt of hybridization, heterospecific breeding is exceptional.

We saw in Chapter 1 that *Mus musculus domesticus* and *M.m. musculus* share a long common north–south frontier, stretching from Denmark to the Tyrol, but only form hybrids in a narrow band along the region of contact whose width never exceeds about ten kilometers (Chaline and Thaler, 1977). We may well ask whether these two subspecies should not rather be considered as true species.

We also discussed the baboons, tailed monkeys living in Africa and representing a complex zoological ensemble. A long time ago this group invaded areas with very different ecological conditions and split into several forms, previously considered by zoologists as true species although able to interbreed. Two other groups, the drills and the mandrills, living separately in the equatorial rain forest, must have broken away from their common ancestral lineage very early and now form two autonomous species. As to the true baboons, living on the fringe of the

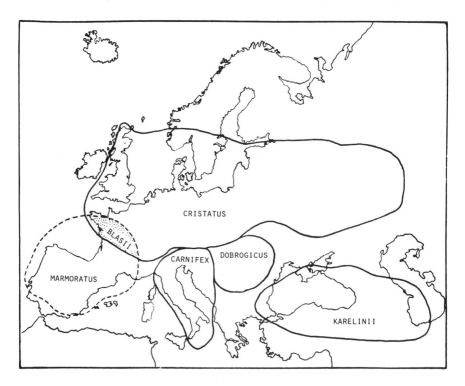

Figure 8. Distribution of *Triturus marmoratus* and *T. cristatus* (with subspecies *cristatus, carnifex, dobrogicus,* and *karelinii*). The natural hybrid *T. blasii* occurs in the zone of overlap. (From Gasser, 1977, after Spurway, 1953.)

wooded and partly wooded savanna that encircles the African forest on the north, east, and south, they split into five different types, according to their geographic localization. Each represents a group that is giving rise to a new species and each preserves its own characteristics, although spontaneous interbreeding is possible in the zones of contact. Mechanisms must exist to inhibit a too-massive gene exchange that would threaten the biological autonomy of each emerging species and would fuse them into a common genetic pool (Lucotte and Dubouch, 1980; Lucotte and Jouventin, 1980).

HYBRIDIZATION AND ADAPTATION

The examples given so far are not the only ones. We could quote many others in all living groups. In fact, the species barriers are often much less rigid than theory would have it, and the wall of the sexual ghetto

Figure 9. Path of the hybrid zone between the carrion crow *(Corvus corone)* and the hooded crow *(Corvus cornix)* in western Europe. (From Mayr, 1963; after Meise, 1928.)

built up to protect the species can be breached. This is due to the fact that the groups we are studying in nature at the present time have not all attained the same degree of isolation nor reached the same milestones on the way to becoming species (see Chapter 11). In practice we encounter all situations, ranging from a simple reduction in panmixia — reflected in a reduced interpopulational gene flow that results in significant differences in the frequency of certain alleles in distant populations, so that their interfertility is diminished — to cases of absolute intersterility. Between these two extremes one finds all intermediate stages: simple, spontaneous hybridization in a fairly wide area of overlap between two areas of geographic distribution; hybridization reduced to a narrow band along the line of contact of two areas; or the appearance of single, accidental hybrids along frontiers. The spontaneous hybridization found in the field can be enough to ensure sufficient interspecific gene flow to play a considerable role in the maintenance of genetic polymorphism in the population.[4]

In the present state of our knowledge, this flow is not always easy to evaluate. It involves electrophoretic analysis of the enzymatic structure of wild populations living in frontier zones where hybridization is common. In practice such research can pose insoluble problems. Except in very favorable (but exceptional) cases where enzymological study is possible, introgression is usually evaluated by the extent of the zone occupied by morphologically hybrid individuals. But this gives only an approximate vision of reality. In fact, some foreign genes, using interpopulational flow, progress gradually and penetrate deeply into the distribution area of the species while nothing, morphologically speaking, bears witness to this introgression. Not all intruding genes achieve the same degree of penetration; this depends on the selective value of each factor in the genome it enters. Some alleles can spread throughout all the populations in a species, while others, less favorable, will remain restricted to the actual zone of hybridization. Thus the frontier of sympatry acts like a semipermeable membrane. Under the influence of selective forces the diffusion of some factors is favored, while others are

4. We have said that intersterility was an essential factor in speciation. In fact, although intersterility can be easily demonstrated between species occupying the same area of distribution (sympatric species) or neighboring areas with common frontiers or even overlap (parapatric species), it is much more difficult to demonstrate between more remote, allopatric species whose members have little chance of meeting. In such cases we speak of *passive* species. Only laboratory breeding can tell whether there is a strict sexual barrier between groups. Such experiments can be surprising. For example, it has been possible to cross butterflies that belong not only to different species but even to different genera (Guillaumin and Descimon, 1976). It is only when their intersterility is confirmed in the laboratory that two groups can be given the rank of true species. Unfortunately, some animals do not breed in captivity. One sees how artificial is the difference between crossbreeding (of different races within the same species) and hybridization (of subjects of different species). In fact, if the concept of species depended rigidly on intersterility, true hybridization would not exist.

inhibited. But in a large proportion of cases, introgression is a far from negligible factor in the maintenance of the polymorphism of the group in question.[5]

Introgression is rarely very great in higher animals, such as arthropods and vertebrates. By contrast, it is commoner in lower animals and particularly in plants, which generally exhibit less strict specific barriers than animals, something that is evident from the fact that there are about ten times fewer plant species than animal species. This difference represents an adaptive phenomenon and recalls what was said in Chapter 1 concerning intraspecific variations in polymorphism. The same rationale is involved, whose essence is as follows. Intraspecific crossbreeding poses relatively few problems in higher species, such as birds and mammals, that have sophisticated means of physiological regulation, such as homeothermy, and can thus survive changes in weather. They have great potential for seeking and finding a suitable partner. The cold-blooded or ectothermic animals do not have the same degree of physiological regulation, and their temporospatial domain is more limited. Able to move around only over shorter distances and for less time, they cannot be so demanding.[6] At the sexual level they must make certain compromises, essential for the survival of the species. As to plants, their situation is even worse. Most of them are fixed in place. Their fertilization depends on chance meetings between gametes that travel passively, using the wind or insects as vectors. The resulting seed must grow wherever it falls, or disappear. Thus many plants are capable of an astonishing level of hybridization and are able to tolerate considerable ecological variation. Hybridization is responsible for their genetic variability (Mayr, 1963) and ensures them great adaptive flexibility.[7] These characteristics of a large part of the plant world were perhaps also true of the animal world at its beginnings. It is possible that in primitive groups specific barriers were less pronounced than they are now. This flexibility could have permitted them to invade all

5. The tangible result of introgression is *intergradation*. This means that a population on the border between two species can have features intermediate between the two, often "graded" over space (that is, in gradients or clines).

6. Nevertheless, long-distance migration is possible among insects; but, except in certain cases like migratory crickets, the migration is passive, allowing the animal little choice. Many insects are continually transported by the wind over various distances, functioning as a sort of airborne plankton that must have been as important for the development of birds as aquatic plankton was for fish.

7. This rule is rather statistical than absolute. There are many plant species whose form is poorly fixed, such as the brambles of the genus *Rubus*, in which tens or hundreds of intermediate types have developed from a few basic types due to more or less complex hybridization. Other examples are the hawkweeds *(Hieracium)*, types of dandelion, and the willows *(Salix)*. These genetically very varied groups do not have exacting requirements concerning their habitat. They are sufficiently polymorphic to adapt easily. In contrast, other species hybridize very little and are rigidly sexually isolated. Their ecological requirements are strict, and their geographic distribution so precise that they serve as indicators of given climatic zones.

habitable locations. Later, each group could have specialized and come to occupy its own particular ecological niche, thus diminishing interspecific rivalry and producing a selective advantage.

But specialization, which permits an individual to exploit a narrower niche more efficiently, develops along with the appearance of more precise and sophisticated functional systems such as sensory receptors, analytic centers, effector organs, reaction programs, and so on. From being a general *factotum,* an individual becomes an eminent specialist. It becomes enveloped in an ensemble of rigid, integrated genetic information that tolerates little variation. Its genome can no longer be modified without threatening this complex but delicate structure. This makes it impossible to introduce genetic material from another species.[8]

At the present time, if interspecific genetic exchange plays a role in the maintenance of polymorphism in plants, its role is much weaker or even nonexistent in many animals.

MECHANISMS FOR MAINTAINING UNFAVORABLE GENES IN THE GENOTYPE

We have just seen how all populations are always receiving new genes, either spontaneously by mutation, from neighboring populations (the interpopulational gene flow), or more rarely from different species (introgression). When they arrive, many of these genes are not favorable and are threatened with elimination by selection. But mechanisms exist for them to be preserved in the hereditary patrimony to ensure the permanence of genetic variety in the species without in any way penalizing it in the face of environmental constraints. These mechanisms involve preventing undesirable new genes from being expressed in the phenotype — keeping them, as it were, silent.

As selection operates on the phenotype — that is, on those factors that are expressed materially — it has no effect on these "sleeping" genes, which can thus remain in the patrimony of the group indefinitely without causing any serious perturbation. But their presence is not totally without importance in the long term, for they constitute a "reserve" of possibilities that the population can call upon if need be. Let us look at the main ways in which the expression of a gene can be inhibited.

---

8. Nevertheless there do exist in the most highly evolved groups more "generalized" or nonspecialized species alongside the specialized ones. They are often the vectors for evolutionary progression toward new frontiers. Their existence does not invalidate our general concepts and is almost always related to *neoteny.* See Chapter 10.

Dominance and recessivity

Some mutations are not noticeably expressed or only weakly expressed in the heterozygous state — that is, when they are present on only one of two homologous chromosomes, the other chromosome bearing the allelomorphic gene. In such a case the gene that is expressed is called *dominant,* and the one that is silent, *recessive.* New mutations are almost always recessive and are only manifest in the phenotype in subsequent generations in individuals that have them in the homozygous state. But these homozygotes represent only a small part of the population.

Let us take a simple example, such as a closed population bearing a gene A that has no allele. Initially, in the presence of no mutation and no foreign intervention, the individuals in these populations are all of the genotype A / A for a given locus. Their gametes are always A, and after fertilization their offspring will be A / A, identical to the parents. In this population the locus A has an absolute monomorphism. The gene A is also said to be *fixed*; it is the only allele found. One day an individual develops a mutation a at the locus A. It will produce gametes A and others a that, when combined with germ cells of other members of the population, will give birth to heterozygous subjects A / a. From then on the genetic monomorphism is lost, but not necessarily the phenotypic monomorphism, for since a is recessive, the heterozygote A / a will remain identical in appearance to the homozygote A / A. At this stage, a remains phenotypically silent. It is only in the next generation, thanks to the crossing of two A / a subjects, that one may find genotypes a / a and therefore a new phenotype a on which natural selection can act. Figure 10 illustrates this type of crossing.

In actuality things are more complex, for recessivity and dominance are both relative phenomena. Complete recessivity is rare. The first geneticists, working essentially with mutations affecting somatic characteristics in the fruit fly, tended to assume that dominance and recessivity almost always obeyed the *all or none* law. When a factor was really dominant, it alone was expressed in the phenotype. This concept of absolute dominance was a result of Mendel's experiments when he crossed smooth peas *(A / A)* with wrinkled peas *(a / a),* obtaining in the first generation plants with peas as smooth as those of the parents, although they were heterozygous. But this type of absolute dominance is rare. Often the heterozygous subject is more closely related to the dominant parent than to the recessive parent without being absolutely identical. For example, when one crosses normal fruit flies that have light-colored bodies with dark-colored flies, the first generation tends to have a light body. The wild light color dominates the dark mutant.

But the crossbred individuals tend to have a darker color than wild ones. The dark mutation is expressed discreetly in the heterozygote. One speaks of *partial dominance*.

If two alleles at corresponding loci are expressed equally, there is no dominance of the one over the other. They are said to be *codominant*. In this case, the crossbreed phenotype is halfway between those of the two parents. One should not speak of dominance or recessivity at the level of the gene, but only at the level of the features seen in the phenotype. A gene is present or absent, no more. Its mode of expression depends on the conditions in which it finds itself, and in particular on the allele facing it on the other homologous chromosome.

Heterostasis: Epistasis and Hypostasis

Most of the features expressed in the phenotype and that interest zoologists need the intervention not of a single gene for their realization but of several. Thus a given gene cannot manifest itself unless other genes compatible with the same action are also present in the genotype. If this is not the case, the gene will remain silent. The block will be due not to a dominant allele of the same series facing it on the other homologous

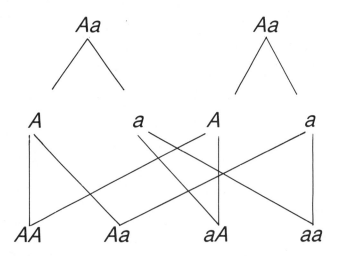

Figure 10. The crossing of two heterozygotes *Aa/Aa* gives offspring of which three-quarters have phenotype *A* (of which one-quarter are homozygotes *A/A* and one-half are heterozygotes *A/a*) and one-quarter have phenotype *a* (all homozygotes *a/a*). These last will be the only ones to express the character *a*, and may be eliminated by selection.

chromosome, but to genes situated at other, different loci and even on other chromosomes. These interrelations between nonallelic genes are called *heterostasis*. We say that a gene is *epistatic* in relation to another gene when it is capable of blocking its action. The inhibited gene is then said to be *hypostatic* in relation to the inhibiting gene. Heterostasis was first described in Japan by Tanaka (1924) in the silkworm. Later many other cases were described, so that today the phenomenon is seen to be a general one.

A classic example is that of the coat color of mammals, which is controlled by several structural genes called color genes that are responsible for the synthesis of hair pigments — essentially black and yellow pigments that, when placed side by side, determine the overall color. In order to be expressed, these color factors must be assisted by other factors called chromogens, situated on different chromosomes. The chromogens can undergo recessive mutations that, when present in the homozygous genotype, cause albinism. In the presence of albino factors the animal remains white, whatever color genes it has in its patrimony. This phenomenon constitutes *cryptometry*. Thus the albino phenotype can be expressed by many genotypes, for its appearance requires only that the animal have a mutation blocking the activity of the color genes in the homozygous state.

Genes necessary for the appearance of a given characteristic but which do not cause it directly have been called *conditional* genes. In fact, the appearance of most hereditary features depends not on a single gene but on a series of genes intervening in a certain sequence. Often it is the activity of a gene somewhere in the chain that sets off the activity of the gene situated immediately alongside it. In other words the activity of a first gene represents the raw material on which the second can act and often serves as a starting signal. Progressing along a certain sequence, this process constitutes the "program" that will finally result in the appearance of the feature in question. In such a system each gene represents a conditional factor for the one that follows and is itself dependent on the activity of the one that precedes it.

Let us take a simple example. Imagine that a first factor F1 provokes the synthesis of an enzyme transforming a substrate S1 into S2. Then a second factor F2 intervenes to transform S2 into F3, and so on until Fn produces Sn — that is, the terminal phenotypic character. In such a chain any new factor will have an effect only after the preceding factor has acted. For example, F2 becomes operational when S1 has been transformed into S2. Thus any mutation blocking a reaction at a given point will automatically block everything that follows.

Ultimately one could imagine the whole genotype being involved in the expression of a particularly complex character. A gene is not an independent element; it is part of a whole functional assembly. It does not intervene at just any time and in just any way, but rather according to a characteristic program in the genome in which it is integrated. Its action depends not just on its own nature but also on other elements of the genotype. It will be active or silent as a function of them. The existence of these phenomena of interaction between multiple genes explains the fact that some features can pass from recessivity to dominance and *vice versa* depending on the nature of other genes present in the same patrimony.

A striking example is the pepper moth, *Biston betularia,* which has been thoroughly studied for more than a century. The commonest form, found in gardens and woods around Manchester until 1848, was chalk white. Since the insect normally settled on light-colored backgrounds, such as birch trees, stones, or lichen, this provided an effective camouflage to protect it from predatory birds. From time to time very dark, sometimes almost black mutants were observed. This *melanism* acted like a recessive character compared with the natural light color. The melanic morph remained in small numbers, for, being easier to detect than its wild homologue, it was at a selective disadvantage. But at the beginning of the industrial revolution the countryside was covered with a layer of soot due to smoke from factories. The background on which the moth habitually settled became darker. From then on the melanic form was at a selective advantage. The light-colored moth became rare, while the dark form became the common one. Under the influence of a rapid change of environment, there was an inversion of gene frequency.

This phenomenon shows how the respective frequencies of two alleles can vary as a function of the environment; it has been closely studied by population geneticists. They demonstrated that selection did not act only on the mutations responsible for the moth's color but also on a number of other accompanying genes, so that the dark feature, which was recessive when it was an exception and of negative selective value, became dominant when its selective value increased. In reality natural selection was not acting only on color factors but on the whole genotype — that is, on combinations of integrated genes, related heterostatically among themselves.

So in a given ecological context, selection tends to retain not only the most favorable genes but also the most advantageous genetic combinations. In a polluted area these may represent combinations that allow melanism to appear and ensure its transmission as a dominant character.

Penetrance and Expressivity

The relationships between nonallelic genes clarify two well-known phenomena that have for long remained unexplained: *penetrance* and *expressivity.*

There exist factors, considered dominant, that are not expressed in all individuals that bear them in the heterozygous state. They are only seen in a percentage of cases, that is, when the other elements in the genotype permit. Such genes have variable penetrance. Penetrance is a statistical concept and can be defined as the frequency of the expression of a gene in the heterozygous state.

Other dominant genes are also expressed to variable degrees. Once again, expressivity depends on the other genes in the genotype. They are said to have a variable expressivity. Expressivity can be defined as the degree of expression of a characteristic in the heterozygote.

Polymery and Pleiotropy

What we have just said allows us to define two phenomena concerned in the expression of all genetic features: *polymery* and *pleiotropy.*

The term polymery, or polygeny, refers to the fact that the expression of a particular characteristic in the phenotype necessitates the intervention of several independent genes. This is a fairly constant phenomenon, particularly at the level of macroscopic features, whether morphological, physiological, or behavioral. All such features that appear far from the basic source of the genetic information need the intervention of numerous biochemical chains controlled by as many genetic systems. They obey complex programs that are still far from being defined. Between 1925 and 1930 geneticists believed, like Morgan, that every hereditary characteristic identifiable in the phenotype was controlled by only one gene, or rather by two genes at the same locus on both homologous chromosomes. But in fact, even features that seem quite simple, like the color of the fruit fly's eye, are now believed to depend upon at least twenty-eight different loci (Grassé, 1978).

Pleiotropy is in a way the opposite of polymery. It can be defined as the fact that a gene is often responsible for the simultaneous appearance of several characteristics that may involve very different aspects of the phenotype. The albino mutation, for example, is associated with loss of color in the skin and the iris, increased susceptibility to infection, and the ability to be easily trained. Albino animals can be tamed more readily than others and seem in general more intelligent, but they are more fragile and have difficulty surviving outside their artificial breeding colony.

Pleiotropy, in sum, concerns the phenotype. At first sight it may seem to involve features that are in no way related. The same genes can affect such different features as skin color, training ability, and susceptibility to infection! The mechanism becomes clearer when we move from the macroscopic to the molecular level. As we said earlier, genes control the synthesis of peptides often used to make enzymatic proteins. A given enzyme is always active at the same place in a biochemical chain, but the same chain can affect different characteristics, morphological as well as functional. Thus a given gene can affect all those features that are influenced at one time or another by the particular protein synthesized. When seen in the phenotype, these different features may have little in common. At the molecular level, by contrast, there are no pleiotropic genes. All of them act in a remarkably unified way, but their actions can affect very different organs or functions.

VARIATIONS IN THE EXPRESSION OF A GENE IN EVOLUTION

The way in which a gene is expressed always depends on other elements present in the same genotype. In general, a new allele introduced in a population by mutation, gene flow, or introgression acts like a recessive factor. It will remain such if its phenotypic activity is unfavorable or neutral, for natural selection will conserve combinations that keep the new genes silent or almost silent. If in the wake of (for example) an ecological change the newcomer becomes advantageous, selection will conserve those genetic combinations that allow it to be expressed. It will emerge from clandestinity, and from being recessive it will become dominant.

Thus, recessivity and epistasis, penetrance and expressivity, all permit the enrichment of the genetic pool without in any way threatening the population itself. They allow the discreet accumulation of potential riches.

the polymorphism of combinations

VARIABILITY THROUGH RECOMBINATION OF THE GENOME

As we said at the beginning of the previous chapter, the phenotypes observed in all living groups depend not only on the nature of the genes present in the patrimony of the species but also on the way they combine in an individual. What is more, the "objective" polymorphism of a population, into which natural selection can "sink its teeth," depends not only on the constant addition of new genes through mutation, interpopulational gene flow, or introgression but also on the gene combinations that are constantly made and unmade during successive generations, thanks to the phenomenon of sexuality.

We shall see the fundamental importance of this mechanism, especially in higher forms, but before going further it is useful to make a preliminary remark. We have emphasized the size and consistency of genetic polymorphism in all groups, but this polymorphism is only useful for evolution insofar as it is expressed in the phenotype, the target of selective pressures. We have already dealt with the mechanisms that permit the silencing of newly arrived genes. If their silence were permanent, such genes would be of no use. In order to contribute to evolutionary progress they must be exposed to natural selection from time to time, and must therefore be expressed to some extent. This is possible through sexual reproduction, which in each generation creates new combinations, new genotypes, and, therefore, new phenotypes.

The fertilization of an egg is the fusion of two germ cells of opposite sex (male and female), nearly always from two different subjects; thus an individual contains genes that were separate in its parents. When they come together in the same genome, they allow new forms of dominance and, above all, epistasis, so that some that were silent can be expressed.

This emergence of new gene combinations goes on uninterruptedly throughout the generations. Thanks to this perpetual mixing up of genotypes, no gene can remain unexpressed indefinitely. At a given moment, when a favorable combination occurs, even the most cryptic gene winds up emerging in the phenotype, thus exposing itself to the forces of selection.

Selection judges not the gene itself, but the whole genotype to which it contributes. In the end it is not so much the gene that is preserved or eliminated but the combinations in which it is acceptable or not, according to environmental constraints. This mechanism implies a continual reshuffling of the cards, but its importance was long overshadowed by that of the one-shot mutation. It is called *recombination* and plays a fundamental role in the adaptation and evolution of groups. Essentially a process of renewal, it happens during the sexual cycle, which we should now look at more closely.

The Sexual Cycle

We have already spoken about the sexual cycle in Chapter 4, so we need only recapitulate briefly here what happens in man, in whom it is not fundamentally different than in other animals, both vertebrates and invertebrates, or even in plants.

All our somatic cells contain 46 chromosomes grouped into 23 pairs, each pair consisting of two similar elements, the *homologous* chromosomes, except for one pair of *sex chromosomes* that consists of two identical elements in the female (XX) and two different ones in the male (XY). Each pair contains one chromosome of paternal origin and one of maternal origin. We are thus crossbreeds of our parents! Our genotype was established when a spermatozoon from our father with a single, or *haploid,* set of chromosomes penetrated a maternal ovule, also containing a single chromosome set. This *fertilization* was our real biological birth. When they fused, these two germ cells reestablished the chromosome pairs. The first (diploid) cell thus formed, the egg, began to divide, and the daughter cells produced all our tissues and organs by a complex constructional process called *ontogenesis* or *embryogenesis,* lasting nine months in man. All our cells stem from the original egg, and each has the same genetic patrimony, since during normal cell division, or *mitosis,* the chromosomes make copies of themselves and each daughter cell receives the same gene complement as the mother cell had. Later the cells differentiate and specialize according to their position in the embryo. Thus elements that were originally exactly the same in the end provide nerves or liver or muscles. This specialization implies that, at a certain stage of embryological development, the differentiated cells utilize only a part of their hereditary patrimony, the part that corresponds to their special function. The rest is suppressed.

At a particular moment some somatic cells become isolated and form our sexual glands or gonads, the testicles in the male and the ovaries in the female. They then undergo a special form of division, *meiosis,* during which each pair of homologous chromosomes is separated, so that only one element goes to the daughter cell. So the germ cells, or *gametes,* have only 23 chromosomes instead of 46, arranged singly rather than in pairs. These cells are haploid, with *n* chromosomes (unlike the diploid somatic cells, with 2*n* chromosomes). The process that makes diploid cells into haploid cells during meiosis is called *reduction.* This is the stage when recombination takes place. At the moment of fertilization we inherited a series of paternal and maternal chromosomes that combined to construct the homologous chromosomes of our genome. But when we make our own germ cells — that is, when we once again separate our pairs of homologous chromosomes — we do not deliver our whole paternal heritage to one gamete and our maternal heritage to another. On the contrary, this separation occurs independently for each chromosome pair; it is random, and all combinations are possible.

If we call our 23 paternal chromosomes A, B, C, D, etc., and our 23 maternal ones a, b, c, d, etc., our genome would be Aa, Bb, Cc, Dd, etc. Our germ cells will not be reformed as the A, B, C, D or a, b, c, d, etc., that we inherited from our parents but as A, B, c, d; a, b, C, D; A, b, C, D; a, B, c, d, etc., in any conceivable order, of which there are 2^{46} possibilities.

In fact these figures are underestimates, for each chromosome has not just one polymorphic locus *A / a, B / b,* etc., but several — a point we shall take up again at the end of the chapter. Among the billions of spermatozoa that a man can produce there are certainly no two with the same composition. The organization of the hereditary patrimony into a number of independent chromosomes opens the way for a series of recombinations in each generation, thus permitting a constant renewal of genotypes on which selection can act. The extent of the possibilities is difficult to imagine. The independent segregation of characters had already been observed by Mendel, before the existence of chromosomes or the details of the sexual cycle were known. It constitutes his third law.

RECOMBINATION AND NATURAL SELECTION

Let us take a ridiculously simple example to illustrate how the independent segregation of characters intervenes in evolution. We shall use a species whose genome only has two pairs of chromosomes, with the alleles *A / a* at a locus on the first pair and *B / b* at a locus on the second.

There are two populations, one with double homozygotes *ABAB,* the other with *abab.* The first subjects (*ABAB*) all produce gametes *AB* at meiosis. If they breed and there are no mutations or foreign influences, they will produce only *ABAB* offspring. In the same way the *abab* subjects will give only *ab* gametes and offspring with the genotype *abab.* If there is interpopulational exchange so that an *ABAB* subject breeds with an *abab, AB* gametes will fertilize *ab* gametes, and *vice versa.* So from the first generation all individuals will be double heterozygotes *ABab.* They will, in turn, produce gametes, but since *A / a* on the one hand and *B / b* on the other are carried on different pairs of chromosomes, their separation will be random, and all combinations are possible. Thus four types of gametes will be produced instead of the two parental types: *AB* and *ab, Ab* and *aB.*

Since fertilization is also random, all crossings are possible between the four types of gametes, as in Table 3. There are twelve possible combinations, of which only the first and last reproduce the combinations of the grandparents (*ABAB* and *abab*). The independent segregation of the two allele pairs has given rise to a wide genetic polymorphism from the second generation.

It is useful to recall here a basic principle that will be essential to the further development of our subject. If *A* is dominant over *a* and *B* over *b,* all subjects with at least one *A* factor and one *B* will have the *AB* phenotype, whatever other elements are present in their chromosomes. If the *AB* phenotype has a strong selective value, the "best" phenotype for natural selection will correspond to several different genotypes (such as *ABAB, ABab, ABaB, AbAB, AbaB,* and so on). Thus natural selection does not simply preserve a single genotype, but all those that produce the favored phenotype, in our case those that manifest *A* and *B.*

In practice, interbreeding couples generally have more than two chromosome pairs with different alleles, so that the number of possible combinations is always very high. In any case, the most favorable

Table 3. Possible Combinations Among Four Types of Gametes

Gametes	*AB*	*Ab*	*aB*	*ab*
AB	*ABAB*	*ABAb*	*ABaB*	*ABab*
Ab	*AbAB*	*AbAb*	*AbaB*	*Abab*
aB	*aBAB*	*aBAb*	*aBaB*	*aBab*
ab	*abAB*	*abAb*	*abaB*	*abab*

phenotype corresponds to various genotypes. That is why selection does not automatically lead to uniformity. In each generation sexual reproduction challenges preexisting gene combinations and creates new ones. This process maintains a strong, permanent polymorphism upon which selection can act, sorting out favorable phenotypes and those that are simply acceptable—for it is the phenotype, not the genotype, that is the target of selective pressure.

Almost always, the phenotypes that are thus "examined" are based on a large number of different genotypes. Thus, contrary to what the neo-Darwinians thought, strong selectivity is quite compatible with broad genetic polymorphism. But this polymorphism depends much more on genotypic recombinations than on new mutations. This endless "game" allows both "old" and "new" genes to become recessive from having been dominant and to go from epistasis to hypostasis (that is, from being expressed to being silent) and *vice versa*. Although a few individuals are eliminated, the "game" allows certain genes, that if expressed phenotypically might be useless or dangerous, to avoid being totally lost. They become "refugees" in combinations in which they can remain silent, holding themselves in reserve and expressing themselves again when environmental conditions permit it. They make up the *genetic load* found in all natural populations, which we shall discuss in the next chapter.

The advantage of this mechanism is evident, for a species can survive a period of severe constraint, very different from what it experienced previously, without a marked impoverishment of its genetic patrimony. It suffices to conserve the favorable combinations in the new environment. The rest is stored as nonexpressed genes. In such conditions the "load" is enlarged, but the variety of the patrimony is not threatened. The population retains all its evolutionary potential.

MUTATION AND RECOMBINATION AS EVOLUTIONARY FACTORS

The study of living species, from the simplest to the most complex, shows that polymorphism is indispensable to the survival and evolution of all groups. Nevertheless, the mechanisms that ensure polymorphism differ between higher and lower organisms.

In the lower forms of life, represented now by the bacteria, mutation plays an essential role in the maintenance of polymorphism. Bacteria mutate frequently, and the simplicity of their genetic complement, a single chromosome, means that the phenomena of dominance and heterostasis are virtually nonexistent. So any mutation can be expressed immediately in the phenotype. In addition, a favorable mutation can be diffused very quickly through a colony on account of the rapid reproduction of the organisms (one generation requires about twenty minutes in *Escherichia coli,* compared with three weeks in the fruit fly and thirty years in man). Also the elimination of a mutation by natural selection is not at all catastrophic in bacteria, for it may well reappear soon after. We know how easily a strain exposed to an antibiotic will give rise almost inevitably to resistant forms. This is, in fact, one of the most serious problems faced by modern therapeutics. Bacteria demonstrate the efficiency of their mutagenic capacity in the face of adaptive difficulties at our expense. They have neither the necessity nor the possibility of maintaining a load of unexpressed genes. They know nothing of the phenomena of dominance and epistasis, which are really found only in diploid organisms.

However, bacteria can indulge in a form of recombination. Two bacteria can fuse and exchange a chromosomal segment, thus providing for a partial recombination of the genetic material. This is *conjugation,* which is now interpreted as a primitive form of sexuality; but at no time is there real gamete formation followed by fertilization. This is not really sexuality but *parasexuality,* or, perhaps more correctly, "presexuality."

The situation is different in higher organisms with complex genotypes and much slower reproduction. Since generations are longer, mutations are rarer, and they are not immediately reflected in the phenotype because they are held back by dominance and heterostasis. When they occur, they need a long time to diffuse throughout the group. They cannot be relied upon to respond quickly to environmental challenges. In these species polymorphism is ensured primarily by genetic recombination linked to sexual reproduction, which mixes up the whole hereditary patrimony in each generation (Ayala, 1982).

This permanent renewal maintains the adaptive possibilities of the species and guarantees survival. Among the newly emerging combinations, some are favorable or acceptable: they will be preserved. Others are unfavorable and will tend to be eliminated. The sexual process always creates some unadapted types: if they are at too much of a disadvantage they do not survive. As Ernst Mayr has said, "They fall by the wayside." This inevitable waste represents the cost of evolution.

SEX AND DEATH IN THE LIFE CYCLE

Without sexuality, polymorphism would be useless. It would remain rigidly immobile and finally disappear as a result of the progressive elimination of unfavorable combinations that would not be subjected to the mixing process. Since any combination can sooner or later become unfavorable, living groups deprived of their sexual dynamic would not survive long.

Sexuality allowed higher organisms to utilize their polymorphism by constantly alternating, from generation to generation, the diploid phase of the individual with the haploid phase of the gametes in the life cycle described above, and which Theodor Boveri glimpsed at the turn of the century.

The diploid phase represents an *emergent* stage during which the individual takes its place in nature, where it can face the forces of selection. During this time it produces gametes that will be more numerous and better distributed if the subject has found the best responses to the constraints of its environment, thereby increasing its vitality, its fertility, its activity, and its longevity. The haploid stage represents the stage of *elaboration,* during which the gametes unite randomly to give birth to new individuals — that is, to new combinations, generally better adapted because they benefit from prior experience. They in turn will face selection. Thanks to the action of sexuality on a highly polymorphic genetic patrimony, higher organisms can always respond to changes in their milieu. In all groups, the long string of generations represents a permanent battle for better adaptation to an ever-changing environment. This is only possible with the periodic disappearance of old combinations and the appearance of new ones. If the life cycle implies birth, it also necessitates death.

Our biological destiny is to dissociate the chromosomal combinations received from our parents and to spread new ones. Once this is done, we must disappear. Death is written in our sexuality. It is an essential part of the cycle. Immortality makes evolution impossible.

Already predicted at the stage of bacterial conjugation, sexuality then developed rapidly, certainly by the Precambrian era. It was by then a feature of plants and of the oldest animal forms. The sexual process is identical in man, vertebrates, invertebrates, and plants. Evolution, so rich in innovation and change, never threatened it. Sexuality, like death, is one of the most constant attributes of life. They must both be of great selective value. Nature, normally so inventive, never created an immortal species.

THE IMMORTAL AND THE MORTAL: THE GENE AND THE INDIVIDUAL

There is a basic contrast between genetic information in the chromosomal genes, which can reproduce indefinitely, and the individual, formed temporarily and expressing this information for a limited time. It was not always so. At the beginning of time, genetic information and the individual were indistinguishable. Bacteria and blue-green algae are both sources of information and living units. As the information became more complex — that is, as the genotype became richer — the two elements became dissociated into the information, capable of eternal reproduction, and the individual, destined to be born and die. The chromosomes we bear in our karyotype come from our parents, who inherited them from their parents, and so on through innumerable generations. Our first ancestors received them from prehominid species that had obtained them, with a few transformations, from even older groups. If we consider only the genetic structure of living organisms, there is a direct link from bacteria to men. All higher organisms die, but their chromosomes are permanent, just like the lower organisms such as bacteria.

Our DNA sequences have their origin in the depths of time, but we can give a date to our birth and that of our ancestors. From time to time since the beginning of the century, people have tried to decide which is the more important: our immortal genes or the individual who expresses a transient combination of them.

Recently the problem has been brought up again by E. O. Wilson (1975) and the sociobiologists, who see in evolution merely a long process of genetic competition. We shall take up this theme later. In fact, such a concept is difficult to accept as soon as one gets beyond the world of bacteria, for it neglects some of the most remarkable achievements of evolution, those that have permitted the appearance of the integrated systems that gave birth to multicellular organisms, systems that cannot be reduced to a series of isolated factors but that represent assemblies resulting in more and more complex functional units. As soon as a sufficient level of complexity is reached, these units form individuals.

Many fundamental structural genes are more or less identical in all living groups. After a certain stage, reached long ago, evolution was no longer dependent on the appearance of new basic structures but rather on new combinations — that is, new individuals, more autonomous, conscious, "personalized."

This "individualization" was what made evolution so original. Evolution based uniquely on competition between genes would probably never have progressed beyond the bacteria.

The Selective Advantage of Pleasure

Polymorphism is one of the fundamental characteristics of living matter. In higher organisms it is maintained by sexuality. Sexual reproduction is neither easy nor economical. Its biological cost is high, on account of the enormous loss of gametes, especially in males. Compared with the "lucky" spermatozoon that succeeds in fertilizing an ovule, millions or billions of others are simply lost. This type of reproduction leads to a form of ontogenesis that is very exacting in terms of living matter and very energy-consuming. What is more, it involves many random factors. It is first necessary to find a suitable conspecific partner, something that often poses many problems. Once the partner has been seduced, it is convenient to keep it as long as it can possibly be of use, which is not always easy. We can judge this from the fact that divorce has never been as common as it is now!

Other risks accompany the development of the egg or the embryo, which can be very complex, necessitating fundamental ecological changes. A tadpole is an aquatic herbivore, but a frog is a terrestrial carnivore. So the developing animal must leave the pond for the field: it must find the correct milieu according to its stage of development, or it will not survive. Nevertheless, sexual reproduction has an enormous selective advantage, for, in spite of its dangers, its incertitudes, its constraints, it has become established in all living groups. It has taken on innumerable forms. If one considers all structured organisms, animals or plants, one is struck by the luxury of detail, morphological, physiological, and behavioral, "dreamed up" by nature to attract the opposite sex and ensure gametic fusion! It is in the realm of forming couples that one sees the most astonishing behavior and the most esthetic phenomena: nuptial plumage and parades, amorous invitations, dances, songs, often with recourse to the most unexpected strategies. Nothing is left out in attracting the sexual partner and consummating the union. Above a certain level of consciousness, the desire provoked by the proximity of the sexes, and the pleasure that is its ultimate gratification, are such powerful advantageous factors that they have been permanently preserved by natural selection. Except in a few very special cases it is hardly conceivable to make love out of duty! A world without pleasure would not only be abominably sad, but it would lose one of the essential stimulants to polymorphism, the basic condition for evolutionary progress.

On the human level these features are intellectualized and may take unexpected forms, from artistic creation to crime. A man in love is capable of the greatest exploits, and the worst misdeeds. Deceived by George Sand, Musset wrote remarkable poetry; the bashful Eupalinos built a temple: "This delicate temple, no one knows, is the mathematical image of a girl of Corinth I happily loved" (Paul Valéry: "Eupalinos or the Architect"). Others murder their neighbor or wage war. Cleopatra's nose was, in the end, merely a pretext for copulation! The history of man, like that of animals and plants, is a long battle for polymorphism.

Sexual instincts play a large role in animal behavior. By declaring them taboo, some religions have developed an easy means of instilling guilt and submission in their followers. No other prohibition would have been so effective.

As we said earlier, sexuality appeared very early and was an essential condition for evolution. Without it variability could hardly reach an effective level, for there would not be really new individuals in each generation. An asexual world would have little innovation — only through mutations that would be of little significance to higher organisms. This is indeed what one finds in practice: asexually reproducing groups can be genetically very polymorphic, if one takes into consideration the alleles at different loci, and the level of heterozygosity may be high; but one finds the same combinations in all individuals. In spite of a varied genetic pool, the population remains very uniform. All individuals look alike, and evolutionary possibilities are virtually nonexistent. Asexually reproducing groups do not represent simpler, and therefore older, forms than others, but rather degenerate branches, blocked in an evolutionary impasse. The worst misadventure that can befall a line, apart from its extinction, is the loss of its sexuality.

THE LIMITS OF RECOMBINATION

We have just seen that the most tangible result of recombination is to furnish, in each generation, new genotypes from which natural selection can choose the best adapted. But there is also the other side of the coin. Let us take a population living where selective conditions have not changed for a long time. Slowly, the best-adapted genetic combinations have been established. Sexual reproduction, and its accompanying recombinations, challenge these environmentally well-adjusted genetic ensembles and may favor new combinations whose selective value could be less. Fortunately there exist several mechanisms that limit the effects of recombination and allow certain particularly favorable gene combinations to be preserved over the generations. We shall consider three of them: asexual reproduction, the reduction of chromosome number, and the blocking of crossing-over, looking only at their phenotypic consequences.

Asexual Reproduction

In asexual reproduction all offspring have the same genotype as their parent, of which they are a perfect copy. They form masses of identical twins. This type of reproduction, whatever its precise form, is found only in lower groups, such as sponges, coelenterates, turbellarians, and bryozoans. As mentioned above, it is always a secondary phenomenon. It appears at a particular moment in the history of a group and enables it to establish a solid ecological niche when a particularly favorable genotype has been developed. Sexual reproduction can reappear, especially if conditions change and become unfavorable. It is the only way a group can escape its immobility and try to find new combinations that will prevent its disappearance.

Reduction of Chromosome Number

We have seen that two pairs of alleles *(A / a* and *B / b)* separate independently when gametes are formed. However, this independence only holds if the two pairs are on different chromosomes. If they are on the same chromosome they are transmitted to the same germ cell together, since, from the point of view of cellular mechanics, the chromosome is the unit of recombination, not the gene. So, except when there is an exchange of segments between homologous chromosomes (crossing-over), the features of a given chromosome are inherited together, at least in the majority of cases. They are *linked* features and form *linkage groups.* There are as many linkage groups in a species as chromosome pairs. When the genetic material is broken up into multiple independent pieces, there can be an infinitely richer polymorphism than if the same material were contained in a small number of large chromosomes.

Let us take five hereditary factors *A, B, C, D, E,* each with a mutation *a, b, c, d, e.*

1. First hypothesis: the five factors are on the same chromosome. When the gametes are formed, if there are no exchanges between chromosomes, all the factors will be transmitted to the same germ cell. Thus the population will have two types of gamete *(ABCDE* and *abcde),* which will form pairs to provide three genotypes:

$$A\ B\ C\ D\ E / A\ B\ C\ D\ E$$
$$A\ B\ C\ D\ E / a\ b\ c\ d\ e$$
$$a\ b\ c\ d\ e / a\ b\ c\ d\ e$$

This species will have very limited polymorphism.

2. Second hypothesis: the factors are on different chromosomes. When the germ cells are formed, the chromosome pairs separate randomly, and the gametes can produce all imaginable combinations, such as *ABCDE* and *abcde, aBCDE* and *Abcde,* or *abCDE* and *ABcde,* and so on. If the number of loci is *n* (five, in our example) and each locus has two alleles, the number of possible gametes is 2^n ($2^5 = 32$). At fertilization all permutations of these combinations are feasible, thus giving a very large number of possibilities for broad polymorphism, according to the formula $(2^n)^2$ ($32^2 = 1024$ in our case). We have assumed only one polymorphic locus per chromosome (A/a for the first, B/b for the second, etc.). In reality things are infinitely more complex, for each chromosome bears a series of loci that can be occupied by different alleles, as we saw at the beginning of this chapter when describing the sexual cycle. This considerably increases the number of possible combinations. Thus, in practice, the conceivable number of combinations is enormous, but the actual number depends largely on the number of chromosomes sharing the genetic material.

As a consequence, the more a karyotype is split into multiple chromosomal elements, the more the species is polymorphic. If this is the case, the most primitive lines will be characterized by a large number of small chromosomes; more specialized lines will have fewer, but larger, chromosomes. Although the karyotypes of many species have not yet been studied, it seems that a reduction in the number of chromosomes is discernable in quite a few groups. This phenomenon seems sufficiently widespread to have merited formulation as Hamerton's law (Buettner-Janusch, 1973). In fact, it is far from constant; frequently during evolution one discovers a series of secondary splits, so that the number of elements in the composition of the karyotype increases again, as for example in *Cercopithecus* monkeys and some carnivores.

How can a species with a large number of chromosomes give rise to other species with fewer chromosomes? In most cases this is brought about by the fusion of two chromosomes at a special region identified

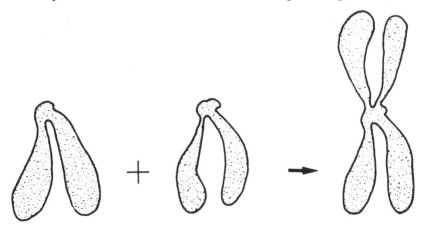

Figure 11. Diagram of Robertsonian fusion, whereby two acrocentric chromosomes become metacentric. We now know of many cases of fusion of acrocentric chromosomes to form metacentric chromosomes that are twice as large but half as numerous. For a review, see Chiarelli and Capanna (1973).

by a constriction, the *centromere*. As we shall see later, the centromere plays an important role in chromosome kinetics, during both mitosis and meiosis. Fusion nearly always happens between chromosomes with terminal or *acrocentric* chromosomes, which can fuse end-to-end to produce a single, large element with a median centromere — a *metacentric* chromosome. Such reorganization was described by W. R. B. Robertson (1916), and is called *Robertsonian fusion* (Figure 11).

In fact, many other forms of chromosomal recombination exist. In all cases they are random, unforeseeable accidents that may be preserved if they are advantageous. The most important in the present context is *translocation* in the strict sense of a reorganization during which a fragment of a chromosome breaks off and fuses with a chromosome of another pair.

Let us take a simple example, a large chromosome I with loci *A, B, C, D, E* and a small chromosome II with loci *F, G, H* belonging to another pair. Let us suppose that I splits between *C* and *D* and that the upper fragment *ABC* is liberated and fuses with *FGH*. We thus have two new chromosomes, a small one I′ with only the *DE* segment, and a large one II′ made by the fusion of *ABC* and *FGH* (Figure 12).

The translocation involves a modification of the linkage groups. At first, *ABCDE* formed one linkage group and *FGH* another, but after translocation the two linkage groups are *DE* and *ABCFGH*. If the

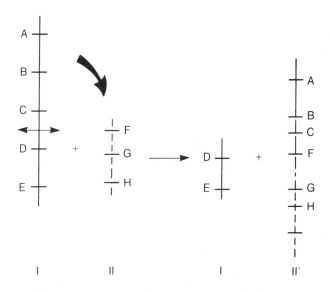

Figure 12. Example of simple translocation between two chromosomes I *(ABCDE)* and II *(FGH),* which become I′ *(DE)* and II′ *(ABCFGH).*

simultaneous presence of *B* and *F* in the genotype is indispensable to ensure the expression of a character of high selective value, the transfer of both factors to a single chromosome — that is, to the same linkage group — represents a considerable advantage, for this arrangement increases the chance of finding both these factors inherited together, or (put another way) reduces the probability that they will be separated during recombination.

Reduction or Blocking of Crossing-Over

Nonrecombination of factors carried by a single chromosome is only guaranteed if the chromosome remains physically intact. This is not always the case. As we saw earlier, during meiosis the two homologous chromosomes making up a pair are aligned side by side and can exchange corresponding segments. Usually the segments exchanged are spatially identical: they bear the same loci. But these loci can be occupied by different alleles.

Let us return to our previous example, a chromosome with factors *A, B, C, D, E* and its homologue with five recessive mutations *a, b, c, d, e*. When crossing-over occurs, it can produce daughter chromosomes of different types according to the site of the break, such as *Abcde* and *aBCDE*, or *ABcde* and *abCDE*, and so on, as in Figures 13, 14, and 15.

Crossing-over occurs randomly at any point on the chromosome, so the chances of two loci being separated will be proportional to the distance between them. This relationship, called Sturtevant's law, has been used to ascertain the relative position of each locus on a given chromosome and thus produce *chromosome maps.*

Since crossing-over involves an exchange of segments between chromosomes that do not necessarily bear the same alleles, it splits linkage groups that would otherwise have remained inviolate, and thus increases genetic polymorphism. Therefore any mechanism that hinders crossing-over will have a stabilizing role. It will be advantageous insofar as it allows a gene combination with a strong positive influence to become fixed. The frequency of crossing-over can be reduced in two ways.

Proximity of Loci and Supergene Formation

According to Sturtevant's law, the frequency of crossing-over between two loci is a function of the distance between them, so logically, loci that are close together should hardly ever be separated, as Figures 16 and 17 indicate. The genes they bear should form an indissociable entity, always inherited as a whole. If these genes cooperate in some useful or vital activity, such a mechanism would be highly favorable.

A functional unit of closely linked genes on the same chromosome was called a *supergene* by Ford.

This means that any chromosomal reorganization, such as translocation or crossing-over, that results in genes with complementary actions coming close together will be advantageous and preserved by selection.

The supergene behaves like a perfectly trained surgical team or orchestra, all of whose members act when and where necessary to complete a common task. If a single member is missing, the activity of the whole team may be slowed or stopped, and their goal never attained.

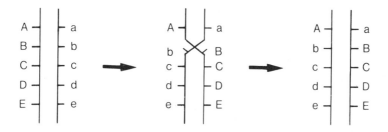

Figure 13. First possibility: crossing-over gives chromosomes *Abcde* and *aBCDE*.

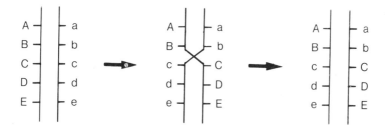

Figure 14. Second possibility: crossing-over gives chromosomes *ABcde* and *abCDE*.

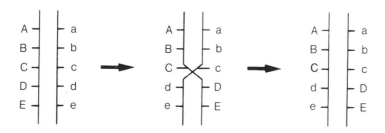

Figure 15. Third possibility: crossing-over gives chromosomes *ABCde* and *abcDE*. A fourth possibility would give *ABCDe* and *abcdE,* and double crossing-over would give, for example, *ABcdE* and *abCDe.*

Many complex hereditary processes that demand the intervention of different factors are regulated by supergenes. Known in all species, they have been especially studied in butterflies, where they are responsible for the astonishing phenomenon of mimicry, thanks to which a given species can take on the appearance of another species, including the other's shape and color, that is repellent to its predators. Through its camouflage the butterfly has a greater chance of escaping destruction. Supergenes are common in the plant kingdom. For example, in tomatoes the factors that code for type of inflorescence are grouped in a supergene. In many species, the alloenzymes that contribute to a single cycle often depend on the same functional supergene. We could cite more examples. These functional ensembles, not dissociable by crossing-over and made up of several genes with coordinated activities, have also been called *linkat*. One can consider "linkat" and "supergene" as synonyms.

Not all supergenes necessarily have the same cohesion. It depends on the proximity of the different elements making up a supergene on the same chromosome, and also on the selective value of the character expressed in the phenotype. As long as selective pressure is high, disso-

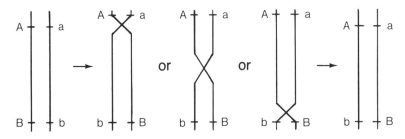

Figure 16. First possibility: the two loci *A/a* and *B/b* are at widely separated parts of the chromosome. Any crossing-over, whenever it happens, has a high probability of separating them. The recombination frequency is high.

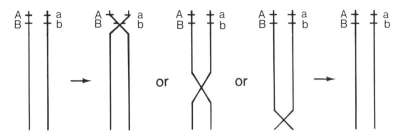

Figure 17. Second possibility: the two loci *A/a* and *B/b* are close together. They are unlikely to be separated by crossing-over. The recombination frequency is low. For contiguous loci the frequency can be almost zero. Such loci constitute a *supergene*, which behaves like a single functional unit. Here *AB* and *ab* form supergenes.

ciation of a supergene may be penalized by the elimination of the individual. If selective pressure lessens, a supergene may dissociate rapidly. Its component elements reclaim their independence and subjects appear that bear only a few fragments of the dissociated supergene, expressing phenotypes that have long since disappeared. If they are no longer eliminated by natural selection, they can spread and even become the common type. The exquisite functional ensemble of the supergene has collapsed like a house of cards. Mimicry in butterflies, mentioned above, is a good example of this complexity and fragility. Populations of butterflies reared in captivity, and therefore protected from their predators, undergo a disruption of their supergenes and the replacement of the wild morph by more or less varied types virtually unknown in natural surroundings.

Every intermediate type of linkage exists — from supergenes (absolute linkage factors), through closely linked genes that dissociate only rarely, to weaker linkage and total absence of linkage.

The species we know today are doubtless the fruit of multiple chromosomal reorganizations during their history. They tend to have stable chromosomal structures that become less and less open to change without a profound modification of selective pressure (see Chapter 8). Some sequences that form supergenes must be very old, unchanged for millions or tens of millions of years, probably because they play a fundamental and indispensable role in the physiological equilibrium of the individual.

A very good demonstration is provided by the major histocompatibility complex, called the HLA system in man and discovered by Jean Dausset in 1958.[1] (An equivalent system, called H[2], had already been described in the mouse by Peter A. Gorer.)

The segment of chromosome number 6 in the human karyotype that carries HLA and its associated mechanisms comprises:

1. The genes that control the synthesis of HLA factors. These define the immunological "personality" of an individual and permit the recognition of self and nonself. They are controlled by four independent loci: A, B, and C, which give rise to surface antigens that can be identified serologically; and D, which controls antigens identified by cellular immunological methods (allogenic reactions of lymphocytes in mixed culture). Other products of the same locus or a nearby D-related locus can also be recognized serologically. All these genes are doubtless those that control the immune response, humoral as well as cellular, at least to a great extent. Since each locus can bear a whole series of alleles, the number of possible combinations is very high, thus providing a rigid immunological definition of each individual.

1. See the text of his Nobel lecture (Dausset, 1981).

2. The genes that code for the markers of the different components of complement close to locus C. These are *C2, C4, C3 (B1),* and perhaps *C8. Complement* is a substance present in serum and indispensable for various aspects of the reaction between antigen and antibody.

A major histocompatibility complex appears to exist, without much modification, in most mammals studied so far (rats, guinea pigs, rabbits, horses, pigs, cattle, sheep, goats, and nonhuman primates such as macaques and chimpanzees). It is found in the chicken with a basic structural and functional organization similar to that seen in higher mammals. An equivalent system, but seemingly based on a genetically simpler model, has been observed in the only anuran studied so far, the *Xenopus laevis* frog (see Vaïman, 1980, for a general review). This shows that the major histocompatibility complex was established very early in vertebrate evolution, for it already exists in amphibians whose ancestors lived at the end of the Primary era. Subsequently, on account of its strongly positive selective value, the system has become more and more perfected.

It is likely that the multiplication of a single chromosome segment has played an essential role in this evolution (Bodmer and Bodmer, 1978). S. Ohno (1980) has stressed the importance of this mechanism in the evolutionary process, and there are many arguments in favor of this point of view. All immunoglobulin molecules are formed of two heavy chains and two light chains, the former composed of four segments and the latter of two, which one can consider as basic units (Figure 18). The glycoproteins corresponding to HLA antigens have striking structural homologies with these immunoglobulin units, especially the $\beta 2$ microglobulin, a relatively simple peptide that is involved in the formation of HLA antigens. Such homologies could not be accidental: rather they bear witness to a common origin. It is probable that certain components of HLA, and in particular the $\beta 2$ microglobulin chains, descend phylogenetically from the same basic information.

A single sequence must have existed in the oldest groups. It could have been responsible for the appearance of membrane antigens. Later it may have undergone multiple duplications, some of which became devoted to antibody formation, an essential step in the development of the immune system, which, dealing initially with the recognition of nonself, improved with time. This phenomenon is a specific example of regulatory mechanisms that were laid down when higher organisms emerged and gradually liberated living beings from the narrow constraints of their milieu. This development was particularly noticeable in the vertebrates, which adopted a terrestrial habitat and became subjected to many more forms of aggression than aquatic species living in a vastly more stable environment.

There doubtless exist many other supergenes representing ancient sequences with considerable selective value. They can be identified almost unchanged in numerous species. Unfortunately, the precise inventory of the karyotypes of living organisms, which would enable us to construct chromosome maps and thus determine the spatial relationships of different known genes with each other, is still very fragmentary.

The *X chromosome* of vertebrates has been best studied, on account of the ease with which we can identify the loci involved, for all bear *sex-linked* characters. This chromosome seems to contain a number of sequences common to phylogenetically very distant species. They have not changed since very ancient times and provide an excellent example of conservatism, whose origin we shall discuss below.

Inhibition of Crossing-Over in the Whole Karyotype

For crossing-over to occur, the two homologous chromosomes comprising each pair must have the same structure, with the same loci lined

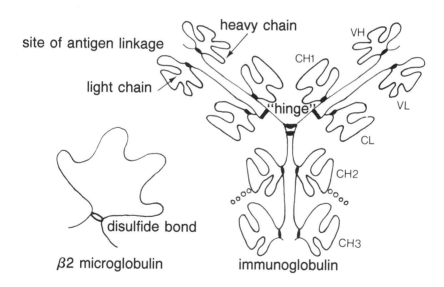

Figure 18. The structure of an immunoglobulin molecule with its heavy and light chains, each made up of segments: four for the heavy and two for the light chains. A molecule of β2 microglobulin is also shown. This is a protein associated with the molecules that determine the immunological identity of tissues (HLA antigens). Its amino-acid sequence is closely related to that of the component parts of immunoglobulin. This suggests a probable phylogenetic relationship between the two systems.

up in the same sequence. If the two chromosomes of the same pair are not identical, crossing-over is inhibited. This is seen frequently in species in which the two sex chromosomes are very different in one of the sexes (the *heterogametic* sex). In most groups the male is heterogametic, with one X and one Y chromosome, and does not undergo crossing-over. The *homogametic* sex, the female, has two X chromosomes and maintains the ability to exchange chromosome segments.

According to C. D. Darlington, the two sex chromosomes, the *heterochromosomes,* must originally have been comparable to ordinary chromosomes and thus more or less identical. They formed a homogeneous pair. At that time crossing-over would have been equally frequent in both sexes. Later, the sex genes, or at least a considerable number of them, gathered in a region that was different on the X and the Y chromosomes, thus forming a series of "sex supergenes." Concurrent with an equally vital phenomenon, the development of sexuality, this grouping together reduced the risk of recombination and thus of threats to a functional ensemble that was indispensable for the survival of the species; it was thus of great selective value.[2]

The region bearing the sex characters evolved differently in the X and Y chromosomes. Thus, the development of sexuality was accompanied by a progressive differentiation of the sex chromosomes, and therefore the reduction and even disappearance of crossing-over in the heterogametic sex.[3]

This phenomenon resulted in a limitation of polymorphism, allowing the retention, at least in a certain percentage of cases, of combinations that were favorable for the population—that is, combinations whose adaptive value was particularly high but that would have been threatened in every generation if crossing-over had taken place.

2. Today this seems a general principle. For example, in vertebrates the difference between the sex chromosomes and the others (the autosomes) is least in the oldest groups (fish and amphibians) and most pronounced in more recent ones (reptiles, birds, and mammals).
3. Recently various authors (for instance, Miklos and Nankivell, 1976; Cassagnau, 1977) have noted that the frequency of crossing-over seems, at least in part, to be linked to the amount of X heterochromatin in the chromosome. Heterochromatin is that fraction of the chromosome that remains visible when the chromosome is resting, when the euchromatin is in the form of a very fine filament, invisible by ordinary microscopy. The chromosomes become visible when the euchromatin develops tight coils at the start of cell division. The heterochromatin forms certain parts of the chromosome such as satellites, the centromere region, and often the extremities of the arms. This fraction seems to be relatively inactive genetically and may represent redundant zones where there is a large amount of repetition of the same sequences of *repetitious DNA.* Its relative amount varies among species and, within a species, from one pair of chromosomes to another. It is at a maximum in chromosomes composed mainly of heterochromatin, but the amount is similar in all cells for the same pair of chromosomes. Thus species with a high proportion of heterochromatin will be genetically stable and evolve little, while those with little repetitious DNA can benefit from more intensive recombinations from within their gene pool and thus have more opportunity for speciation when faced with varied environments. However, this "rule" seems not to apply to some groups, including the primates.

Plants, whose sexual "tolerance" (many species accept introgression by foreign genes) we have already noticed, almost always have poorly differentiated sex chromosomes; the rate of crossing-over is high in both sexes, thus maintaining a broad genetic polymorphism. The same is true in invertebrates, except for insects, and in fish. Almost all other groups possess very differentiated sex chromosomes, which inhibits or blocks crossing-over in the heterogametic sex. But it is especially by means of minor anomalies, such as inversions, of the *autosomes* (nonsexual chromosomes), preserved in the heterozygotic state, that crossing-over is blocked. It has been observed for a long time that these chromosomal heterozygotes are mainly encountered in populations situated in the center of a species's geographic area of distribution, where environmental conditions are "moderate" and the selective pressure is the least severe. In such regions environmental constraints are best met by the adaptive abilities of the group. We saw in Chapter 3 that this situation permits quite wide genotypic variation. In central areas one finds the most genetically polymorphic populations; peripheral zones, where the milieu is harsher, are less "tolerant." In central areas supergenes appear and are fixed by the blocking of crossing-over subsequent to chromosomal reorganization. Once established, these new combinations will be transmitted by gene flow and reach peripheral areas where environmental conditions, and therefore selective pressures, are much tougher. They will become established permanently if they have a selective advantage, and can then replace existing combinations. This mechanism contributes to the protection of the adaptive ability of peripheral populations living at the frontiers of the species's area of distribution, that is, at the limits of its genetic possibilities.

THE RICHNESS OF MECHANISMS OF VARIATION

What we have discussed so far demonstrates the richness and multiplicity of the mechanisms that maintain biological variety in all groups, even if certain forms of inhibition — such as supergenes or blocking of crossing-over — exist to allow the conservation of some acquired selective features.

Polymorphism is a fundamental law of living organisms. It is expressed not only in the nature of the hereditary patrimony by the presence of several alleles at many loci, but also at the level of combinations among genes that are challenged in each generation.

A population is a system in a permanent state of reconstruction, and the potential of its patrimony greatly exceeds what the naturalist has occasion to observe in the ridiculously limited space and time in which chance has placed him.

But it is possible that variation goes further. We have spoken so far as if genes always maintained the same fundamental action, continually delivering the same message — as if their expression depended uniquely on the "genetic environment" in which they found themselves, such as the nature of the allele on the other homologous chromosome, the nature of other nonallelic factors present in the genotype and linked by epistasis, and so on. In reality, it is conceivable, at least in eukaryotes (cells with a true nucleus containing a sophisticated chromosomal complex), that genes do not always pass on all their information. In higher groups many genes are made up of two types of fragment: the *exons*, which express the hereditary message, and the *introns*, which represent, in general terms, zones of silence. These zones are perhaps not constant; and it is possible that the way in which the gene is read varies according to the circumstances and the respective proportions of introns and exons, so that it can code for different proteins in different cases. Indeed, one cannot exclude the possibility that several "polymorphizing" mechanisms are active at any one time, such as recombinations plus differential gene read-outs.

polymorphism in the face of selection

THE PARADOX OF INTRAPOPULATIONAL POLYMORPHISM

If the origin of polymorphism seems clearly established, the mechanisms whereby it is maintained are at first sight less obvious. How does one explain why individuals freely interbreeding, living together at the same place and the same time, and therefore subject to the same selective pressures, do not have the same genetic background? The polymorphism seen in the hereditary patrimony of all populations is contrary to the neo-Darwinian assumption that there should be one particular genetic combination, and one only, that responds best to a particular environmental condition: the "optimal" genotype seen in the species's or race's characteristic holotype, the inevitable fruit of natural selection. Intrapopulational polymorphism is a paradox to neo-Darwinism. And yet, the fact that it has been conserved so consistently during the course of evolution suggests that it corresponds to the requirements of natural selection and confers some sort of advantage. Its universality even leads one to suppose that it might constitute a condition *sine qua non* for the survival of the group.

This paradox may be resolved in two possible ways. First of all, it is conceivable that all mutations are neutral as far as natural selection is concerned, or that all alleles at a given locus have exactly the same selective value, which amounts to the same thing. In this case, the distribution of polymorphism within a population would be random. This theory, put forward by Motoo Kimura and his pupil Tomoko Ohta of the National Institute of Genetics at Mishima in Japan, is supported by a certain amount of experimental data.

It is also possible to think, like many modern evolutionists, that if individuals always show a tendency to heterozygosity and populations to polymorphism it must be that ecological conditions, and hence the forces of selective pressure to which a given individual or group may be subjected, are variable in space and time. In the face of this environmental "mosaic" a very varied genetic complement would confer certain advantages. Let us examine these two theories.

The "Neutral" Explanation

The *neutral theory*, widely published in the last fifteen years, denies or minimizes the influence of selective pressure, at least in the majority of cases. If natural selection did not exist, or if all genes, or almost all, had a neutral selective value (which would have the same effect), natural populations would be spontaneously polymorphic. Suppose that a locus bears a gene A or its mutation a. If A and a have the same selective value, or if they have no value at all, their distribution in the population would be random, no more ordered than that of playing cards dealt from a well-shuffled pack.

Kimura has expounded his theory in a series of publications over the last fifteen years or so (Kimura, 1968–83; Kimura and Ohta, 1973). It has been taken up by Jack King, Thomas Jukes, and others, beginning in 1969. We have discussed it in detail elsewhere (Ruffié, 1976) and will restrict ourselves here to a review of its essentials.

For Kimura, selection can exist, but basically as a negative force capable of eliminating certain really deleterious genes. This is Mullerian selection, which differs from positive Darwinian selection with its concept of an advantage for the best characteristics, as we said earlier. But this negative selection would operate only in a small number of cases, for the immense majority of mutations would remain neutral. The neutralists think that when selection acts, its role is more one of conservation than of diversification. It is, as Susumo Ohno wrote in 1980, an "agent for the maintenance of order."

Kimura presents a series of arguments derived from observations of natural populations.

1. Mutations are more numerous in molecules that play only a secondary role in cell physiology than in molecules involved in a fundamental function. In other words, the most important molecules, those subject to the strongest selective pressure, are those that have fewer mutations and thus tend to evolve less. Histone IV, for example, has a chain of 102 amino acids, and its role is to compress the DNA filaments, several centimeters long, in the nucleus of eukaryotes. In doing this the total length of the histone molecule is involved. The entire chain is active. The histone IV found in the sweet pea and that found in the calf differ only by the substitution of two amino acids. The ancestral molecules common to both animals and plants existed at least one and a half billion years ago. Thus during all this time, selection has allowed only two mutations to become fixed — which seems quite insufficient to explain the evolutionary change from flowering plant to mammal.

On the other hand, fibrinopeptide A from horses and donkeys (two closely related species, since they can hybridize) is made up of a much shorter sequence of only 16 amino acids. The fibrinopeptide A of the two species differs by four substitutions, an enormous number if one considers the shortness of the chains and the fact that the donkey and the horse must have separated quite recently, perhaps a few tens of millions of years ago. Unlike histone IV, fibrinopeptide plays only a secondary role. At the required moment it merely leaves the molecule of inactive fibrinogen by fragmentation at an arginine group. As a result of this amputation, the molecule becomes active as fibrin in the coagulation process of the blood. In its whole length, fibrinopeptide A has only one active site, at the arginine that indicates the site where the break must be made.[1]

In the same way, the C chain of the proinsulin molecule, the precursor of the hormone that lowers blood sugar, which must at a particular moment break off from the molecule to produce active insulin, has many more mutations than insulin itself.

Recently, the introduction of particularly sensitive electrophoretic techniques has made it possible to demonstrate that the structural proteins of a cell do not mutate; nor do the membrane enzymes. Both are influenced by steric relations with neighboring molecules. On the other hand, soluble enzymes with less constraints mutate more frequently (see Jones, 1980).

2. If we consider relatively complex active molecules, such as enzymes or hemoglobin, mutations mainly affect zones that are functionally relatively unimportant, where modification of the structure of the peptide chain would have no particular consequence on the molecule's activity. The active sites remain intact.

In higher vertebrates hemoglobin is formed of a tetramer with four globin chains wound around a heme nucleus containing iron atoms that can fixate oxygen in a relatively labile way. The deep regions that contain the heme play a major role in the functional activity of the molecule. More superficial regions are less important and have mutated ten times more during evolution.

Another important example may be cited. We now know that the unit of genetic information coding for an amino acid is a sequence of three purine or pyrimidine bases situated side by side on the DNA chains that make up the chromosomes. They are called codons or triplets. In this system it is essentially the first two bases that are rigidly

1. See Russel F. Doolittle (1980) and A. Henschen *et al.* (1980) for a discussion of the evolution of fibrinogen and its components.

coded, the third being capable of replacement by a synonym, which means it can be changed without the semantic content of the codon being modified: the same amino acid will always result. It is this third, "neutral" base that shows the greatest frequency of mutations.

Thus almost always, mutations are much more frequent in molecules or fractions of molecules of little functional importance than in those that have an important role in cell biology. Mutations are very rare indeed in molecules or fractions of molecules that are physiologically very active.

3. This observation can be transposed to the ensemble of populations making up a species. If mutations were of selective value, groups living in a more or less constant environment, such as in the depths of the oceans, would be much more monomorphic than those in more variable environments, such as tidal zones subject to constant change. But the first are just as polymorphic as the second, if not more so.

In reality, Kimura and his colleagues think, polymorphism increases as conditions become more neutral. Functional constraints maintain monomorphism. When these constraints decrease, the rate of preservation of mutations in a group tends to be at its maximum. This may explain the apparent paradox observed in primates: although of recent origin, they gave rise very quickly to the hominid branch that led to man.

Evolution of the human race, such a vast phenomenon in biological terms, took only a few million years, probably between five and seven. This very rapid evolutionary process was marked by a remarkable lack of change at the molecular level. Very few mutations can be found between man and chimpanzee.

Kimura's discoveries are of fundamental significance. We shall see in Chapter 7 how — at least starting from the organizational level of the eukaryote — innovation is achieved more readily by new combinations and arrangements than by new mutations. Thus for higher organisms, mutations cannot explain species transformation.

The Selective Explanation

Selectionists, who represent a considerable number of naturalists, work from a different point of view. For them evolution, and thus genes and genetic combinations, have an essential adaptive role. Environmental variations thus act on genetic polymorphism in populations and play the essential role in evolution. We saw in Chapters 1 and 2, when speaking of ecogeographical rules, adaptation and acclimatization, and clines and gradients, that the hereditary patrimony of each population varies as a function of the environment. There can be little doubt that for many factors or combinations of factors such variations result from a "choice" brought about by a selective process.

Thus the appearance of new groups implies a continual role for the forces of selection. A fish has all the genetic makeup that will allow it to live in the water, and the ancestors of the first amphibians that began to live on the land had to modify their genes as a function of their new habitat. The conquest of the land happened through a progressive adaptation to different environmental conditions. The descendants of the first sea creatures that invaded the land colonized shores and estuaries first and then, little by little, spread inland into the heart of the continents. A similar adaptive progression occurred when certain reptiles conquered the air and gave rise to birds. These different evolutionary levels involve the adoption of new habitats and the exploitation of new niches. They are in perfect accord with the concept of populations.

Certain genes are doubtless neutral at the moment they are being studied, but we can say of neutralism what Dr. Knock said about health: "It is a precarious state that promises nothing good."

There are serious arguments in favor of the nonneutrality of many factors. Let us consider several widely separated populations, each living, however, in a geographic area with similar ecological constraints. If we look at the frequency of different alleles at several polymorphic loci by electrophoresis, we almost always find the same frequencies in each population. If their isolation is complete, gene flow cannot explain this equality. It can be related only to selective pressure. Certain species of crickets show *chromatic polymorphism,* characteristic variations in their body color. However, the distribution of morphospecies is constant for a given location from one year to another (Dreux, 1977).

The same phenomenon has been observed in laboratory colonies. It is common to raise fruit flies in *demometers,* cages where the environmental conditions, such as temperature, humidity, lighting, and food, can be varied. These are also factors of selective pressure. By studying the evolution of gene frequencies in the same population during successive generations, we can judge the adaptive value of certain genes to changes in environmental parameters.

1. When fruit flies of different varieties, and thus with different allelic frequencies, are raised in identical conditions but in separate cages, one finds that the allele frequencies tend to become uniform after a number of generations. In other words, there must exist certain frequencies that represent an optimal adaptation for particular ambient conditions, and all populations faced with the same selective factors tend in that direction (Lucotte, 1978).

2. J. R. Powell (1971) provided another proof of the relation between polymorphism and selective pressure. He varied a certain number of ecological parameters in his colonies over a period of time. For example, in the first demometer, conditions remained stable. In the second, one parameter was varied during the experiment, while in the third, two were changed, and so on. After a number of generations he found that genetic polymorphism was maximal in populations that had been subjected to the most numerous and most drastic ecological variations, while it was minimal in those that had been reared in stable conditions. There was a linear relationship between ecological variation and genetic polymorphism.[2]

In general, the enzymes used for the metabolism of sugar, an internal substrate that is remarkably constant in all species, are much less polymorphic than others acting on less specific external substrates, such as esterases, phosphatases, or hydrolases. The first are critical enzymes, the second peripheral enzymes. This observation was first made in the fruit fly, but was later confirmed in other animal species and even in man (Lucotte, 1978; Lucotte and Lefebvre, 1980).

These experiments show that genetic polymorphism in natural populations seems to be the response of living groups to constant changes in environment. Thus, far from being unfavorable, this polymorphism confers a powerful selective advantage on the individuals that possess it. In a constantly changing environment there is no "optimal" genotype nor holotype. This could only be imagined in a rigidly stable milieu, that is to say an environment of constant selective pressure. A world in which ecological constraints were always the same in all places and at all times would probably harbor genetically monomorphic groups incapable of progress. Evolution is born of genetic variation and the selective variation of fluctuations. It is inconceivable in an invariable system. The conditions in which we live are just the contrary of stable.

We may add here that this mechanism is effective because at the molecular level there are no totally recessive genes. Even the genes that seem not to be expressed in the macroscopic phenotype—the only one taken into account for so long by Mendelian genetics—are not completely silent. Their activity can be revealed by sophisticated techniques such as electrophoresis, but may remain undetected by direct observa-

2. One should not confuse *variations* in selective pressure with the *type* of pressure. In general, frequent violent variations lead to a high degree of polymorphism, whereas severe but *permanent* conditions tend toward relative genetic "poverty," with only acceptable combinations surviving and even these becoming rarer as the pressure increases. This is why in the center of a geographic area where relatively moderate conditions exist, polymorphism is often very noticeable in spite of the small amplitude of the variations. There is no contradiction between the two phenomena. See Chapter 3.

tion. And this action is sufficient to enable any gene to be used for the purposes of natural selection at the proper time.

Selection will not act to eliminate or generalize a gene, but rather to favor certain combinations and discourage others. Almost always a gene will be "judged" in relation to the functional ensemble to which it belongs, not in relation to its mere presence, which in itself is not very significant. So with the exception of highly unfavorable mutations, which must be rare, the adaptation of the hereditary patrimony to environmental fluctuation is expressed much more by variations in gene frequencies than by the loss, or the total fixation, of certain alleles.[3]

THE INSTABILITY OF THE TERRESTRIAL ENVIRONMENT

The terrestrial environment is not stable; it is subject to incessant spatial and temporal changes. Let us consider these two types of variations.

Spatial Variations

The earth's surface is not an ecological "continuum." It is subdivided into an incalculable number of microclimatic zones, often interdigitating with each other and changing over short distances, particularly in areas where the subsoil is irregular and the relief complicated. There are no homogeneous zones. Almost all are subdivided into smaller zones, each demanding different forms of adaptation.

The relationships between the genetic polymorphism of a group and the nature of the subsoil where it lives have been known for a long time, particularly concerning certain macroscopic features such as coloration.

We saw at the beginning of this book the role played by food chains in the equilibrium of living species. Predacity is the drive behind these chains. It is one of the most important selective forces and tends to favor individuals that are best adapted to their background and are thus less visible. They are *homomorphic*.

In Chapter 2 we spoke of the banded snail studied by Maxime Lamotte. It has numerous genetically determined morphs, so distributed that their camouflage helps them avoid predators, in particular the song thrush. This adaptation is so subtle that one can sometimes find different morphs separated by only a few meters — for example, one popula-

---

3. Fixation can be considered the elimination by an allele of all others for a given locus so that it appears in all members of a population. The factor has become "public," and the locus monomorphic. If the eliminated allele still survives in a few rare subjects, such as in a single family, it is a "private" factor.

tion living in a field and another in a bush. There are nevertheless exchanges between the two to maintain the homogeneity of the species and conserve a permanent reserve of variability in each population. Similar examples can be found in other groups. Homomorphism between animals and their background is a widespread phenomenon. It is the origin of what have been called *substrate* races, local populations that adopt the best camouflage for their particular environment. Homomorphism is of obvious adaptive value. It is not the background that creates the selective pressure, but the chances of being seen by a predator. When the background variation is continuous — an ecological gradient — one can observe a similar continuous morphological variation or genetic cline. This relationship is compatible with the ecogeographical rules discussed in Chapter 1.

Obviously, spatial variations are not restricted to predatory factors but are related to all ecological parameters that constitute selective elements, such as climate, attacks by microbes, parasites, or viruses, food supply, or competition.

Spatial ecological variations are not only found over large distances. An insect or a lizard that moves from the sunny side of a rock to the shady side changes its world. These variations are most obvious in terrestrial species. It would be tedious to list more examples. We may simply say that in one way or another this "ecological mosaic" applies to all species.

Temporal Variations

Ecological conditions, and therefore selective constraints, vary with time. According to their duration we can distinguish three types of variation.

First of all are those due to the rotation of the earth, lasting twenty-four hours, and corresponding to the alternation of day and night: *circadian* or *nycthemeral* variations. There are also *seasonal* or *annual* variations related to the movement of the earth around the sun. Finally, there are *secular,* or better, *millennial* variations, such as warm periods and ice ages, with a much longer time constant and whose origin is still unclear.

It is possible that such changes affect the whole universe. Worlds in which conditions are constant are difficult to conceive in our perpetually moving cosmos. Such variations obviously play an important role as selective factors. Before going further we might examine them in more detail.

1. Daily variations have short cycles. Throughout the day there are changes in temperature, sunshine, availability of food, the dangers of aggression, and so on. At different moments, different genetic combinations will be at an advantage.

2. Seasonal variations are more prolonged. In those species that produce several generations each year, there will be periodic variations in the frequency of different morphological features such as color or size. These variations are reflected at the level of the genes themselves; the frequency of different alleles changes regularly throughout the season.

In 1940, N. W. Timofeeff-Ressovsky pointed out that the red variety of the ladybird *Adalia bipunctata* predominated in winter, whereas the black variety was mainly found in summer. Other examples have been described in butterflies. Populations of *Colias eurytheme* found in the cold season tend to be black and are active at low temperatures. In the warm season the adults are lighter in color and more active at higher temperatures. When both varieties are placed in similar temperatures, the former variety seeks shade and is relatively inactive.

In some cases the dissimilarity between seasonal types is sufficiently large that they have been described as belonging to independent species and even different genera. For example, the variety of *Hestia assimilis* found in the dry season is very different from that found in the wet season (Guillaumin and Descimon, 1976).

These modifications are related to changes in gene frequency from one season to another. They reflect the optimal diffusion of the alleles giving the most advantageous combinations for environmental conditions. Such variations can easily be followed by observing morphological features, but it is probable that they also (and perhaps mainly) involve physiological and behavioral characteristics whose adaptive value is certainly greater.

Other temporal variations are related not to alleles but to chromosomal modifications, so they are easy to observe (Cassagnau, 1977). They were described by Dubinin and Tiniakov (1946) and Dobzhansky and Epling (1944) in the fruit fly. *Drosophila pseudoobscura* has several types of inversion on its third pair of chromosomes. Their frequency changes with the season, according to a very regular cycle. This phenomenon was described by Dobzhansky in populations of fruit flies at Piñon Flats in California (Figure 19).

3. Those variations that stretch over much longer periods are called secular. They are, in fact, "millennial" and may be related to climatic changes, such as ice ages, or geographic variations, like continental drift or orogenetic movements. They are generally progressive, but can be related to more sudden modifications due to natural catastrophies like volcanic eruptions, landslides, or floods. Their periodicity is less regular than that of nycthemeral or seasonal modifications.

Like spatial variations, temporal variations affect all environmental factors, such as climate, predacity, nutrition, and pathology.

THE SELECTIVE ADVANTAGE OF POLYMORPHISM

All the species that surround us are, therefore, subject to permanent modifications in selective pressure, both spatially and temporally.

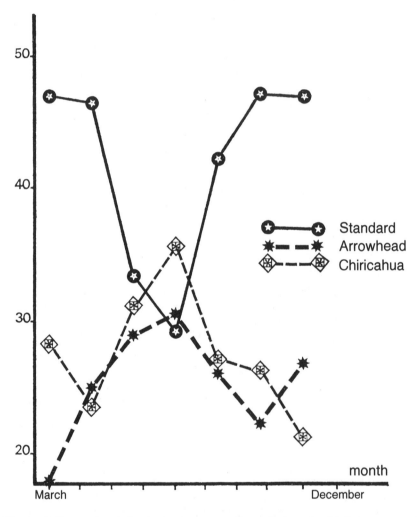

Figure 19. Seasonal variation in the frequency of inversions on the third chromosomal pair in three variants of *Drosophila pseudoobscura* at Piñon Flats. (From Dobzhansky and Boesiger, 1968; data from Dobzhansky and Epling, 1944.)

Some are periodic, succeeding other variations that may have acted in the opposite direction. Thus an individual responds to environmental heterogeneity by heterozygosity, just as a population responds by polymorphism. In an environment subject to multiple variations, individual heterozygosity and population polymorphism confer immense advantages.

The Advantage of Polymorphism at the Individual Level (The Advantage of Heterozygosity)

We saw in Chapter 1 how electrophoretic techniques have shown that heterozygosity is a permanent feature in all individuals, whatever their species and whatever their habitat. As early as 1926, long before electrophoresis was known, the importance of heterozygosity in wild populations — including those that seem at first sight very monomorphic, at least in their morphological characteristics — had been emphasized by the Russian geneticist S. S. Chetverikov (trans. 1961); it was later developed by his pupil, Timofeeff-Ressovsky (1940). Subsequently many other observations were reported, both in natural populations and in the laboratory (Dobzhansky and Boesiger, 1968; Pasteur, 1974; Lucotte, 1977). They concern all hereditary features and demonstrate that heterozygotes are in general much less sensitive to environmental variation than homozygotes. Apparently better equipped to respond to the constraints of the environment, they are the ones who always win the battle of natural competition.

East and Jones pointed out early in the century (1919) that plant heterozygotes were particularly vigorous, and this was exploited in agriculture. From 1930, the Americans utilized hybrid corn with a much higher productivity than the "pure" forms. Later experiments on animals, including insects (particularly fruit flies), rodents, and birds (particularly the Japanese quail), showed that this rule was universal (see, for example, B. Boesiger *et al.,* 1975; E. Boesiger, 1974a, 1974b, 1975; Ayala, 1977, 1982). In general, individuals born of crossings between very different, distant populations with a high degree of heterozygosity are in better physical condition and have greater vitality and resistance to the aggression of the environment than their parents.

It is impossible to cite all the experiments of the last few years, but we may quote the particularly demonstrative work of Francisco Ayala as an example. He used two varieties of *Drosophila serrata.* The first population (I) came from a pure line, very endogamic and therefore weakly polymorphic, for all individuals carried certain sequences of identical chromosomes coming from their common ancestors. The second population (II) was obtained by mixing two wild strains living in two widely separated areas. It was therefore highly crossbred. Electrophoresis showed that group II was twice as polymorphic as group I. Ayala put both populations into a demometer and provided the same ecological conditions in terms of limitation of food and space in order to increase the selective pressure. After twenty-three generations, the demographic growth of the second, more polymorphic population was twice that of the more endogamic population I, as can be seen in Figure 20.

These observations agree with others that have been made in human half-breeds, who are often at a biological advantage compared with their ancestors. This phenomenon has been called *luxuriance* or hybrid *heterosis,* although it would be better to say *half-breed* heterosis, for, as we have seen, the term hybrid should only be applied to crossings between subjects from different species.

By contrast, consanguineous lines become more and more fragile and often end by degenerating. Consanguinity leads to *depression* or *antiheterosis.* In general, consanguineous individuals are genetically less polymorphic than others. All have common ancestors from whom they receive at least a part of their genetic patrimony, so they have a certain number of genes representing the same chromosomal segments and transmitted without much change from generation to generation. By interbreeding they inevitably finally produce individuals with many

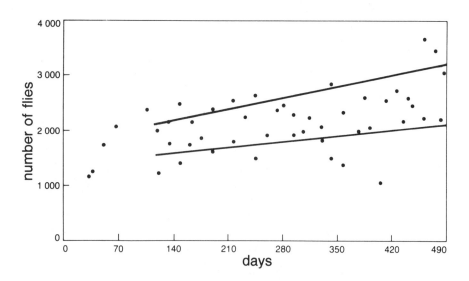

Figure 20. The effect of genetic variation on the ability to respond to selective pressure. (From Ayala, 1977.)

homozygous loci. They are sensitive to external variations, for their adaptive responses are relatively restricted.

The Mechanism of Heterozygous Advantage

What are the causes of the advantages enjoyed by heterozygotes? They are certainly rather complex. The first element to be considered is *genetic homeostasis,* a term coined by I. M. Lerner in 1954 to explain the luxuriance of the crossbreed. It can be formulated as follows: Heterozygotes have a much wider range of genotypic responses than homozygotes and are thus at an advantage both during development and during adult life. They are better able to respond to the constraints of a constantly changing environment.

A simple comparison can explain how heterozygosity represents an advantage. Take three tourists at the border between France and Spain, able to pass freely from one country to the other, but unable to change their money. The first has French francs in both pockets. As he walks along the border, he will sometimes be in France, where he can eat without difficulty. When he ventures into Spain, however, he will risk dying of hunger. The second tourist has Spanish pesetas in both pockets. When in Spain, he can buy whatever he wants, but when in France he suffers all forms of privation. The third person has pesetas in one pocket and francs in the other. He can meet all his needs whichever country he finds himself in. This polymorphism, represented by the possession of two currencies, has considerably enlarged his ecological niche and the resources he can exploit. He has an advantage over the other two. His chances of survival and reproduction are greater.

The adaptive capacity of heterozygotes is expressed phenotypically in a curious way. D. S. Falconer (1960) and others have demonstrated that when subjected to strong environmental constraints, heterozygotes show fewer phenotypic modifications than homozygotes. They can maintain their biological constants within acceptable limits more efficiently. They are less perturbed. The variety in their genetic background permits them almost always to respond by physiological mechanisms that do not threaten their physical structure. They have a greater biological inertia. Thus the genetically most polymorphic populations are phenotypically the most stable, which at first sight may seem paradoxical. In fact, the more heterozygous loci an individual has, the more it is protected against environmental changes and the aggressions they provoke.

Let us take a very schematic example that reflects the story of our three tourists but in biological terms. An enzyme A allows maximum physiological activity between 15° and 20° C. Let us suppose that there has been a mutation a, with the result that the best activity is now between 20° and 25° C. A homozygous subject A/A will work best between 15° and 20° C, and a homozygote a/a between 20° and 25° C. On the other hand the heterozygote A/a will have a range from 15° to 25° C. Individuals with A/a will be able to be active for longer periods during the day or the year than individuals with A/A or a/a, each of which will be restricted to a narrower range of temperatures.

Similar advantages can be had in terms of space. Heterozygotes may have access to a wider territory, an obvious advantage over a homozygote restricted to narrower frontiers. Let us suppose, for example, that the two alleles A and a are found in a population living on the slopes of a steep mountain, where there are wide changes in temperature with altitude (Figure 21). Subjects with A/A will remain near the summit, while subjects with a/a will remain on the lower slopes. However, A/a individuals can live at any level. Not only can they cross the top of the mountain just like A/A, but they can also colonize the lower slopes on both sides. Genetic polymorphism produces a pioneer spirit. In natural populations, four principal elements intervene constantly in selection.

1. Climate. Animals and plants must adapt their physiological functions to a series of climatic givens such as temperature, humidity, and wind, and also to variations in these factors with time, such as diurnal or annual cycles. This adaptation is particularly important for reproduction, which is often related to a seasonal rhythm, particularly in cold-blooded animals. The greater the genetic polymorphism, the wider the territorial frontiers and periods of activity.

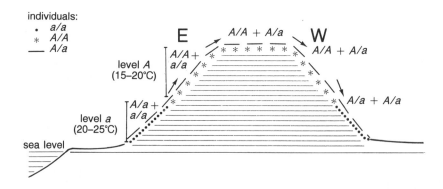

Figure 21. Schematic diagram of a mountainous region where mean temperatures decrease with altitude. Individuals with a/a live on the lower slopes, while those with A/A live near the summit, on the same side. Heterozygotes can live at all altitudes and even colonize the other side of the mountain. They ensure gene flow throughout the species and thus its cohesion.

2. Predation. We have already seen its importance as the motive force of food chains, channeling energy from plants to carnivores via the herbivores. All animals are both hunters (or harvesters) and hunted. Their chances of survival are greatest if they possess a large number of strategies for attack, defense, and flight. Many of these behavioral patterns are innate, at least in the beginning, and depend on the variety of the genetic patrimony.

3. Food. The less a species is fastidious in its choice of food, the more it is likely to survive unfavorable seasons or periods of shortage and the more likely it is to be able to exploit territories forbidden to others.

4. Diseases. Pathogenic complexes, particularly parasites, play an essential role in selection. Individuals equipped to defend themselves on all fronts, particularly by the possession of a diversified immunological arsenal, are the best armed.

In all these cases genetic equipment that is the most varied furnishes the broadest possible response and confers an obvious advantage. In general, heterozygotes dispose of a much greater temporal and geographic latitude than do homozygotes; they thus have access to a wider variety of natural resources and are better armed to face life. The probability of their reproducing is greater. How can one doubt that heterozygosity represents a highly favorable situation in terms of natural selection? Much evidence is available to prove this hypothesis.

We may mention two examples, based on the work of Lucotte *et al.* (1978), Frelinger (1972), and others. Lysozymes are proteins found in eggs. In the quail there are two alleles. Electrophoretically the first shows a slow migration (type *L*), while the second is more rapid (type *R*). Some subjects are heterozygous, and during electrophoresis, two bands are formed by migration of the two proteins *(L / R)*. One of the functions of lysozymes is to dissolve the membranes of bacteria. Thus normally these enzymes have an important role to play in protecting the egg against infection. It is technically possible to measure precisely lysozyme activity during incubation. The result demonstrates that the activity of the alleles *R* and *L* varies with the pH of the medium, and that the optimum pH for the enzyme R is not the same as for L. As the pH of the egg white decreases during incubation, the two enzymes are not effective at the same time. Thus the heterozygote is at an advantage because the functional range of the enzyme is wider over a given time. There will be a greater antibacterial protection than in the case of the two homozygotes.

Conalbumin is an egg protein that transports iron, whose structure and function are related to those of haptoglobin. It protects the white of the egg against bacterial infections in a very different way from the lysozymes, by reducing the iron content of the medium and thus preventing bacterial growth. There are two allelic forms, A and B. Heterozygotes A/B have a lower iron content than the two homozygotes A/A and B/B, thus offering a greater resistance to bacterial infection.

One could cite many examples in both animal and plant kingdoms. Nevertheless genetic homeostasis alone cannot explain the advantages of crossbreeds. In certain cases other mechanisms must intervene.

The two following experiments by Lucotte and Kaminski (1978) and Lucotte *et al.* (1977) are a good demonstration. They concern hybrids of chickens and quails. Alcoholic dehydrogenase in the serum of the chicken can be identified on account on its rapid electrophoretic migration. In the quail it is slower. In the chicken / quail hybrid the enzyme shows up as three bands, one corresponding to the paternal form and another to the maternal form, while the intermediate band is made up of the two halves of the two molecules combined in the form of a polymer. Alcoholic dehydrogenase is sensitive to high temperature, which greatly reduces its activity. The form of the enzyme found in the chicken is more fragile than that of the quail. However, the hybrid enzyme is more resistant than those of the chicken and the quail added together.

In the chicken, pseudocholinesterase has a quicker electrophoretic migration than that of the quail, the hybrid variety being intermediate. In particular doses, eserine can progressively inhibit pseudocholinesterase. This progressive inhibition of the enzyme by eserine can be represented by curves that are comparable for the quail and chicken. On the other hand, the inhibitory curve of heterozygous pseudocholinesterase is different. It demonstrates that eserine acts in a much more limited zone of concentration than in the two other cases, thus showing that the hybrid enzyme is much better buffered than those of the two parents. Thus the capacities of heterozygotes can exceed the sum of the capacities of the two homozygous parents. Their luxuriance is often much greater than a simple additive effect would suggest.

This phenomenon is found when crossbreeding leads not only to the coexistence of two parental enzymes in the hybrid but to the appearance of a polymer. The fact that a complex molecule will have new properties is not surprising. An increase in the semantic value of a molecule — that is, in its intrinsic properties — as a function of complication seems to be a fairly generalized phenomenon.

G. B. Johnson (1976) suggests that the advantages possessed by hybrid molecules could be due to an *allosteric* phenomenon — that is, to the capacity certain proteins have to change form and function under the influence of specific molecules, which means according to the nature of their surrounding medium. At the enzymatic level allosterism is expressed by the possibility that a molecule can receive "signals" and modify its activity in consequence. Allosterism always implies a certain molecular complexity and must be the basis for many regulatory phenomena.

Whatever the nature of the mechanisms, the advantages of heterozygotes seem clear. They have a more varied genetic background and are capable of facing more serious modifications of the environment than homozygotes. They suffer less from change and adapt better. They survive more easily the multiple aggressions of the environment. Homozygotes, on the other hand — and particularly consanguineous individuals — can respond only to a small number of alternatives. Obliged to live in a constantly changing environment, they come up more quickly against barriers they cannot overcome.

Is this advantage consistent? Even if the heterozygote cumulates the advantages of homozygous parents, does it not also cumulate their disadvantages? According to Bruce Wallace (1968), the heterozygote does not have a permanent selective advantage over the homozygote, which in particular circumstances can be in the more favorable situation. But as was said earlier, natural environments are not constant, and all living beings are subjected to incessant variations in their selective constraints. The advantage of heterozygosity is related not to the nature of the constraints but to their variations. For a general review of this problem, see Christen (1976).

All these observations lead to the conclusion that many loci have a greater selective value in the heterozygous state than in the homozygous. This property has been called *superdominance* or *overdominance,* and the relevant locus is said to be overdominant when it is occupied by a different allele on each homologous chromosome. In the example above, with two enzymatic mutations A and a that react differently at different temperatures, one would say that the genotype A/a is overdominant compared with A/A or a/a. We will take this question up again at the end of the chapter in studying variations of the selective value of a gene as a function of other elements of the genotype.

The Advantage of Polymorphism at the Level of the Population

We have just seen that polymorphism represents an advantage to the individual, but it is also advantageous to the population, for it enables the whole group to adapt. When ecological conditions change, a heterogeneous population can modify its gene frequencies over several generations, thus putting into circulation those combinations that respond best to the new constraints. This adaptive plasticity is possible thanks to the reserve of genes conferred by the heterogeneity.

We have seen that groups with several generations per year can adapt to seasonal modifications by changing certain features such as color, a phenomenon that must reflect periodic changes in the frequency of certain alleles.

Let us look again at our example of the enzyme A that mutated into a so that the optimum temperature for A was between 15° and 20° C and for a between 20° and 25° C. Let us imagine what would happen in a species having several generations annually and living in a limited geographic zone. In the warm season, the commonest populations would be those with *a / a* and *A / a*. The frequency of *a* would be high, and that of *A* much less. The opposite would be true in the cold seasons, when we would find mainly *A / A* and *A / a*. Now the frequency of *A* would be much greater than that of *a*. Thus the species would survive by continually adapting the genetic structure of its populations to the constraints of the environment. This phenomenon is well known to zoologists. Among the Culicidae exist a number of varieties of the common gnat, *Culex pipiens,* many of which disturb our nights both in town and in the country. Some forms are ubiquitous. In California there are large numbers of the subspecies *C.p. fatigans,* found mainly in tropical and subtropical regions. In the same areas one finds also *C.p. pipiens,* which can live in harsher climates. During cold winters *C.p. fatigans* is eliminated and only *C.p. pipiens* is found, together with a related variety that has been promoted to an autonomous subspecies *C.p. comitatus* but is really only a hybrid of *pipiens* and *fatigans.* Thus in the cold, which is presumably a highly negative selective factor, *fatigans* can disappear temporarily as a subspecies, but its genetic pool survives in heterozygous forms in *comitatus.* This "shelter" will permit it to reappear when climatic conditions are more favorable. It will simply be necessary for the recombinants derived from the cross-bred *comitatus* but having features of *fatigans* to be once more at an advantage.

A similar phenomenon has been described in Asia as far as Japan, where one finds *C.p. pallens,* the Asiatic hybrid of *pipiens* and *fatigans* (Grjebine *et al.,* 1976). This back-and-forth from one group to the next is simply the phenotypic expression of variations in the frequency of factors in the same gene pool.

The zoologist may see in *comitatus* and *pallens* independent species or subspecies. For the geneticist the two forms are simply different ways of assembling ancestral genetic pools. Their role is important because they permit the conservation of the entire hereditary patrimony of a group, but in new combinations that are "tolerated" by natural selection during difficult times. Selection has transiently eliminated one subspecies — that is, one sort of recombinant — without having much effect on the fundamental, hereditary patrimony, as can easily be seen in the spring, when the vanished species reappears.

This process has been seen in all groups, including vertebrates, even if it does not go so far that the changes in gene frequencies actually create new seasonal species or subspecies (Pasteur, 1977a). During these transformations, rare alleles become common and common ones rare, but none of them actually disappear completely. The population's survival depends on a new genetic equilibrium. Even if the frequency of the factors is sometimes modified fundamentally, they are never completely eliminated. This new equilibrium manifests as a changed gene frequency — not as an impoverished patrimony, as neo-Darwinism would require.

In similar selective conditions a monomorphic population might disappear, as can be seen in some domestic species that are very sensitive to environmental variations. But polymorphism in populations has another advantage. In a monomorphic group all individuals are tied to the same genetic programs and tend to do the same things — such as choosing food or sexual partners — at the same place and time. This form of monomorphism creates fierce internal competitions — an obvious disadvantage. In a polymorphic population, on the other hand, the times and places of various activities are much more widely dispersed. Polymorphism reduces competition and increases resources. It is a favorable selective factor.

We have often said that the ecological niche in which a species develops is simply the material reflection of its genetic capacity. The greater the degree of polymorphism — that is, of heterozygosity — the more an individual or a group is able to exploit its natural environment. Increased polymorphism widens the ecological niche. It adapts a population to new behavioral patterns and new functions. It confers new powers and allows the organization of peripheral *subniches*.

THE MECHANISMS LIMITING GENETIC POLYMORPHISM

Most biological processes involved in evolution are like balance sheets, with a series of positive elements set off against other, negative elements. For the balance to be positive, it is sufficient that there are more positive than negative elements (Ruffié, 1976). These actions and reactions, whose comparative importance will vary from time to time, give great flexibility to living organisms. They permit constant adjustment.

Genetic polymorphism is no exception to this fundamental rule; thus, in addition to processes that tend to increase, or at least maintain, genetic heterogeneity within groups, there are also mechanisms that limit it. There are three of these, which we shall describe briefly.

Natural Selection

We saw in Chapter 5 that selection rarely acts on an isolated gene, which in itself has no intrinsic selective value, but rather on combinations of genes, that is, on the whole genotype. We also said that there does not exist a single well-adapted genotype for particular environmental conditions, but several. These are the ones that selection favors. Selection acts constantly but in moderation, except on highly unfavorable monstrosities that are very rare and quickly eliminated. In reality, natural selection never has the rigid, pitiless character of the selection imposed by man on domestic animals, which is designed to eliminate radically at birth any individuals slightly different from the chosen model. Natural selection is indulgent. It operates by a host of small deviations in a probabilistic way. In statistics, improbable events sometimes do happen; selection also lets improbable events happen from time to time.

It has its limits, of course, but its range is wide. Only what is outside those limits will be selected out. This is why no population contains all the theoretically possible genotypes but only a fraction of them, those that are compatible with the environment. Every natural population has considerable polymorphism, but only within the limits of what is tolerated by natural selection.

Chance

In Chapter 3 we saw that chance plays a role in biological evolution, and earlier in this chapter we saw the importance that Kimura and his colleagues give to random phenomena in the process of evolution. Chance affects all stages of the life cycle, from birth to death. In many species one individual will produce millions and even billions of gametes, but often only a few of them will be fertilized. All the rest will be lost. Chance plays an essential role in small populations, where the total loss of a gene is not uncommon.

We said that if necessary, a whole population could be born from a single couple and even from a single fertilized female. In such a case the genotype of the founder or founders constitutes, at least in the beginning, an important factor limiting polymorphism. This is the founder effect mentioned in Chapter 3. Certainly the role of selection should not be underestimated, for it is responsible for species adapting to their environment. The stability of these adaptations, which form part of the characteristics of the species, proves that it is of genetic origin. But this in no way diminishes the importance of chance, which is superimposed on natural selection and attenuates its effects. Chance arrives from time to time to disorganize the biological order that selection has established. It periodically shakes up the too-restrictive barriers and allows evolution to change course. Chance is "anticonservative."

Isolation

We saw in Chapters 3 and 4 how gene flow plays an essential role in the maintenance of the heterogeneity of populations. Any obstacle to these exchanges almost always manifests as a decrease in polymorphism and a tendency for a population that has lost contact with its group of origin to become a subspecies and even a new species. Sedentary birds have twice as many species as migratory varieties (Mayr, 1963), and similar examples can be found in many phyla. Such phenomena are mainly found in isolated groups with few members. When acting on small populations, isolation amplifies the effect of chance.

THE FRONTIERS OF POLYMORPHISM

Genetic polymorphism acts between two limits. The lower limit occurs at the point where it would be disadvantageous to have too many homozygotes in natural ecological conditions, which, as we have seen, vary a great deal over space and time. Only domestic animals, created by man using particularly strict artificial selection techniques and living in the very stable human environment, can survive with a very high degree of homozygosity. Many domestic animals would have little chance of survival if returned to nature. Their genetic background is too poor to allow them to defend themselves against all the constraints. There is also an upper limit, for only those mutations that can be integrated in an ensemble giving the individual good environmental adaptation are preserved. The others are eliminated. Genetic polymorphism is liberal, but never overpermissive.

THE GENETIC LOAD

In all populations there exists a percentage of deleterious genes that can produce by segregation a number of individuals in each generation homozygous for these mutations and thus at a great disadvantage. As we have already seen, they are threatened by selection and risk "falling by the wayside" as the group progresses. These genes represent the *genetic load,* a concept suggested by H. J. Muller in 1950 and already referred to in Chapter 5. It can be defined as the "difference between the actual fitness of the population and the postulated value of a genotype with maximal fitness" (Mayr, 1963) or as "the difference in selective value between the real population and a hypothetical population made up of only the best genotypes" (Petit, 1977).

Based on the notion of the best genotype, the idea of the genetic load is directly related to typological thinking. The concept must be reexamined in the light of data from population genetics, because its foundation, the "best genotype," is so vague a notion. In any given context one can certainly assume that at any polymorphic locus one particular allele may be favored, which will therefore enjoy a greater probability of reproduction. But generally this does not last, in view of the instability of the terrestrial environment and its perpetual variations. So we retain the term "load," now in common usage, but use it to refer to the mass of genes that are either not expressed or very little expressed, but nevertheless remain present in genotypes in a heterozygous state. If the circumstances change they can "come out of hiding." All that is needed is for the combinations in which they take part to become favorable.

VARIATIONS IN THE SELECTIVE VALUE OF A GENE

In its original definition, the genetic load that characterizes a population at a given moment implies the invariance of the selective value of each gene. Attached to the gene like an indelible label, this value would depend uniquely on the gene, remaining constant whatever the other elements of the genotype might be. This concept postulates that the overall selective value of a genotype is no more than the sum of the selective values of the different genes that make it up. But that hardly corresponds to reality. Selection does not act on each gene separately and could in no way be defined as the sum of the selective values of each gene, added up one by one, for the simple reason that no gene acts in isolation but always in an integrated way. Natural selection acts on complex, functional ensembles of which each component part is capable of multiple interactions (see Chapter 4).

As we have seen, the selective value of a gene varies with (1) the other genes in the genotype and even in the population, and (2) the external conditions that confront it — that is, the selective forces the individual is called upon to face. A mutation can remain silent in certain circumstances and express itself fully in others.

It would be useful to explain now the possible variations in the selective value of a gene as a function of the individual's genotype; as a function of the genotypes of the whole population; and even, finally, as a function of the biological environment.

Variation in Selective Value as a Function of the Whole Genotype

The selective value of a gene depends first of all on the other elements of the genotype.

1. We might recall the general case already mentioned, that of functional *complementarity*. Many enzymes and proteins are functionally correlated — for example, those that work together in the same metabolic pathway. It is obvious that the selective value of one enzyme depends on the presence of the other, without which it would be useless. If an enzyme E1, highly favorable in a given milieu at a given time, is under the control of two genes, *A1* and *B1*, these two factors have a powerful selective value when they are present simultaneously in the genotype. If *B1* is replaced by a mutation *B2* with less synthetic power, the value of *A1* could be considerably reduced and even abolished, for in reality selection "judges" the functional group *A1 / B1*.

In the case of a mutation that perturbs the whole system, the organism must find a replacement solution such as a *vicariant* metabolic pathway (as is often the case). If not, it is condemned to disappear.

2. Balanced polymorphism and overdominance are other very common phenomena that contribute greatly to the maintenance of genetic variety. We have seen how two alleles with complementary actions can give a considerable selective advantage to the heterozygote (overdominance, described earlier); but it is also common for a gene that is deleterious and even lethal in the homozygous state to be favorable in some heterozygous situations. This explains the persistence of lethal genes in many wild populations. In 1930 R. A. Fisher demonstrated that genes that are very unfavorable in the homozygous state can remain indefinitely in a population if they confer an advantage in the heterozygous state.

A well-known, classic example is that of hemoglobin S in man and its relationship to resistance to malaria. Hemoglobin S is characterized by the replacement of glutamic acid, the amino acid in the sixth position on one of the beta chains of the globin, by another amino acid, valine. This substitution transforms hemoglobin A (Hb A) into hemoglobin S (Hb S). Homozygotes for this mutation (*Hb S / Hb S*), with only hemoglobin S in their red blood cells, have a severe congenital anemia often incompatible with prolonged life. This is drepanocytosis, also called sickle-cell anemia, because the red blood cells of these patients have a deformity that makes them look like a sickle or a crescent.

In the heterozygous state *(Hb A / Hb S)* this mutation is quite compatible with life, there being a sufficient quantity of normal hemoglobin in the cell to preserve its functional integrity. Indeed, the presence of hemoglobin S effectively protects against malignant tertian fever, the plasmodium being unable to develop properly in the presence of Hb S. Thus in a zone of endemic malaria, homozygotes *Hb A / Hb A* are threatened by the malaria, which is often fatal. Homozygotes *Hb S / Hb S* suffer from severe congenital anemia and often do not reach reproductive age. Only heterozygotes *Hb A / Hb S* remain in satisfactory health, showing no signs of anemia and highly resistant to malaria. This mechanism was described over thirty years ago by Allison (1954), who demonstrated the correlations between the distribution of the *Hb S* gene and that of *Plasmodium falciparum* malaria. Since then many other cases of fixation of deleterious genes have been described — for example, deficiency in glucose-6-phosphate dehydrogenase — that illustrate the phenomenon of balanced polymorphism.

In the great majority of cases, apart from frankly pathological mutations, a gene is neither "good" nor "bad": all depends on what accompanies it. The discovery of balanced polymorphism condemns the concept of the genetic load. In addition, it shows how fallacious careless eugenic campaigns can be. While it might be legitimate to want to avoid certain phenotypes, it can be dangerous to eliminate genes whose role in the heterozygous state might be irreplaceable.

Let us look again at the example of sickle-cell anemia. Black Africa and a part of the Indian subcontinent pay a heavy toll to this disease, which represents an enormous burden for families and for nations. Steps have even been proposed to prevent the birth of *Hb S / Hb S* homozygotes, whose often short life can be a real trial for their families. Such a prophylactic campaign is now conceivable. It would involve the diagnosis *in utero* of *Hb S / Hb S* homozygotes and the proposal of voluntary termination of the pregnancy. But eradication of the *Hb S* gene must certainly not be attempted. Its loss could provoke a catastrophic plague of malaria in a totally unprotected population.

The phenomenon of balanced polymorphism, which is simply an accentuated form of overdominance, must account for the persistence of many mutations that in the homozygous state would be unfavorable or even incompatible with life — the lethal mutations. Thus polymorphism is related not only to constant variation in the environment but also to the phenomenon of overdominance, which ensures that heterozygotes will always have an advantage over homozygotes.

Variation of Selective Value as a Function of Gene Frequencies in the Population

As we have seen, the selective value of a gene varies as a function of other elements in the genotype of the same individual; but it also depends to a great extent on the frequency of other genes present in the population. This fairly new idea was put forward by Claudine Petit (1951, 1958) and has been confirmed on several occasions since then (Anxolabéhère and Périquet, 1972; Anxolabéhère, 1976a). The most striking manifestation of this phenomenon is the *advantage of the rare type.* In a wild population the rarest gene almost always has less selective value than a more frequent allele. That is indeed why it is rare. But from time to time the rare allele has a distinct advantage, at least as long as it remains present at low frequency. This advantage disappears when the frequency increases. For a given ecological situation two alleles each reach a certain value, establishing a sort of equilibrium at which they tend to stabilize.

The advantage of the rare type, which at first seems paradoxical, has been described in three situations.

1. Sexual selection. In colonies of fruit flies, males with rare genotypes are more successful in coupling than those with common genotypes. This phenomenon has been observed in other species, but its precise mechanism is not understood; nor do we know how the sexual partner, a female in this case, identifies a rare genotype. It may be that the nuptial parade is different and more attractive for the females.

2. Larval competition. When two forms of larvae compete in the same milieu, the percentage of hatching is greater in the less frequent form. This is probably due to the fact that the two types do not exploit exactly the same microhabitat. Thus the rare variety is less subject to competition than the common, will have more food, and will be less susceptible to losses and the inhibiting restraints of overpopulation.

In general any variation that causes the subject to live on the edge of a crowded territory is advantageous, because competition is less in this area. The same conclusion was reached by Ken Ichi Kojima concerning genes coding for several enzymes in the fruit fly — in particular, esterase-6 and alcoholic dehydrogenase. Similar results have been obtained recently for other enzymes (Ayala and others).

We shall see in Chapter 8 that this mechanism must explain the tendency to progressive separation seen in populations that have begun to diverge. The same phenomenon of "reciprocal separation" has also been seen in sibling species. Selection conquers rather than eliminates. It tends to occupy empty spaces rather than destroy populations that already occupy full spaces. This is why mutants that nibble away at the frontiers of a species often have a strong selective advantage. They are less troublesome to the others and are less troubled by them. The future is always bright for pioneers.

3. To these mechanisms that give advantages to rare genes we can add the observation that the normal functioning of selection is weakened for exceptional morphologic forms in species that are victims of predators, such as insects, snails, herbivores, mammals, and rodents, among others. We have already stressed the fundamental role that predacity plays in natural selection. Sometimes a predator will retain in its memory a "blueprint" of the commonest form of the species it is hunting. The rare form may thus not be identified and escape. This was called *apostatic* selection by B. Clarke (1962).

Nevertheless, all these balances can be upset if the frequencies of the different alleles vary. As soon as initially rare larvae become common, they in turn encounter the limiting factors of overpopulation. In the same way a formerly exceptional morph that has become common will be exposed to the hunting instincts of the predators. The advantages discussed above last only as long as their rarity value is maintained. Thus any variation in the frequency of an allele in a population can change its selective value.

This phenomenon seems quite generalized. It applies even to chromosomal inversions, whose selective value varies with their frequency in the population under consideration (Kojima and Tobari, 1969). At present, many people think, like Petit (1971, 1973), that the advantage of the rare type is just as important, if not more so, than heterosis in maintaining genetic polymorphism in natural populations. Several authors have proposed models of this *frequency dependence*. This issue is beyond the scope of this book, however, and we can only mention some of them here, such as Petit (1966), Petit and Nouaud (1976), Anxolabéhère (1976b), Ayala (1971a, 1971b, 1972, 1982), and Kojima and Tobari (1969).

Variation of Selective Value as a Function of the Biological Environment

By biological environment we mean the nature and frequency of other gene ensembles belonging to other species in the same biotope. We need no longer stress the fluctuations in selective value of genes as a function of the environment, since we have already discussed this problem in detail. Nevertheless, we should emphasize the selective relationships uniting the genetic patrimonies of different species living in the same spatiotemporal unit. At the beginning of Chapter 1, the importance of food chains in the equilibrium of the living world was discussed in detail. They represent the energy flow originating in the sun, which is made available by plant photosynthesis to herbivorous animals and subsequently to first- and second-generation carnivores. These chains imply strict and precise relationships between participating species. Weakness or loss of a single link could endanger the whole system, although there are often many alternative pathways in the more complex food chains that can bypass defects in the primary pathway. In order for the system to function, each population in a given food chain must live in perfect relationship with those that precede it and those that follow it. This equilibrium between populations of different species is really an equilibrium between the genetic pools of all species in a particular ecosystem. We can see how a modification in a particular link — such as a change in color or in herbivorous or carnivorous habits, for example, or perhaps a change in the local flora — could modify the selective value of genes belonging to other groups in the same network.

Let us consider an extremely simplified food chain made up of a series of elements such as plants (P), herbivores (H), first-generation carnivores (C_1), and second-generation carnivores (C_2). Let us suppose that as the result of an event such as an accident, an epidemic, or a climatic change, C_1 becomes rare or extinct. The chain is in danger of being broken beyond stage H unless C_1 is replaced by a new species C_1' capable of playing a similar role — that is, eating H and being eaten by C_2 — and still in balanced proportions so that the general equilibrium is maintained. The vicariant species (C_1') will always differ from C_1 in several features, such as its times of activity, its color, and so on. Thus population H on one side of it in the chain and especially population C_2 on the other side required to adapt to a new situation, may undergo genetic modifications — for example, a rare allele that so far has been of no particular use may become common in the new context because it is now advantageous.

Paleontology has demonstrated that in spite of numerous modifications during the history of life on earth, no biotope has ever remained overpopulated or deserted for long. All changes are followed by regulatory phenomena that reestablish the chain, sometimes with new species that are either born of a process of evolution or immigrate from elsewhere to occupy an area that is becoming deserted (May, 1978). The most striking (or caricatural) example is doubtless man's appearance on the earth. First a particularly efficient predator and then an enthusiastic farmer, he has profoundly modified the genetic equilibrium of many ecosystems. The arrival of a new species is never devoid of consequences for other elements in the biotope. We should think of evolutive genetics no longer in terms of independent gene pools belonging to particular species, but in terms of patrimonies linked by multiple interactions.

POLYMORPHISM AND EVOLUTION

Genetic polymorphism thus appears to be a very generalized phenomenon. We meet it in all natural populations, and we are beginning to glimpse the mechanisms that cause and maintain it as well as those that limit it. Fundamentally, it seems to be related to chance, to permanent fluctuations of the environment, and to ecological constraints.[4] Such fluctuations mean that a group of living organisms is never in permanent equilibrium with its milieu. It must constantly readapt in the face of the variations it encounters. This explains the usefulness of polymorphism and the multiplicity of the mechanisms that tend to preserve it and extend it. These mechanisms can be found at all levels, individual as well as populational. Certain of them, such as recessivity and heterostasis, allow unfavorable genes to survive by ensuring that they are not expressed in the phenotype, where they would fall victim to natural selection. This is a "passive" defense. These passive mechanisms, known for a long time only to geneticists, justify the concept of genetic load. Other, more important mechanisms are based on the fact that most genes, even if disadvantageous in the homozygous state, may be useful to the heterozygote. This "active" defense constitutes *balanced* polymorphism and *superdominance,* whose effects seem very widespread.

The diversity of the systems designed to maintain genetic polymorphism leads to great security. If one of these "polymorphizing" systems, such as the interpopulational gene flow, were neutralized, the others would intervene to save the population from falling into the fatal trap of monomorphism. A wild population that is accidentally or artificially brought to a state of monomorphism by the loss of many of its members or by domestication is very likely to disappear.

A world made up uniquely of domestic animals is inconceivable, just as is a humanity made up of pure races. The only way a monomorphic group suddenly plunged into natural conditions can survive is to reestablish its polymorphism very quickly. If it cannot, it is condemned, unless, like domestic animals, it enjoys particularly protected artificial conditions such as those ensured by the human milieu.

The universality of genetic polymorphism makes it one of the fundamental laws of living beings. It must be of very positive selective value because it allows each population to adapt constantly to the variations of its environment, to enlarge its niche and increase its resources. When two populations of the same species are remote enough to be subject to very different selective pressures, they can diverge sufficiently to be-

4. Thus we can assume that the gene frequencies observed in a population at a given moment depend on: (1) The nature of selection. The frequencies vary up to a level that gives them a maximum value in the particular selective context; for as we have seen, this value changes with the frequency of the allele. When this optimum is attained the population is in equilibrium, a state that will not last due to the constant changes in environmental conditions. (2) Chance, which can act in all ways. One can never say initially, concerning the establishment of a particular gene frequency, what is related to chance and what to selection. Attention has recently been drawn to the multiplicity of factors underlying polymorphism (see Dubois, 1980).

come intersterile and give rise to autonomous species, each exploiting a particular niche. Such a movement is only conceivable in the context of a widespread genetic polymorphism. This is the basis of the pioneering spirit of populations, which has given rise to evolutionary movements. The origin of species is founded on such permanent, multiple variations. We can well understand the enormous effort made by nature to preserve and extend it.

Only the populational model is compatible with an advancing pattern of evolution such as we see in real life. Polymorphism is involved at all stages of the history of life because it is a highly advantageous characteristic, indispensable to the progression of evolution.

the evolution of integrated systems

STRUCTURAL GENES AND REGULATORY GENES IN EVOLUTION

All structural genes are backed up by regulatory systems responsible for the differentiation of tissues and their division into organs during embryonic development. The importance of these phenomena is not lost in the adult, for they ensure that bodily integrity is maintained and that the component parts remain coordinated. We should stress the contrast between the multiplicity of programs of differentiation, each characteristic of a given species, and the uniformity of the basic materials that they have at their disposition. This uniformity is seen at the cellular level. In the whole vast family of vertebrates, all cells can be classified into less than two hundred types that are almost identical in all groups. The same uniformity is found at the molecular level. Apart from a few minor details, structural and enzymatic proteins are the same everywhere.

It is true that when we look at related species we can recognize certain alleles that exist in some but not in others. But this difference, at most the substitution of a few amino acids, is almost always without any great effect on the phenotype.

Let us take two complementary examples that illustrate the relatively minor importance of structural genes in speciation. Georges and Nicole Pasteur (1980) point out that the six sibling species of the marine polychete worm *Capitella capitata* differ in all the alleles studied so far, whereas morphotypically they are indistinguishable. Man and the anthropoid apes of Africa apparently have 98 or 99 percent of their proteins in common — that is, they have the same structural genes at 98 or 99 percent of their loci.

Thus two species differ more through phenomena of regulation than by mutations, which, in fact, have a rather modest role, often without great significance to the phenotype, as Kimura and the neutralist school have clearly shown. From the biochemical, histological, or even genetic point of view, man is simply a differently regulated chimpanzee!

The same phenomena of resemblance or identity are seen in phylogenetically distant groups. All the basic structures of the human being can already be found in coelacanths, whose ancestors appeared at the beginning of the Devonian period about 380 or 390 million years ago.

Many neo-Darwinians considered speciation as a result of "noise," which is to say errors occurring at the moment the DNA chains are replicated; these errors are then sorted out by selection. But this picture is not legitimate: the genetic keyboards on which a population is designed are active much earlier. Error does not in fact play much of a role in the dynamics of evolution; error is indifferent or deleterious. Evolution is based rather on new combinations involving basic materials that have changed little since time immemorial. At present speciation seems fundamentally like a regulatory phenomenon.

If we restrict ourselves to a biochemical view of living organisms, nature seems rather miserly. In this domain evolution shows little imagination. All species, including those living, or surviving, at the extremes of our globe's ecological conditions and thus showing very particular forms of adaptation, are made up of the same, or closely related, materials.

The fundamental proteins of our cytoplasm — those that form the cytoskeleton, cytochromes, and respiratory enzymes, for instance — differ not at all, or very little, from those found in unicellular organisms; and the way in which we derive our energy from the metabolism of glucose is hardly different from that used by aerobic bacteria. The same can be said for all basic physiological processes, such as the stocking or utilization of energy sources, the elimination of waste products, and so on. Certainly if we consider proteins globally, there are species differences in the great majority; but when we look more closely we often find that in these more or less complex molecules there are a certain number of similar sequences that suggest a former common origin. For example, insulin, a relative simple hormone of molecular weight around 6,000 and made up of two chains, the A chain of twenty-one amino acids and the B chain of thirty, only became functional quite late in vertebrate evolution. However, sophisticated immunological methods demonstrate the existence of insulinlike sequences in bacteria. Throughout natural history evolution has progressed, but in doing so it has used a number of basic elements, from which it proceeded to construct in a more and more complicated "combinatory" way. In the beginning there would simply be the polymerization of monomers already present in ancient groups — for example, the hemoglobins, the immunoglobulins, and certain hormones such as growth hormone. But this polymerization was later accompanied by a divergent evolution of the redundant parts, thus giving every species its own biochemical identity. Almost always, molecules precede function, for the latter emerges when new relationships are established between preexisting molecules — such as when physiochemical changes occur that are needed to permit functional interactions. But the actual molecules do not change. It is the combinations that are evolutionary, not the structures.

This can be illustrated by the phenomenon of lactation, which is peculiar to mammals but involves hormones containing molecules that are the same as, or closely related to, molecules already found in the reproductive cycle of much older vertebrate groups. A similar example concerns the dozens of neurotransmitters that are constantly being identified in the human brain. Among them have been found many hormones already known for their activity in very different organs — the stomach, pancreas, gall bladder, or vascular system — and active in many other living organisms. Enkephalin and somatostatin can be found in the digestive system, whereas gastrin, cholecystokinin, vasoactive intestinal polypeptide, insulin, glucagon, and related molecules are found in the brain. Morphine has its origin in plants of the poppy family, but can replace cerebral enkephalins and endorphins at neuronal membrane receptors.

Indeed, all living organisms have an almost identical cellular chemistry. Jacques Monod's aphorism "What is true for the bacterium is also true for the elephant" (1970) can be understood in this way; but it should not be applied to the diversification of species. The appearance of new combinations — that is, new developmental programs characteristic of each group — constitutes the very foundation of evolution. If we consider only fundamental biochemical structures and elementary reactions, everything is already established in the bacteria. The variety of the living world is due much more to regulatory phenomena than to structural phenomena.

Measurements of DNA in many species show that the increases that can be found at various levels of the phylogenetic tree can be correlated with the appearance of new relationships between metabolic chains and the formation of functional ensembles whose complexity exceeds our powers of imagination.

We should look for the motive force of evolution not in structural genes, but rather in the way they combine their action through ever more complex systems. The basic materials have not changed significantly for hundreds of millions or even billions of years, but the construction methods have been modified profoundly. At the present time evolution must be reconsidered in terms of relationships between structural genes rather than in terms of the appearance of new structures.

Let us take an example based on morphological criteria, which are very useful to the systematist. Experiments in which heterospecific grafts are performed in vertebrate embryos demonstrate that the morphogenetic substances responsible for the formation of the limbs, for example, are the same in all groups and are distributed in a basically identical way. All vertebrates have a forelimb composed of a humerus whose head articulates with the scapula, and of an ulna and a radius, followed by a series of small bones, the carpal and metacarpal bones and the phalanges. Precisely the same elements are found in all groups, but the relative development of each element varies according to the species. In the end, variations are due to the way in which morphogenetic substances act differently in space and in time to produce a fin in a fish, a leg in a lizard, a wing in a bird, or an arm and hand in a man. There is no specific protein for a fin or an arm, for a leg or a wing, but rather a series of regulations acting on each tissue responsible for the organ in question.

All are derived from the same genetic model; their specific variations depend on differential regulations, each with a particular adaptive value. It is useless to look for differences in basic nature. Beyond a certain level of complexity that is rapidly attained, most evolutive modifications are based purely on new arrangements of preexisting structures, not on the appearance of new structures.

We shall see in Chapter 11, dealing with transphyletic evolution, that this is true even for groups that today seem phylogenetically widely separate. But in these cases the relationships are simply more difficult to discern, for the common ancestors lived long ago and have already disappeared.

The evolutionary process is like an enormous Lego set with enough pieces to construct the most varied objects — a wheelbarrow, a crane, a clock. All these can be made out of the same elements, the only difference being their arrangement and interrelationships.

"Morphological innovations in evolution can doubtless be explained essentially, if not exclusively, by a reorganization in the relations between the level of synthesis of preexisting proteins, each relationship being tied to a spatial and temporal localization in activity. One can assume, with a good chance of being right, that the majority of evolutive processes are due to the reutilization, under conditions of different regulation, of a genetic patrimony already acquired" (Petit and Zuckerkandl, 1976).

"At higher grades of organization, evolution might indeed be considered principally in terms of changes in the regulatory systems" (Britten and Davidson, 1969).

Ernest Mayr said, "The day will come when it will be necessary to rewrite the greater part of population genetics in terms of interactions between regulatory genes and structural genes" (1963).

This day has come, even if we still do not know the details of the mechanisms that regulate structural gene activity in eukaryotes. Henceforth, naturalists must abandon the traditional neo-Darwinian plan and assume that, apart from rare cases of highly deleterious mutations, evolution is not so much a sorting out of new alleles but rather the appearance of new combinations. This allows us to have a clearer idea of the mechanisms responsible for the birth of species. We must now define a certain number of concepts utilized by Dobzhansky, Mayr, Ayala, Lewontin, and most populationists that are indispensable to a global genetic approach to speciation.

COADAPTED SYSTEMS

The genotype of a species is a functional entity made up of a limited series of structural genes supported by a large number of regulatory systems. The latter coordinate the former as a function of a program whose final result is the construction of the phenotype characteristic of the species. This phenotype must be acceptable in the given environmental conditions: this is the only thing required of it.

The program does not correspond to a real material object like a musical score or a recipe; it represents the nature and number of more or less complex interactions between structural genes and their supporting regulatory systems. A more and more complex series of "logical" circuits is thus formed as we climb the phylogenetic tree, resulting in the establishment of a host of integrated, sequential functions with multiple relays and possibilities for vicariance, as we have seen earlier.

As we said in Chapter 6, all genes in a given patrimony participate in an efficient and harmonious ensemble. By their interrelationship they achieve a high selective value. But this selective value depends on their simultaneous presence in the same genome and may disappear if a single element is missing. Such an ensemble was called a *coadapted system* by Dobzhansky (1950).

"Good Mixer" Genes

There exists a category of genes that play an important role in maintaining the cohesion of coadapted systems. They were called *good mixers* or "jacks of all trades" by Ernst Mayr (1963) and *sociable genes* by Georges Pasteur (1964). For a given locus, a gene may function favorably when the locus is occupied by one allele and much less favorably when it is occupied by a different allele. The good mixers are genes that are in the privileged position of working well in the presence of very varied alleles. They thus favor intrapopulational polymorphism which, as we saw in Chapter 6, confers a considerable advantage to the whole group. It is obvious that the selective value of a good mixer depends on the nature of the patrimony of the whole population. It can be very high in one population, but much less in another. In general, good mixers are acknowledged to be related to branching enzymes located at sites where feedback mechanisms can originate. They are like marshaling yards: If necessary they can trigger a process of a vicariance.

These good mixers usually have a higher selective value in open populations that can receive many foreign genes than in very closed populations tending to have a high frequency of homozygotes. In the latter case the good mixers lose much of their importance and can even disappear.

Inertia in Coadapted Systems

Coadapted systems have a considerable inertia when faced with new mutations or environmental modifications. The inertia of a coadapted system maintains the perenniality of the species in spite of multiple internal and external changes that constantly threaten living organisms. We should look at this phenomenon in more detail.

The Buffering Power
of Coadapted Systems Against Variations in the Genome

Let us take the case of a foreign allele, arriving in a system by either mutation or external gene flow. This newcomer can cause a metabolic disturbance at a certain level, such that substances that should be metabolized no longer are. They accumulate and, through a series of feedbacks, collateral chains, thus far repressed, are now opened up. The metabolism of the cell is restored, cost what it may, through different pathways. In the end, thanks to these vicariant systems, the final phenotype is achieved. Thus, unless one of the fundamental functions of the cell is threatened drastically, most metabolic modifications brought about by the arrival of a new allele are taken care of by regulatory mechanisms that operate at all levels and constitute something like safety locks.

Lerner (cited by Lucotte, 1978) summarizes this mechanism in a very simple diagram (see Figure 22). Let us suppose that the indispensable phenotypic feature t is the product of the simultaneous action of

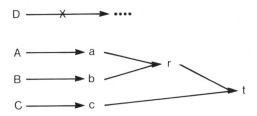

Figure 22. Unmodified metabolism.

three metabolic chains $A \to a$, $B \to b$, $C \to c$, such that a and b interact to give an intermediate factor r that acts in turn with c to give the terminal character t, as seen in Figure 22.

If a mutation blocks B the product b is no longer synthesized. A continues to function alone but to no avail. It makes a, which cannot be used because there is no b, and it accumulates in the cytoplasm of the cell; t no longer appears. When its concentration reaches a certain threshold, the cytoplasmic a can activate a chain $D \to d$, which has so far been repressed, but which is capable of interacting with a and producing a product s that acts on c to form t. The chain $D \to d$ has thus played a vicariant role in supplanting the deficient chain $B \to b$ and ensuring by another pathway the utilization of a and the realization of the phenotypic feature t (Figure 23).

These vicariant chains, opened up when there is a deficiency in one of the stages of normal metabolism, are very numerous in metazoan cells. They add to the richness and complexity of regulatory systems and explain the inertia found in species (that is, well-integrated genetic ensembles). When faced with a challenge to their biological order, after a new mutation for example, individuals often find a way of maintaining their phenotypes within acceptable limits.

More than twenty years ago Albert Vandel (1963) recognized this phenomenon and called it *autoregulation*. In his opinion an isolated

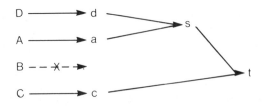

Figure 23. Modified metabolism.

mutation had little evolutionary significance. Everything depended on the way it was received in a coadapted system and how it was used.

Vicariance makes possible the realization, whatever happens, of various essential properties without which an individual would lose its specific character and perhaps its chances of survival. It explains the buffering power of genotypes and the existence of developmental channels postulated by Conrad H. Waddington and confirmed by Mayr, channels through which all developing organisms must pass and that play a fundamental role in the stability of species.

Buffering Power and the Neutral Theory

In Chapter 1 we saw how, for a given population with more or less homogeneous phenotypes, many loci related to enzymatic synthesis could be polymorphic. In other words, individuals living side by side, apparently similar and perfectly interfertile, can have enzymes that differ slightly at the molecular level. But these differences, detectable by electrophoresis, do not usually have an appreciable phenotypic expression. The mutations that are responsible for them lead to the formation of similar individuals. This means that a modification in a DNA sequence can be "cushioned" by vicariant systems in the genotype that possess a certain buffering power.

These observations have led many geneticists to adopt the neutral theory of Motoo Kimura, already discussed in Chapter 6, according to which the majority of mutations concerning structural genes have no selective value. This theory is accepted by many workers and agrees quite well with the inertia of coadapted systems proposed by Dobzhansky.

We can now begin to see how we might reconcile Kimura's neutral model and the evolutionists' selective model; it should be possible to end the apparent contradiction between the neutral theory and the selective theory of evolution that we spoke of in Chapter 6. The neutralists almost always argue on the basis of particular mutations such as the substitution of an amino acid or a peptide chain. Thanks to the buffering power of coadapted systems, most such mutations have little effect on the phenotype. They escape natural selection; they are practically neutral. One can see how they appear regularly over a period of time, independently of the evolutionary "shocks" that can be observed at times in all lines. Kimura does not deny the place of selection, but he thinks it is only important in the case of highly deleterious mutations that are exceptions and quickly eliminated. In reality the selections that constrain the living organism to adapt to its environment act on the phenotype, which is to say on the material expression of a combination of genes corresponding to a coadapted system; and such a combination could not remain selectively neutral. Selectionists take into consideration just these adaptive characteristics — whether morphological, physiological, or ethological — whereas the neutralists concentrate on particular isolated mutations. But while the replacement of an amino acid in a peptide is often without any particular value, there can be little doubt that the genetic combinations responsible for making a fish also adapt it for swimming, or that the combinations that give rise to birds adapt them for flying. It is difficult to imagine that the creation of such wonderful swimming or flying machines should be the fruit of chance rather than selection.

Neutralists and selectionists do not speak the same language. The former concentrate on genes and their polymorphism; the latter on genetic organization, translated into phenotypes. As we saw in Chapter 5, the birth of a new species depends much more on new organizations than on new mutations. Setting these two models up against each other has created confusion between selection of genes and selection of combinations — that is, of phenotypes. If, as Kimura quite plausibly proposes, selection intervenes in evolution only to eliminate very unfavorable mutations (a form of Mullerian selection), it constitutes an essentially conservative process, opposing change. But above the level of the eukaryote, it is the selection of combinations, and thus of gene frequencies, that has played the essential role in evolutionary movements.

In the present state of our knowledge, neither the neutral theory nor the selective theory, taken in isolation, can explain all the known phenomena. In particular the survival of rare alleles in all populations suggests an absence of selection, while the existence of monomorphic loci cannot be explained by the neutral model. In fact the two mechanisms are not exclusive and may well act simultaneously, but not at the same level: the first acts on genes and the second on phenotypes. The contradiction between the two theories now seems more theoretical than real.

Buffering Power and Environmental Variation

Not only does the phenotype remain stable in most cases of new mutations; the same inertia is observed in relation to environmental modification. Confronted with new ecological constraints, a sufficiently rich and complex coadapted system can utilize vicariant pathways in response and maintain the phenotype within acceptable limits.

Let us take the very simple case of the synthesis of enzymes for the metabolism of lactose in the bacterium *Escherichia coli*. Their synthesis is blocked if the medium contains glucose, a sugar that can be utilized directly by the bacterium. As soon as there is a lack of glucose, the catabolic chain for lactose ceases to be repressed, as long as there is lactose in the medium. Thus, according to the circumstances, the bacterium can use different catabolic chains with the same final result: energy production from glucose either directly or after its liberation from lactose.

Vicariance involving much more complex mechanisms certainly exists in eukaryotes. Population geneticists have long emphasized the buffering power of the genotype of most species when faced with ecological fluctuations. This buffering power increases with the degree of heterozygosity. It must be similar in nature to the phenomenon of heterosis of crossbreeds already discussed in Chapter 6. "The homeostasis of development allows us to understand why in natural populations we find a predominance of similar phenotypes — the wild phenotype — in spite of the fact that no two genotypes are identical" (Boesiger, 1969).

GENETIC HOMEOSTASIS

Thus we see that the realization of indispensable phenotypic characters can follow several pathways. This is why new mutations, just like environmental changes, do not cause noticeable modifications, at least in many cases.[1] This inertia in coadapted systems faced with change, either internal (mutations) or external (ecological variations), constitutes homeostasis or *genetic homeostasis* — a term created by Lerner about thirty years ago and adopted by Dobzhansky, Boesiger, and many others.

Many experiments have demonstrated the ubiquitous nature of genetic homeostasis, which in spite of modifications imposed on the genotype tends to stabilize the phenotype so that it can best respond to the constraints of the environment. For example, by appropriate crossbreeding in a strain of fruit flies it is possible to select two extreme features, a long thorax and a short thorax, thus creating two populations that differ fundamentally. As soon as the selective pressure is relaxed by allowing free breeding, the length of the thorax tends to average out so that the two groups end by merging and the two populations become indistinguishable. Figure 24 demonstrates this evolution schematically.

Homeostasis must also explain the morphological constancy of certain species introduced into new territories. We have already talked about *Mus musculus,* the common mouse, found in both our fields and

1. As E. Boesiger says, different genotypes often give more or less identical phenotypes, but the genotypes are sufficiently different that the individuals bearing them are intersterile. They thus form two sibling species.

our houses, which has split into a number of types. *M.m. brevirostris* is a commensal species of *M.m. musculus* that lives around the Mediter-ranean. Its northern limit is southern France. Imported to Central and South America by the Spanish invaders, it spread widely, and is now found in the southern part of the U.S.A. The Old and New World races have remained very similar in spite of the obvious differences in their environments (Britton-Davidian et al., 1978).

To illustrate the relationship between the polymorphism of the geno-type and the monomorphism of the phenotype, we can take as a model the phonetic construction of a word. The phenotype is the spoken word; "orchestra," let us say. Each letter represents a gene with a particular value; the arrangement of these letters forms an integrated whole — the word — with a meaning — its semantic content. But the spoken word could be written in several different ways ("aurkestra," "orkestra"), any of which when read out loud means the same thing, the phenotype. Its semantic content has not changed. If it is a password needed to enter a fortified town to escape a relentless enemy, the sentries guarding the gates will allow anyone who can *pronounce* the password to enter, however it is *written*. Natural selection works like the sentries. It judges the spoken word (the phenotype) without considering the letters (the genes) that make it up.

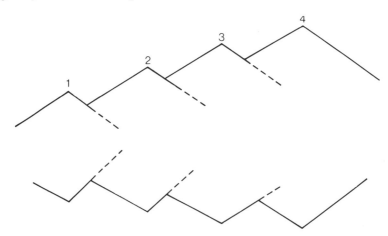

Figure 24. Evolution of thorax size in two groups of fruit flies deriving from a single population following rigid voluntary selection. This is done by sorting the shortest thoraces into one line and the longest into another, eliminating intermediates at each generation. If this strict selection is relaxed (at points 1, 2, 3, and 4) the thoraces of offspring of both lines tend toward a mean value, probably representing the best "adaptive" size for the animal. (This simplified diagram is adapted from those of Robertson, 1916.)

DEVELOPMENTAL CHANNELS AND EPIGENOTYPE

The notion of *developmental channels* is derived from those of co-adapted systems and genetic homeostasis. All species, and more generally all taxonomic groups whatever their level (genera, families, orders, classes), possess a certain number of characteristic invariant features, whose permanence is probably due to the fact that they are indispensable to the survival of the individual and the group. Their presence is necessary to ensure a minimum of adaptation to the environment.

Birds' wings, their porous bone, their sensory system, their feathers, are all features specially adapted to flight. It is difficult to conceive of species of birds without them. The same is true of more general attributes covering wider taxonomic units, such as placentation, the single-chambered eye, or mammalization.

All these characters have a remarkable stability. Their realization depends on coadapted systems that group structural genes together to form "a cybernetic network of reactions so interrelated that any perturbation in the early stages leaves the final state unchanged" (Lucotte, 1978). They correspond to what Waddington called the *epigenotype*.

In other words all coadapted systems imply by their very nature a certain number of obligatory channels that the development of the embryo must follow and that lead to the laying down of characters that are indispensable to the species. Thanks to the inertia of coadapted systems and the multiple vicariant pathways that are possible, these channels enjoy great stability. They ensure that development does not deviate from a principal pathway leading to the best-adapted phenotype, the fruit of a long selective "experience." Any potential deviation immediately brings into play regulatory phenomena that redirect the ontogenetic processes along the principal pathways.

The rigorousness of some developmental channels might explain the survival of very ancient species, some of which really merit the name "living fossils" and have not varied from time immemorial despite continual new mutations and periodic ecological changes (Delamare-Deboutteville and Botossaneau, 1970). Some modern bacteria that already existed in the Algonkian era hold the record for longevity.[2]

Developmental channels are an ontogenetic expression of genetic

2. We might also mention the astonishing discoveries of H. Dombrowski (1963), who described bacteria in a state of "hibernation," trapped in salt crystals from the Zachstein Sea for 250 million years. They included specimens of *Bacillus circulans* apparently identical to those that are commonly encountered today. Other bacteria in a similar state were described in Middle Devonian salt in Saskatchewan and in Silurian salt in New York State. If these observations are confirmed, they would represent, to use the expression of Delamare-Deboutteville (1970), "absolute living fossils."

homeostasis and represent a conservative, antievolutionary process. For evolution to progress it must free itself of them. This can be done by a "genetic revolution" that upsets the coadapted systems; without it, species would enter an evolutionary cul-de-sac. This problem will be further considered in Chapter 8.

COMPLEXITY AND INNOVATION

For the geneticist there is a striking contrast between the invariance of structural genes that have not changed since long ago except in a few details and the innumerable variations in the combinations they form, which as we have seen are really responsible for evolutionary movements.

In vitro genetic recombination is now widely practiced. It is well known that one can introduce a DNA sequence from any animal or plant species into a bacterium or a eukaryote cell, and that this sequence will provoke exactly the same protein synthesis regardless of the nature of the host cell. Thus a bacterium can be made to manufacture a hormone found only in vertebrates. Once it has received the information, it will carry out the synthesis even if the final substance is of no use to it.

A *cistron* (a DNA segment that codes for a particular substance) coding for histidine always causes histidine to be synthesized, whether it is in a bacterial chromosome or a human karyotype. A given sequence of DNA makes the same protein, whether in *E. coli* or the fruit fly. But in the former it is participating in the construction of relatively simple structures with limited possibilities, while in the latter it forms part of particularly complex assemblies with many applications.

An increase in the complexity of biochemical combinations that form living matter leads to innovation. Every extra degree of complexity opens the door to hitherto unknown possibilities. It creates new qualities from preexisting basic material that is assembled in a different way. This is one of the most astonishing properties of life. It matters little that very few new structural genes have appeared for millions or hundred of millions of years, for at each stage of evolution countless new functions have emerged due to new combinations. It is what François Jacob (1977) called *bricolage,* or making a new object from already existing ones. Unlike the industrialist the *bricoleur* does not make really new things, although functionally he is innovating. "Novelty derives from new associations of old materials. To create is to recombine" (Monod, 1970). For Jacques Monod innovation consists almost always of "new utilizations of universal metabolic sequences, first employed in other functions."

We recognize the important role played by duplication or multiplication of a given DNA sequence in the laying down of complex molecules like hemoglobin, immunoglobulin, or enzymes. This redundancy is not really a new "invention" but a form of polymerization often involving diversification of an already existing chain designed for simpler functions. A related phenomenon is the grouping together of hitherto isolated molecules, which thereby become modified and acquire new properties.

One of the most striking examples is provided by the vertebrate endocrine system. Very often, the organic molecules incorporated in hormones or their immediate precursors appeared in phylogeny long before they were functional. These molecules became "revolutionary" when they were reorganized and coordinated in a system that permitted them to interact. *The innovation was not the existence of the proteins but their capacity to interact.* The appearance of new interactive abilities probably constitutes a large part of functional innovation at the molecular level (Petit and Zuckerkandl, 1976).

At the present time this concept has taken on considerable epistemological importance for all life sciences. Tomorrow's physiology will have to study correlations rather than analyze functions, and genetics will form an integral part of this study.

THE TWO PHASES OF THE EVOLUTIONARY MOVEMENT

Most naturalists since Lamarck and Darwin who have become interested in the phenomenon of evolution have studied only the late phase of the emergence of species, particularly the metazoan species that concern us directly.

For the geneticist evolution begins much earlier. Seen overall, the history of evolution covers two major periods.

1. An archaic period during which nucleotide and protein complexes were formed, involving essentially structural genes and their immediate products. At this stage almost all the peptides that would serve as a basis for living matter were constructed. By its conclusion, elementary macromolecules had formed and the first cellular organisms appeared, typified by the archibacteria that are now quite well known. These were species living three or four billion years ago, neither plants nor animals. They appeared before the divergence of the two kingdoms and were well adapted to conditions on our primeval planet, with its atmosphere containing low levels of oxygen. Three modern groups may well be related to these species: the halophiles, in very salty water; the thermoacidophiles, in warm sulfur springs; and the methanogens, in marshes. Their regulatory powers were probably extremely rudimentary. This archaic period, during which mutations must have played an important role and at whose beginning there was no life, was, at least biochemically speaking, the only real period of creation. At this stage relatively high-energy aerobic systems replaced the anaerobic systems that had been the only ones possible in an environment with low oxygen levels. All this happened only once, which demonstrates the essential unity of the molecular structure of the living world.

2. A more recent period during which new mutations played a secondary role and that is characterized essentially by the appearance of combinations of preexisting basic information and by their integration into more and more complex regulatory ensembles. These multiple "constructions" were made possible by a considerable enrichment of the genome, which characterized the transition from prokaryotes to eukaryotes. Although much shorter than the archaic period, this stage represents the largest part of the evolution of species and the only one studied by zoologists and botanists. It is the more spectacular of the two. It leads directly to us. At its beginning almost all the molecular structure of living matter was already created; what followed was rearrangement, construction, or assemblage.

The Relationship Between Complexity and Semantic Content

The second phase of evolution thus consisted of the construction of more and more complex edifices from the same elements and the appearance of new relationships between these elements. The semantic value of a system — that is, the information it contains — depends not just on the basic sequences, or structural genes, that make it up but also on the way they are arranged and their interrelationships. As the amount of information increases and becomes organized into a number of programs, the semantic value of the whole increases, not only quantitatively but also qualitatively. Nature has climbed a certain number of steps in integration, always based on the same elements — the sixty-four triplets of the genetic code — but forming more and more complex ensembles, so that each step reveals hitherto unknown new qualities compared with the preceding step. The principal steps may be regarded as the cell, multicellular organisms, social development, and so on. This is an essential concept, which we shall discuss in greater detail in Chapter 15.

The alphabet can be taken as a simple, if rather approximate, example to illustrate the links between complexity and increase in information content. The alphabet is made up of twenty-six letters, each with a phonetic value. For example the letters *e, f, o, r, t,* taken singly, represent merely a series of sounds; their semantic content is very small.

However, we can group these letters into combinations with different implications. We can form the words *fort, effort, offer, free,* and many others whose semantic value is richer because each word signifies not only a series of sounds corresponding to the sequence of the letters but descriptors such as "fort" or concepts such as *effort, offer,* or *free.*

From the same alphabet of twenty-six letters we can increase the semantic content of the system by using more complicated combinations. We can construct sentences such as "It was an effort to free the fort," but we can also go much further and with the same twenty-six letters write St. John's Gospel, Darwin's *Origin of Species,* or Karl Marx's *Kapital.* These are texts whose semantic value is even greater because they contain ethical or philosophical treatises. If we analyze the letters used in each we find approximately the same proportion of a's, b's, and c's or of x's. It is not the nature of the component parts that differentiates these words, but the way in which they are grouped. We can also add that spelling mistakes or printer's errors found from time to time, like isolated mutations, very rarely alter the general sense of the text — the "phenotype."

The genetic code can be compared to an alphabet of sixty-four letters — or rather sixty, if we count the codons that are used as signals. A letter would be a codon (or rather, the amino acid that it codes for), which itself has little significance. Words correspond to peptides, amino acid sequences whose functional value is limited. Sentences are represented by enzymes made up of several peptides that have a precise function. Texts are much more complicated programs containing the organizational plans from which a living being will arise.

Evolutionary movements stem from these associations. The examples we have already given of hormones, hemoglobin, and immunoglobulin, are very indicative. Phenotypic modifications are due less to new mutations than to new combinations. They are epigenetic. Most evolutionary changes that we can analyze at present are such epigenetic phenomena.

TELEOLOGY AND TELEONOMY

This model is too simple to be rigorously exact, but it permits us to imagine how combinations of basic units with little semantic value can give rise to more complex units whose semantic value is richer and even different in nature. Such a model could lead to the formation of species. A problem that came to light very early was whether speciation was based on a preestablished program or depended on the laws of chance. We shall come back to this important question, which is still being debated, in the next chapter. Suffice it to say here that this problem is related to the general question of "finality" in biology. To describe the persuit of a goal by living organisms we use the words *teleology* and *teleonomy,* whose meaning is not agreed upon by all biologists; many see in them the final avatar of finalistic thought (Mayr, 1976, 1981).[3]

Determinism has no place in the life sciences. Evolution may not have a particular goal, but it has reached the stage of hominization, which we tend to take as the ultimate one. Since man has existed, many consider that the only justification for the universe was his creation. But no arguments except religious ones support this vision of natural history, so flattering for us. If a tapeworm could think, it would doubtless consider that man existed to give it a suitable shelter in his intestines. Man's pretentiousness is just as ridiculous as that of the taenia.

For the moment, there is no solution to the contradiction between evolution's lack of a goal and the incalculable number of possibilities it has before it. How did evolution "choose" which pathway to take? Can selection alone explain this choice? Or should we assume a total absence of choice and invoke pure chance, or almost pure chance, as the supporters of the neutral theory would have it?

Even if it were possible to construct very simple words by taking letters of the alphabet at random, one could hardly hope to write St. John's Gospel or Marx's *Kapital* in this way. Even if we had time available on the geological scale, the probability of such a result is zero.

However, two sets of factors substantially reduce the number of possible alternatives and thus the role of chance.

1. Evolution took place in a series of steps, each involving more complex groupings. When we write a text, we draw not on isolated letters but on prepared sentences or even complete paragraphs.

2. In addition, as discussed earlier, vicariance allows the use of a certain number of words or sentences that are synonymous. This reduces the random factor considerably, but is not enough to explain the coherence of the phenomenon of evolution entirely.

We must admit frankly and modestly that we find ourselves faced here with one of the most mysterious frontiers of knowledge.

COMPLEXITY AND LIBERTY

A simple genetic system such as that of the phage or — to a certain extent — that of bacteria imposes its program on the individual, who must obey it to the letter; the alternatives are few or none. On the other

---

3. Mayr's comment was also made in a seminar at the Collège de France, March 30 to April 4, 1978. Jacques Monod, in his inaugural lecture at the Collège de France on November 3, 1967, remarked, " 'Teleonomy' is the word we can use if, through objective modesty, we prefer not to use 'finality.' "

hand, a complex genetic system provides many possibilities for regulation and vicariance. The individual can choose several pathways, and a certain margin of variation is open to it. As we have seen, this margin increases with greater structural complexity and greater polymorphism. The more genetically polymorphic an organism, the more alternatives it has at its disposition. It can adapt to new situations without threatening the whole structure. The organism becomes freer of the constraints of the environment. Its various levels of response allow it to face a host of different situations *actively*. This tendency is seen in all animal groups but particularly in higher vertebrates — birds and mammals — which have developed homeothermic mechanisms. Such individuals can maintain their own internal "climate," which frees them to a great extent from external climatic problems. At a later stage ethological responses were added to these physiological ones. The animal was able to invent and perfect adaptive behavior; learned behavior replaced innate behavior and was better adapted because it was conscious and voluntary. Able to spread rapidly, it led in man to the establishment of culture.

This growing complexity of the genome corresponds to the three response levels discussed in Chapter 1: *genetic* adaptation by combinations of favorable genes; *physiological* acclimatization by the introduction of physiological regulatory mechanisms; and *cultural* adaptation by the acquisition of behavioral patterns that effectively respond to new situations.

For example, reptiles can do little when faced with a major cooling of the climate. They can only change their geographic localization or prolong their hibernation. When their period of activity becomes too short to ensure their reproduction, the species disappears. In similar circumstances a mammal uses its thermoregulatory mechanisms. Its heart beats faster and its basal metabolism increases. It seeks richer food to lay down more fat as a protection against the lower temperature. Some dig deeper holes in the earth. As to man, most of his reactions are cultural. He puts on more clothes, constructs better shelters, and builds fires.

Thus the enrichment of the genetic patrimony provides an escape from rigid programs and opens alternative pathways. It is a step towards liberty. Albert Vandel (1968) called it "this essential period in the history of life during which was established the association of thousands and millions of cellular elements into a solidly organized and regulated whole. This history is nothing other than the birth and the development of liberty in the world."

PART TWO

the
evolution
of
populations

speciation and transspecific evolution

THE DIFFERENT ASPECTS OF EVOLUTION

In the first part of this book we studied the importance, consistency, origin, and significance of the polymorphism found in all living groups and showed that our traditional typological way of thinking does not correspond to reality, since species are made up of heterogeneous natural breeding groups, the populations. In the light of this information we can now examine the precise nature of evolutionary movements.

Fundamentally, evolution is marked by the appearance of new species. Charles Darwin was not wrong to call his book *The Origin of Species.*

Based on observations of those groups that are accessible, zoologists and paleontologists have long distinguished two types of phenomena.

1. Microevolution, which could also be called "specializing" or "regressive" evolution, corresponds to true speciation. A primitive nonspecialized species is split into several daughter species, each of which tends to occupy a different niche. This involves multiple adaptations (thus the term "specializing" evolution), which means that in each splinter group we can find animals that swim, jump, crawl, or fly; we can find herbivores, carnivores, diurnal and nocturnal animals, free-ranging animals, parasites, and so on. But in spite of these differences we find in each species the same basic construction, inherited from the common ancestor. Thus speciation corresponds to the divergent pathways taken by the descendants of a given line to exploit multiple resources. We shall see the selective advantages involved a little later.

This evolutionary model can be represented by a tree whose more or less numerous branches dichotomize. This is *adaptive radiation* or "branching" evolution, whose mechanism will be studied later. Such a model would seem necessarily to force adaptation ever further so that the life-style becomes more and more specialized (or, to put it another way, the niche becomes narrower); from this idea comes the term "regressive" evolution.

The first arthropods, the trilobites, are very ancient, for they appeared in the oceans toward the end of the Precambrian era about 600 million years ago. In the course of time their descendants colonized all habitats as they became available, including the water (where they are mainly found as shellfish) and the land (the spiders and particularly the enormous group of insects). Their dynamism was such that certain shellfish invaded the land, and some insects returned to an aquatic environment. The relationship between a given species and its habitat is often so tight that, for instance, a particular insect can sometimes live only on a particular part of one plant species (buds, stems, leaves, or flowers).

This evolutionary movement is irreversible. No line returns along the pathway it has followed. An animal that has lost an organ never regains it in its old form. If it must face new constraints and needs an organ that it has already lost, it will invent a new one. This is Dollo's law of the irreversibility of evolution. Nature innovates, but does not repeat.

2. Macroevolution is a "progressive" evolution or *typogenesis*. While microevolution multiplies species, it was thought for a long time that it was unable to create new types of organization. It is relatively easy to see that a lobster, a fly, a spider, a scorpion, and a centipede are built along similar lines and that it is possible to pass from one group to another by modifying segments or appendages. All have an approximately similar construction. There are also transitional forms, some of which can still be found today: the *relict species,* or living fossils. But if we consider paleontological data we find many more, because many key species that were transitional forms disappeared quickly and can only be identified as fossils. We shall see their importance later. Comparative embryology also allows us to trace the phylogenetic origins of a group, for the development of the embryo preserves certain characteristics of the ancestral groups. Ernst Heinrich Haeckel formulated a *biogenetic* law — ontogeny recapitulates phylogeny. Even though a little exaggerated, this formula has a certain value.

Nevertheless, microevolution as defined until recently could not account for the appearance of new types of organization. Even if it could conceivably explain the relationship between myriapods, shellfish, and insects, or (in a different domain) that between a toad, a lizard, a gull, and a man, it could not explain the transition from invertebrates to vertebrates. Thus the terms macroevolution (Vandel), typogenesis (Schindewolf), and even "progressive" evolution were coined to define the birth of new types. This is the only form of evolution that is really important for the biologist.

We now know that there is no basic difference between these two types of evolution. They are part of the same process. Their artificial separation was due to the fact that as time passes intermediate forms are lost; but this does not mean that they never existed, and we find evidence of many of them. Before going further we might consider the importance of macro- and microevolution to polymorphic groups that have a populational structure.

THE GENETIC NATURE OF THE SPECIES

Fundamentally, what we have defined as microevolution now seems to be the birth of daughter species from an ancestral species. For the geneticist the species is an ensemble of coadapted genes (as described in Chapter 7) — that is, a series of structural genes supported in an orderly way by regulatory systems ensuring multiple interactions. These interactions are the program for the particular species and prepare it for a well-defined niche. In principle the program is invariable. It ensures that flies always give rise to flies, and camels to camels, each with its specific habitat.

This coadapted functional ensemble has various requirements.

1. It must be harmonious, equilibrated, and well integrated to avoid the splitting up of the system under the influence of selective constraints. Faced with constant environmental changes, the system must have a certain inertia: the genetic homeostasis that we have already discussed. It ensures the perenniality of the phenotype in an ever-changing milieu. This homeostasis depends on the existence of multiple vicariant pathways within each genotype, thanks to which different polymorphic genotypes can result in approximately equivalent phenotypes that are adapted to the selective constraints.

In addition, homeostasis allows the preservation of all possible genotypes, thus maintaining a certain polymorphism in the group, the nature of which we saw in the preceding chapter. If this homeostasis did not exist the species would collapse like a pack of cards at the first severe constraint it met.

2. This functional ensemble must be in some way isolated from others making up related species. We shall see later that there are many mechanisms for isolation that prevent a permanent exchange of genes between groups. Such generalized introgression would hinder the mechanisms of speciation and would constantly disturb acquired adaptive features.

3. This functional ensemble must correspond to an ecological niche. As Albert Vandel (1963) wrote, "What we call a species constitutes a state of equilibrium between the physiology of the animal or the plant and the milieu in which it lives."

This fundamental notion, already mentioned in Chapter 2, cannot be overemphasized. The ecological niche defines a species just as rigidly as its hereditary patrimony. It is simply a material projection of the patrimony. To change its niche a group must modify its patrimony and thus create a new species. There is constant adjustment between species and niches.

In a farmyard, a population of worms living in the compost heap has genotypes adapted to this environment. It cannot live in the nearby pond, which represents part of the niche of the ducks; and the ducks cannot live in the compost heap.

This adjustment seems rigid as we observe it. But considered over large biogeographic distances and long paleontological time scales, species are seen to be on the move and, as we have seen, can be transformed by modifications in their genetic program.

THE ORIGIN OF SPECIATION

The nature of the prime mover of speciation is still a subject of debate. We have seen that it cannot be the appearance of a new mutation or even of a series of new genes. This saltationist theory was defended for many years by the mutationists — in particular Hugo de Vries, beginning in 1901, Richard Goldschmidt in 1940, and later by many others — but cannot be accepted in its original form. It is incompatible with data from population genetics. It has been brought up to date by cytogeneticists, particularly Jean de Grouchy (1978, 1980), who see the origin of every species as an isolated individual with a chromosomal rearrangement. We shall come back to this hypothesis later in this chapter when we consider the role of chromosomal changes in speciation as a factor in sexual isolation.

In truth, we can hardly envisage speciation today in any other way except through a populational model implying a challenge to the functional ensemble formed by the genotypes of the whole population, (its coadapted systems) and its progressive replacement by a new system. In most cases, this transformation is not brutal but rather slow, spread normally over many generations. It affects all individuals in a reproductive group, that is, all those participating in the same genetic pool. It is the whole of this pool that evolves, not an isolated individual.

What can explain this evolutionary "paradox" that a species, genetically stable by definition, should become unstable to create a new species? In other words, why should a genetically coadapted system in a state of equilibrium embark on a different pathway? Two explanations are possible. For some workers, the change has an endogenous origin. For an unknown reason, modifications take place in the genotype itself. Speciation would thus be a spontaneous process. This possibility was mentioned in the previous chapter when we discussed teleology and teleonomy. Like embryogenesis, evolution would follow a program, one that would depend not on the individual but on the phylum to which it belongs. In fact this model extends Haeckel's law to the level of the phylum. It is possible to imagine several motive forces behind such a model. First of all there could be an irresistible tendency toward "creative evolution," as Henri Bergson had it, with one aim in mind: man, or the omega principle postulated by Teilhard de Chardin. In France this theory of endogenous evolution has been supported more or less implicitly by P. P. Grassé. In fact, no one has ever proved the existence of such a program, nor its nonexistence. If we admit it we accept creationism, although perhaps replacing God by the "program of programs." This does not resolve the problem, but only delays it.

According to other workers, the origin of speciation could be in changes of selective pressure related to ecological variations, aided and abetted by chance. It would not be the coadapted system making up the species's hereditary patrimony that would take the initiative for evolution; rather, the selective constraints would first be modified (climatic changes or migrations, for instance). Since the adaptive possibilities of a species are not unlimited, if they are exceeded by environmental constraints the group would have only two possibilities: to disappear, which would be the more frequent solution, or construct a coadapted system capable of responding effectively to the new environmental constraints — a sort of "genetic revolution." This would be the function of speciation.

Most geneticists favor the second hypothesis.

PHASES OF SPECIATION

In populational terms, we can break down the process of speciation very schematically into three phases.

1. The passive isolation in space or time of populations subject to different selective pressures.

2. A genetic revolution consisting of the replacement of one coadapted system by another more suitable to the different ecological conditions, achieved by the combined effects of isolation suppressing gene input from neighboring populations and of selective pressure.

3. Permanent reproductive isolation protecting the group's acquired characteristics and preventing its return to an earlier form by gene exchange with parent populations.

These three factors can imply either that an entire species A is transformed into species B quite homogeneously (*anagenesis*) or that several populations of the same species that are far apart (particularly on the periphery of a geographic area) might be affected heterogeneously. In this case the mother species A might divide into several daughter species, A_1, A_2, A_3 . . . An, each adapted to a particular niche. This is *cladogenesis*. Both phenomena are gradual and involve the whole of a population. They are similar in nature, differing only in the way they are expressed.

Parcellation of the Species and Change of Selective Constraints

The parcellation that occurs during a phase of passive isolation ensures that gene exchange between populations is stopped, or at least decreased. It may be quite prolonged. Ambient factors in a given area may change spontaneously. When a certain level is reached, the population will have exhausted all the adaptive resources offered by its genetic polymorphism and can only survive if its genome is profoundly restructured. This restructuring will give rise to a new species that will replace the old one. The two groups will thus succeed each other in time, overlapping geographically but not temporally. They are said to be *allochronic*.

Study of geologic strata has revealed many examples of this type of successive speciation. The caballine line that led to our domestic horse began in North America. It is marked by a whole series of species succeeding each other in time and tending to adapt to a changing environment. Their first ancestors must have had five digits like typical vertebrates. The *Hyracotherium (Eohippus)* at the beginning of the Eocene were as big as a dog and still had four digits on their forelimbs and three on their hind limbs. They lived in forests, and their teeth show that they ate leaves; their feet allowed them to move around easily on the soft soil of the undergrowth. *Eohippus* was found in both Eurasia and North America. In the Oligocene it was replaced in America by *Mesohippus,* a larger animal that had lost its smallest digit. It still lived in wooded areas. A related form, *Anchitherium,* reached Europe, where horses had been absent for a long time. In the Miocene a relatively dry climate developed, and in many regions plains replaced the forest. The major caballine line was once again modified to adapt to a new niche. Tough, silica-containing grass replaced leaves as food, and the teeth were modified to have high crowns (hypsodonty), to withstand the increased wear. The lateral digits became even more reduced, while the middle digit became longer and stronger to provide better purchase on the hard soil of the savanna and allow the horse to run more quickly —an obvious advantage in fleeing from predators. This was *Pliohippus* of the Miocene, the size of a pony. A related type, *Hipparion,* reached Eurasia and spread as far as Africa. Our modern horse appeared in America at the end of the Pliocene: *Equus* had lost what remained of its lateral digits and was a remarkable runner. Like its predecessors *Equus* spread from America to the Old World, invading all of it except for Australia. It formed permanent colonies but disappeared from America, its continent of origin, after the last Pleioquaternary glaciations, a few tens of thousands of years ago (Romer, 1970).

Another, equally remarkable case is that of the mole rats of the Pliocene and Pleistocene, which, over four million years, gave rise to seven successive species *Mimomys occitanus, stchlini, polonicus, phocoenicus, savini,* and *Arvicola cantiana* (which for the first time manifested a feature frequently found in modern rodents: continuous growth of their teeth). They produced two modern species *A. terrestris* and *A. sapidus.* These various species, distributed over a long period of time, are related by intermediate populations (Chaline and Thaler, 1977). There was never a sudden change that could represent a mutational process, chromosomal or otherwise.

Many other examples of allochronic and sympatric evolution are found in most groups. Often evolution is *diachronic:* several species appear simultaneously—at least on a geological scale—from a common stem, usually following the occupation of new niches (cladogenesis).

This quite common type of evolution was discussed in Chapter 3, when we examined the future of populations living at the periphery of a geographic area. Let us recall the essential point. When a mother species is distributed over a wide but heterogeneous area (is eurotopic), the peripheral populations are relatively isolated and subject to severe selective constraints that vary from one place to another. Each population will try to adapt as best it can to local conditions. They can sometimes do this only by giving rise to new coadaptive systems that represent newly emerging species becoming specialized for marginal niches. If such a movement extends in several directions it may become adaptive radiation or branching evolution, mentioned earlier. This can frequently happen in populations at opposite poles of the distribution area, which may therefore be subject to divergent selective pressures.

For some modern species this process of evolution is going on before our very eyes, but too slowly for us as humans to be able to follow it from beginning to end. Nevertheless, by observing a series of related groups it is possible to find different stages of speciation in certain lines that are being transformed. A striking example is fruit flies of the *willistoni* complex, studied by Dobzhansky, Ayala, and others. These flies are found in Central America and form about fifteen groups possessing the characteristics of all stages of isolation: from true species, perfectly intersterile, through semispecies, to subspecies that have a certain degree of interfertility (Ayala, 1978). This situation is similar to that of the baboons mentioned earlier.

Many other examples can be found in frogs, reptiles, birds, and others. Recall the case of the herring gull, *Larus argentatus,* mentioned in Chapter 2. It has a circular distribution around the North Pole in both Old and New Worlds. Neighboring populations of this group are interfertile, but those at the two extremities of the area, now living in contact because as they widened their distribution the two ends eventually met, are intersterile. Thus they became two independent species *L. fuscus,* the lesser black-backed gull, which represents the ancestral species and is found in the Old World, and *L. argentatus,* the herring gull, which differentiated in America beginning with migrants of the preceding species and is now secondarily established in Europe after crossing the Atlantic.

The apterous carabid and tenebrionid beetles have a higher level of speciation than the winged forms. Any form of philopatric behavior — that is, attachment to a given area — tends toward speciation. This is already clear in some invertebrates like butterflies, and even more so in higher vertebrates like birds and mammals that have social behavior related to the care of their offspring that brings them back, at least for part of the time, to the same nest. Some reptiles, amphibians, and even fish (like the stickleback) show this type of attachment to a territory.

The role of geographic isolation as an essential factor in speciation and thus in evolution can no longer be doubted. There are many arguments in favor of this rule. We may mention a few.

1. Speciation reaches a maximum in insular systems, where each island may possess its specific fauna. Endemic species in archipelagos are consistently much more numerous than those in neighboring mainlands. This is true of the British Isles compared with continental Europe; of Ceylon versus India; of Formosa, the Ryukyu Islands, and the Japanese archipelago compared with the Chinese mainland; of Tasmania and Australia; and of the Mediterranean islands versus southern Europe (Mayr, 1963).

For terrestrial animals the sea forms a barrier that is a very effective factor in speciation. It is not by chance that Darwin and Wallace made their most remarkable observations in archipelagos — the Galapagos and Hawaii for Darwin, and Indonesia for Wallace. It is well known how in 1835 the navigator of the *Beagle* noticed that each island in the

Galapagos had its own species of finches but that each seemed to be derived from a single variety found on the South American mainland. He made similar observations about giant tortoises.

The Pacific Ocean, containing many islands scattered irregularly through it, provides many examples of speciation by isolation, affecting all groups. Land snails of the genus *Partula* are widely established around the Society Islands. They form multiple species, some of which occupy a very narrow territory. For instance, *P. salifana* is only found at the top of Mount Salifan on Guam, and *P. filosa* in a single valley in Tahiti. Six species are known in the Marquesas, eight in Samoa, two in the Fiji Islands, seventeen in the New Hebrides, eleven in the Solomon Islands, five in the Admiralty Islands, four in the Caroline Islands, three in the Marianas, and a single one in New Guinea. (This work by Crampton, Clarke, Murray, and others is cited by Franc, 1977).

Also in the Pacific, another equally remarkable example is the mosquitoes of the group *Aedes scutellaris*, found over an enormous area that stretches from Madagascar, the Maldive Islands, and India to Indonesia, the Philippines, Australia, and the South Pacific. Some forms are widely dispersed into a large number of insular species, some of which are very localized. In many cases the mosquito seems to have followed very recent human migrations (Bekin, cited by Grjebine *et al.,* 1976). It seems that the group was fairly homogeneous at the beginning but probably widely genetically polymorphic, thus allowing it to break up rapidly into a large number of species under the influence of geographic isolation.

Lygodactyls are geckos, a variety of small lizard whose feet bear suckers that allow them to walk across ceilings. They are common in Africa south of the Sahara, Madagascar, and the islands of the Indian Ocean. They are also found in America. This group has been closely studied by Georges Pasteur. *Lygodactylus capensis* is an African species, widely distributed and with considerable genetic polymorphism. It is a pioneer group characterized by a propensity to establish peripheral colonies that give rise to many daughter species by cladogenesis. Two species live on the African continent; two others, on the way at least to becoming species, live in the small islands of the Indian Ocean; but in Madagascar there are about fifteen species, some of which form another genus.

The case of Madagascar is typical; although this large island is thirty-five times smaller than the African continent, it has seen a rate of speciation of lygodactyls four to five times that of Africa. This is due to the wide variety of possible niches and is partly related to the absence of predators, which allows these small reptiles to live in conditions that would not be possible in Africa. But it is also due to the ability of the lygodactyls to adapt to a variety of niches because of the richness of their genetic patrimony. There are also two groups in South America (Pasteur, 1977b).

On continents the situation is very different from that seen on islands. Speciation is only found when there are geographic barriers or, according to Mayr (1963), in "regions that are insular in one way or another," (assuming that remoteness can be an effective barrier even if there is no natural obstacle). Such is the case of the herring gulls discussed earlier. Forest clearings, mountain lakes, caves, geographic zones based on rivers, all can produce an archipelago model even in the middle of an enormous continent and thus give rise to rich forms of speciation.

2. A second, just as striking form of proof of the importance of geographic isolation in speciation is derived from paleontology. If we consider geological eras, we find speciation most active at the time of great ecological upheavals. At the end of an ice age the sea level rises, continents are submerged, and the summits of chains of mountains appear as rows of islands; this has almost always led to the appearance of numerous new species.

As an example let us consider a large area occupied by a series of

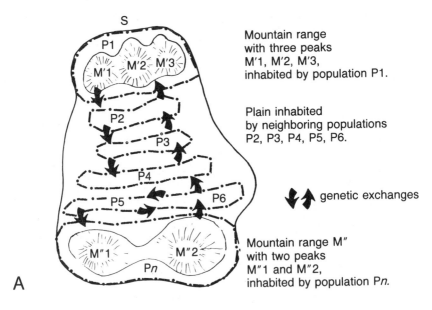

Mountain range
with three peaks
M'1, M'2, M'3,
inhabited by population P1.

Plain inhabited
by neighboring populations
P2, P3, P4, P5, P6.

genetic exchanges

Mountain range M"
with two peaks
M"1 and M"2,
inhabited by population Pn.

Figure 25. Progressive separation of populations living on high ground as a result of a rise in sea level.

A: Stage 1. The whole island is above water. Several populations (P1 to Pn) inhabit the available space. They exchange genes by interpopulational gene flow. Only species S exists.

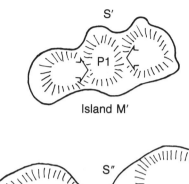

Island M′

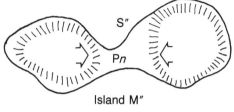

B

Island M″

B: Stage 2. The water level has risen, and the two mountain ranges now form two islands, between which there is no gene exchange. Through chance (drift) and different selective pressures the two populations, after long separation, can give rise to two new species, S′ and S″.

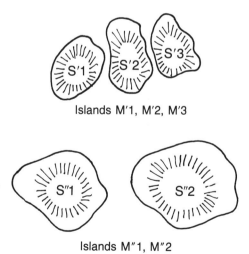

Islands M′1, M′2, M′3

C

Islands M″1, M″2

C: Stage 3. The water level has risen even more, leaving only the peaks above the sea, so that S′ and S″ are further isolated and may eventually form five emerging species S′1, S′2, S′3 and S″1, S″2.

populations P1, P2, P3 . . . P*n,* all belonging to the same species S. Let us imagine that the area is made up of a large plain limited at the north and south by two mountain chains M' and M" (Figure 25). Initially the species is composed of a series of populations living between the two mountain areas. In spite of ecological differences between M' and M", the species is relatively homogeneous, for numerous gene exchanges are possible between all the populations, which are still interfertile. When the climate becomes warmer, the sea level rises. The central plains are progressively submerged, so that all the populations P2 to P6 disappear. Only P1 and P*n* survive, taking refuge in the mountain chains, which are now islands. Gene exchanges are interrupted. P1 and P*n* will diverge more and more through chance and genetic drift, but also because the environmental conditions, and thus the selective pressures, are different in M' and M". After a certain time P1 and P*n* may become intersterile. They have become independent species S' and S". If the sea continues to rise until only the peaks are above sea level, there will be three islands in M' (M' 1, 2, and 3) and two islands in M" (M" 1 and 2), so that the species will form five isolated groups, S' 1, 2, and 3 and S" 1 and 2. If the interpopulational gene flow is weak or nonexistent, and the situation lasts long enough, the five groups could become intersterile themselves and give rise to five new independent species.

The Pleistocene ice ages, the ones we know best, were times of intense speciation, particularly for mammals and birds in Europe and North America (both of which were modified topographically by the climatic changes more than were the tropics, because of the advance and subsequent retreat of the glaciers). Still, the warm regions of the earth experienced periods of heavy rain, separated by droughts, that increased the areas of the deserts and created xeric barriers between groups that had been in contact.

Another series of transformations is related to continental drift. The continents move on enormous tectonic plates in the lithosphere (the more or less viscous external layer of the earth's crust) at a speed of a few centimeters per year, which on a geologic scale is quite fast. As a result of these movements, continents that were once locked together break up and create new frontiers. Others that were once separate come together and collide, often causing mountain chains to rise at the site of their collisions. All of this causes topographic and climatic modifications that favor the breaking up of groups and speciation.

So the evolutionary history of the animals and plants that surround us is closely related to the paleogeographic and paleoclimatic history of the regions where their ancestors lived.

The Genetic Revolution

This essential part of speciation has already been referred to. We shall simply summarize its essential features here.

An isolated population subject to constraints that are at the limit of what it can tolerate uses all its adaptive capacities. It reduces the frequency of certain genes that have become useless or dangerous, and calls upon others that were somewhere deep in its genetic load as soon as it seems they may be useful. Hitherto unexpressed genes come to the surface; while others that were once in the forefront become recessive and are no longer expressed, according to the mechanisms described in Chapter 4. A small number of genes will be eliminated: undesirable elements are "purged" when the genetic load can no longer keep them. There is a continual "ballet" involving the selective value of factors that depends on their frequency, as we saw in Chapter 6.

The population that is being reformed liquidates those combinations that are incapable of conferring sufficient adjustment to the environment and only preserves the rare "acceptable" genotypes. At this stage, the isolated population has become a race or subspecies of the parent group from which it derives; it differs in that its gene frequencies are profoundly modified. This, however, may not be sufficient to adapt it correctly. It will become even poorer under the combined influence of isolation, which suppresses gene flow from the outside, and severe selective pressure, which continues to demolish former combinations and imposes a drastic sorting out so that the group's variety diminishes even more.

Because it is monomorphic, such a population is fragile. Its homeostatic capacity is reduced and its genetic inertia feeble. It is very susceptible to selective forces and chance events. Either it will disappear, which happens frequently, or it will end by reconstructing a coadapted system that is better able to respond to the selective constraints than the former system. New relationships are built up from those factors that have been preserved. A new program develops, gradually enriched by factors that have been introduced by mutations or, more rarely, by introgression and that are useful in the new context. Adaptive modifications make themselves felt. With this genetic revolution the group passes from being a race to being an emerging species.

Thus, the genetic revolution that leads to speciation implies: (1) a different distribution of alleles, with perhaps the disappearance of some and the acquisition of others; and (2) the establishment of new relationships between these alleles, with dominants becoming recessives and epistatics hypostatics, for instance.

Regulatory systems play an essential role in the changes. Former vicariant pathways may become main channels, and vice versa. Other pathways that have not even existed are organized and become functional.

Attempts at speciation are not an exceptional phenomenon, but rather a normal tendency that manifests itself continually on the ecological frontiers of most groups. What is exceptional is for them to succeed and advance the evolutionary movement.

The genetic revolution is nothing like the storming of the Bastille. There is nothing violent or bloodthirsty about it. The reconstruction of the genome is slow and progressive, occurring piece by piece, each new acquisition causing a series of feedbacks that in turn provoke new transformations. It takes place over many generations and needs much time. This explains why one cannot hope to observe it in the laboratory. It is much easier to modify gene frequencies, as breeders often do to obtain new races, than to restructure a coadapted system.

Domestication has never created a new species. Speciation implies the passage from the quantitative to the qualitative. We can only hope to understand some features of this revolution in a few groups that we see around us that are really emerging species. Such is the case of the *willistoni* complex of the Central American fruit flies that we mentioned earlier as an example of isolation in evolution.

The population model, implying a progressive transformation of a whole reproductive group by successive steps over a very long time, is incompatible with the typological view, which follows a species back to a single mutant ancestor or a series of mutants that appeared suddenly.

We have already seen that most mutations responsible for the extensive genetic polymorphism of all wild populations are usually expressed only weakly in the phenotype, if at all. However, a few very rare mutations are exceptions to this rule and show spectacular changes. The drosophilists work on such cases. They can be such that the mutant seems to belong to a different species, whereas it is perfectly interfertile with other members of the group from which it is derived. A genetic revolution is just the opposite, for at the beginning it can be unnoticed at the level of the phenotype and yet involve sexual isolation that gives rise to an emerging species—a sibling species of the ancestral group from which it is derived, and virtually indistinguishable on traditional morphological grounds.

In Chapter 1 we saw that *Drosophila melanogaster* and *D. simulans* are very difficult to distinguish phenotypically and yet are two independent species totally incapable of interbreeding. Although morphologically almost identical, they differ in their system of coadaptation, and their differences must be related to thousands of genes. They represent two independent programs. On the other hand we mentioned two simple mutations of *D. melanogaster* that profoundly modify the phenotype. The first, the white mutation, removes the normal red color from the eyes, and the second, the vestigial mutation, reduces the wings to stumps. Any fly with both mutations would never be classed as *melanogaster* by a nonspecialist. And yet, in spite of the spectacular morphological differences, the two types are interfertile, for they bear the same program and only differ in their mutations, which are certainly spectacular but do not implicate the overall coadapted system. They are unquestionably members of the same species.

Speciation is related much more to the relationships between genes than to mutations.

Reproductive Isolation

A peripheral population, cut off for many years from other populations of the same species, finally succeeds in its genetic revolution. We are then in the presence of an emerging species, still quite modest but with its new system of coadapted genes, its program, and its own particular niche. It is separate from its ancestors, which occupy a different niche. From now on selection will act on two distinct groups that do not share precisely the same ecological conditions and will tend to adapt more and more efficiently to their respective environments. Thus they will diverge more and more. What is important is that each should conserve its acquired adaptive features and subsequently improve on them. This can only be achieved if sexual isolation is strict; otherwise new interpopulational flows — which are now interspecific — will be set up if (for example) geographic barriers are removed. This would threaten the new structure that has been built up with such difficulty. Thus any organic modifications that tend to ensure sexual isolation will have a positive selective value. From now on active isolation, related to the nature of the partners themselves, will tend to replace the passive isolation, usually geographic in origin, that as we have seen generally constitutes the prime mover of speciation. This active isolation ensures that two species with a common origin that have been separated for a long time and then come into contact again no longer have the capacity to hybridize. This was the case of the herring gull we spoke about earlier. The acquired features are preserved. Speciation will not have been in vain.

The mechanisms that ensure sexual isolation are very numerous. There are two types: *postzygotic* and *prezygotic* isolation, depending on whether the reproductive barrier exists before or after fertilization.

Postzygotic Sexual Isolation

In this case copulation and fertilization are possible, but the process stops there. The incompatibility is thus beyond the level of the partners, who are quite capable of mating. Postzygotic isolation is often the first to develop and is seen as a more or less marked reduction of fertility. In its first stage hybrids can be born but are relatively infertile. They have far fewer offspring than conspecific pairs. This is why the zone of hybridization between two neighboring species is often limited to a narrow band that does not tend to become wider. As we saw in Chapters 2 and 4, examples can be found in many groups. At a later stage one of the sexes becomes sterile, and then finally both. In the most rigid form of this isolation, fertilization takes place but results in an abortion sooner or later.

The Mechanism of Postzygotic Isolation

These difficulties in fertilization are doubtless due to the divergence of the two genomes, which can appear at two different, but not exclusive levels. The first is the molecular level and consists of a cytochemical incompatibility. There are often important differences between the two coadapted systems. The modification of the hereditary patrimony by the new species during its genetic revolution may be too great for it to be able to interact with that of the related species. An egg resulting from such a fertilization will not be viable, for the hereditary material received from the father—the literal "patrimony," cannot interact with that of the mother, the "matrimony." There can be no development.

Such cytochemical incompatibilities can have many causes. In some cases the cytoplasm has a lethal effect on the heterospecific nuclear material, as C. L. Gallien (1970) showed when he transplanted the nucleus of one species into the enucleated cell of another species. Sometimes the spermatozoa have a lytic action on the cytoplasm of the ovum. In other cases the incompatibility is immunological and is manifested by a local or generalized reaction of the mother organism against the spermatozoa or the fertilized egg. This form of immunization of the mother by the fetus can be found within a single species. It is called alloimmunization, and is well known in man in relation to the rhesus factor.

But the incompatibility can also be chromosomal. We saw in Chapters 4 and 5 that when the germ cells are formed by meiotic division the two chromosomes forming a pair must be closely apposed before separating again at reduction division. During this phase *pairing* or *synapsis* can occur, permitting the exchange of homologous segments by crossing-over.

When the karyotype of the daughter species has been modified by translocations or inversions, pairing becomes difficult and the two genomes cannot align their corresponding loci except by complicated loops and other figures that make meiosis difficult.

Speciation is often, though not always, accompanied by chromosomal changes. Related species often have related karyotypes, apparently derived by quite simple modifications: this is Robertson's rule.

From these data it is possible to establish a chromosomal "taxonomy" that retraces phylogeny by a series of successive modifications of the karyotype. A remarkable table of primate evolution has been constructed by Bernard Dutrillaux and his colleagues (1979, 1980, 1981); Chiarelli and Capanna (1973) have also produced inventories of vertebrate groups. This leads us to the role of reorganization of the karyotype in speciation.

Chromosomal Speciation

For a long time cytogeneticists could study in detail only the giant chromosomes that exist in certain Diptera. For this reason the karyotypes of most species of fruit fly were known quite early, and a number of chromosomal rearrangements described. Since about 1960 techniques have become available for cultivating cells from various species, blocking mitosis in prometaphase, and splitting the nucleus by hypotonic shock to liberate the chromosomes and permit their easy identification. In this way good pictures of the karyotypes can be obtained and analyzed, using techniques introduced a few years later to show banding patterns. These are light and dark bands peculiar to each chromosome. They permit the detection of even minor reorganizations, as was done, for example, in the giant chromosomes of the salivary glands of larval Diptera. In this way we can identify corresponding sequences in related species. The chromosomal complements of a large number of animals, particularly the mammals, practically unknown until then, were rapidly established. This work confirmed the general application of Robertson's rule.

This gave the mutational theory of evolution a new lease on life but in a new form, that of chromosomal speciation. Certain cytogeneticists established a causal relationship between the two phenomena and recognized the origin of the sexual isolation of an individual that would engender a new species in a rearrangement that had accidentally involved a karyotype. This hypothesis was supported in France particularly by Jérôme Lejeune and Jean de Grouchy.

Chromosomal speciation might proceed in the following manner.

1. Within a population an individual, the *propositus,* undergoes a chromosomal change. It then carries this anomaly in the heterozygous state and transmits it to half its gametes, that is to say, to half its offspring.

2. If the offspring with the altered karyotype (still in the heterozygous state) are viable and interbreed with the first individual or among themselves — a common situation in small isolated populations that are very endogametic — a quarter of the subsequent offspring will carry the anomaly in the homozygous state, as seen in Figures 26 and 27.

This process has been suggested as one explanation of hominization by Chiarelli (1967), Lejeune (1968), Ruffié (1973), and de Grouchy (1978, 1980). It is now known that the ancestors of the hominids appeared in East Africa about the middle of the Miocene, when they split off from the common stem of the African anthropoid apes. The chimpanzees, the higher primates that are closest to us (our cousins, in a way) have forty-eight chromosomes, of which eight are acrocentric. Man has only forty-six, of which six are acrocentric. It is thus logical to think that members of the ancestral phylum had forty-eight chromosomes. Chimpanzees and gorillas kept to this original plan. The hominids underwent a translocation of two pairs of acrocentric chromosomes that became a single pair of mediocentrics. This was actually not a real Robertsonian translocation but a terminal rearrangement with the loss of a fragment of heterochromatin from the short arms of the two chromosomes that fused but without loss of hereditary material, the euchromatin. Banding studies of the karyotypes of the African monkeys and of man have allowed the identification of the chromosomes responsible for this modification. This "Adamic theory" of hominization is so called because it postulates the existence of a single ancestor, just like Adam of the Bible.

In actual fact, even if everyone agrees about the great importance of karyotypic modification in the isolation of a species, it is not evident that these chromosomal rearrangements are the first stage in speciation in all cases. They may represent a secondary event, consolidating the sexual isolation of a group that has already begun to separate. This hypothesis merits some consideration. Let it be said from the beginning that the appearance of a species from a single chromosomal mutant is not theoretically impossible. Such a mechanism is even the most satisfactory for explaining cases of sympatric speciation, during which a new species appears in the very center of the area of distribution of the parent species and seems to become independent right away, without the intervention of geographic isolation. However, this situation is not common. Studies carried out in the field involving very varied groups

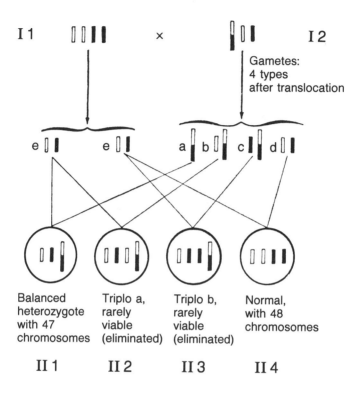

Figure 26. Conversion from 48 to 47 chromosomes by balanced translocation.
First stage: Appearance of several subjects carrying an anomaly in the heterozygous state. A subject has two pairs of small acrocentric chromosomes (I1, to the left of the parental line) and undergoes a fusion (translocation) of one acrocentric element onto the other (to the right of the same line; I2). The subject thus has only 3 chromosomes, one of which, large and metacentric, is derived from the translocation. All the gametes of I1 are identical *(e)*. By contrast I2 will produce four types of gametes *(a, b, c, d)*. At meiosis each chromosome pair must split, and since the translocated chromosome cannot be divided into its constituent elements, it will pass into a gamete in one piece. This may result in overdosage. If fertilized by a normal subject, the translocated individual can produce four types of offspring, as can be seen on the descendants' line:
II1: Balanced heterozygotes carrying the same anomaly as the translocated parent.
II4: Normal homozygotes chromosomally similar to the original population and the normal parent.
II2, II3: Subjects with overdosage (triplo a or triplo b), almost always nonviable.

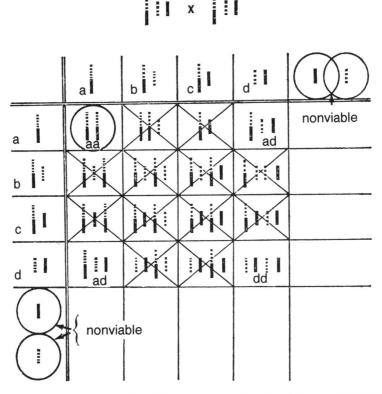

Figure 27. Conversion from 47 to 46 chromosomes from two heterozygous parents by translocation.

Second stage: Appearance of subjects carrying the anomaly in the homozygous state. This diagram illustrates the crossing of two balanced heterozygotes found in the previous generation. Each again produces four types of gamete *(a, b, c, d),* giving sixteen possible combinations. Of these, twelve are nonviable (overdosage) and four are balanced: two with the translocation in the heterozygous state *(a/d, a/d),* one with two pairs of chromosomes identical to the normal grandparent *(d/d),* and one with the translocation in the homozygous state *(a/a).*

This last represents a new balanced karyotype (with an even number of chromosomes), having exactly the same chromosomal sequences as the ancestral type, but arranged differently as a result of the translocation (thus giving it one less pair of chromosomes). If acceptable, this karyotype can spread rapidly through a whole population, which tends to become isolated from neighboring populations from which it was derived. Insofar as this group is in the process of transformation, this isolation can confer a certain advantage, for it protects its acquired characters, especially adaptive ones, against the threat of "foreign" genes. This defense against hybridization provided by the chromosomal rearrangement must, at least in some cases, be an essential stage in speciation, which becomes an irreversible process. Thus, rearrangement that reaches the homozygous state and is stabilized could be the start of a new species.

suggest that speciation almost always implies not a single individual but a whole population that has become physically separated from the parent species. This is as true in fruit flies as in baboons. We have already analyzed this mechanism for collective speciation, which manifests itself first of all by a transformation of gene frequencies and then of coadapted systems — a genetic revolution.

This form of progressive populational speciation explains the fact that all stages of transition — races, subspecies, semispecies, and sibling species — are encountered in all developing lines. Chromosomal speciation, a sudden and individual phenomenon, would present a rather different picture.

In addition, the populational model accounts for the genetic polymorphism that is always present in emerging species. This could not be true of a group recently issued from a single ancestor, for this ancestor would have been capable of carrying and transmitting only a small fraction of the genetic variety of the line from which it came. So we must ask ourselves what the precise role of chromosomal reorganization is in those cases — probably the most common ones — where speciation involves a whole population. Actually the accidental occurrence of such rearrangements is not exceptional. It can be found in all groups and in all generations. Such accidents are well known in man, and since the discovery of the chromosomal origin of mongolism by Jérôme Lejeune and his colleagues in 1959, they have opened a new chapter in pathology, that of the chromosomal diseases. In a normal population a chromosomal rearrangement constitutes a rather unfavorable event for the affected individual, even if it is not expressed in the phenotype, as is common. It represents a handicap for any offspring, as difficulties may arise at the time of the formation of germ cells by meiosis. We saw in Chapters 2 and 4 that this phenomenon is one of the principal ways of inhibiting hybridization and thus a mechanism for protecting the biological integrity of species. In addition, any chromosomal reorganization within a homogeneous population having sufficient genetic exchange with other populations in the same species and in reasonable equilibrium with its milieu, constitutes a disadvantage. In general, this type of anomaly has great difficulty crossing the hurdle of meiosis. It tends to be eliminated more or less rapidly.

However, the situation is different when a peripheral population, undergoing its genetic revolution, "keeping its distance" from the parent species, and attempting to exploit a new niche, is involved. This isolating movement is doubtless of positive selective value insofar as it reduces competition and increases the individual's resources. We shall see in Chapter 9 that reduction in competition is almost always beneficial. Thus any element that favors this isolation tends to be retained. This is true of chromosomal reorganizations that interrupt interpopulational flows and avoid threats to genetic combinations that help in adapting to the niche being taken over. These reorganizations confirm the new system of coadaptation and prevent retreat from it. By abolishing any possibility of sexual exchange with the parent species, the new karyotype protects what has been acquired by the population that is becoming independent.

This is not simply a hypothesis but is supported by many arguments. We may mention a few of them.

1. Many sibling species that are undoubtedly intersterile have identical or almost identical karyotypes. Now, most zoologists believe sibling species to represent groups that have recently split off from a common stem. Sibling species must represent the first stage of true speciation.

2. On the other hand, perfectly homogeneous species can demonstrate a certain genetic polymorphism. Although fairly rare, this has been observed in various groups, particularly in arthropods. We may recall that Theodosius Dobzhansky described chromosomal polymorphism in *Drosophila pseudoobscura,* with the frequency of its different types varying cyclically with the seasons (see Chapter 6).

Chromosomal polymorphism can manifest itself in clines, with regular spatial variations. This is the case of *Jaera albifrons,* a small crustacean found commonly on the Atlantic coast of Europe in the intertidal zone, described in detail by Bocquet (1972). In the subspecies *J.a. syei* the diploid number of chromosomes is at a maximum, with 28 in the male, on the coast of northern Germany and diminishes toward the southwest. It falls to 26 in Holland, 24 in the Boulogne area, and 22 between the Cotentin Peninsula and northern Finistère, reaching 20 in southern Finistère and 18 on the Basque coast. Another regularly diminishing gradient is observed as one goes north from Germany toward Scandinavia. A few hybrids with an odd number of chromosomes can be found naturally in the zone of contact of two groups, but their number is always limited.

Cases of chromosomal polymorphism have been described in insects and particularly butterflies, certain of which, like *Lysandra coridon,* vary in clines. In this species the karyotype changes gradually from 92 chromosomes in northern Europe to 87 in Spain and Italy. There seems to be no mixing between neighboring populations with different karyotypes (de Lesse, 1970). Other examples are found in shellfish and rodents, including the common rat *Rattus rattus,* and in prosimians such as bush babies and lemurs.

These regular variations in time and space suggest that chromosomal reorganization may, in certain cases, have an adaptive value.

Apart from clinal distributions, explained above, which are peculiar to certain groups, other chromosomal mutations giving rise to less regular variations can be found particularly in the center of the area of distribution where environmental conditions are relatively moderate. In this zone, selective pressure is lower than in the periphery and "tolerates" a certain number of deviations of all forms in genetic information that would not be possible on the borderlands where selective pressures are much greater because they mark the extreme limit beyond which the species cannot penetrate. It is precisely in these marginal zones that new species tend to appear, just where chromosomal reorganization is rarest.

3. Speciation is progressive, prepared for long in advance by the variation of certain characteristics of the species along clines, as we saw in Chapter 2. This gradual evolution along a series of gradients cannot have its origins in a reorganization of the karyotype, for this is a sudden, unique phenomenon.

4. Finally, the study of many emerging species that can be found today shows that the real process of speciation usually occurs before any *major* chromosomal reorganization that may lead to intersterility.

Let us look again at the mouse *Mus musculus,* which is splitting into several newly emerging species (Chaline and Thaler, 1977). We have seen that two forms live in non-Mediterranean Europe: *M.m. domesticus,* a subspecies found in the west, and *M.m. musculus,* found in the east — in Russia, Scandinavia, and eastern Germany. They can be distinguished by a number of minor biological details. There is a zone of hybridization between the two groups that cuts through Germany in a north-to-south direction, but that is only a few kilometers wide. Electrophoretic study of enzymatic alleles has confirmed that this is indeed a zone of limited hybridization. In the northern part of their area of distribution the two species have approximately the same karyotype, but this is not the case in the south, around the Alps and in Italy. Normally in *M. musculus* there are 40 chromosomes, all of which are acrocentric, but in the southern part of its territory there are local populations in which the number tends to be smaller, between 22 and 38, as a result of a number of Robertsonian fusions. These chromosomal subspecies are often isolated in mountainous regions or islands, like a group with 22 chromosomes that is found in the center of the Appenines. There exist zones of contact with the standard 40-chromosome variety so that chromosomal hybrids of all sorts can appear; but they are always restricted to a narrow region.

The mole rat, *Spalex ehrembergi,* is an underground rodent. Four races live in contiguous areas from the north to the south of Israel with, respectively, 52, 54, 58, and 60 chromosomes, the first type being probably the most ancient. Thus, contrary to what is usually observed, the division of these mole rats into subspecies is accompanied by an increase in the number of chromosomes, not a decrease. Electrophoresis shows that each race has its own allelic frequency but that none of them possesses an exclusive mutation. Each has certain peculiarities, either physiological, such as basal metabolic rate, or ethological — the males of different races recognize and attack each other, thus provoking a rather rigid sexual isolation. These differences seem to be related to local ecological constraints such as the degree of aridity. The splitting of *S. ehrembergi* into subspecies is linked with an adaptive evolution to the subhabitats of its area of distribution.

We have already spoken of the chromosomal complex of the *Anopheles maculipennis* mosquitoes in Chapter 1. An important vector for malaria, they are widespread throughout the world and made up of a whole series of species. In the Nearctic region there are six, located on the North American subcontinent: *A. freeborni, A. quadrimaculatus, A. earlei, A. aztecus, A. occidentalis,* and *A. punctipennis.* Of these the first two transmit malaria. These Nearctic variants are the oldest and also the easiest to identify individually by their morphological features. Their geographic distribution is fairly rigid, and they all have marked ecological differences. In addition, their karyotypes show major alterations from one species to another. Even when hybridization is possible, the offspring are always sterile. There is a strict sexual isolation of each group: they have long since passed the stage of sibling species in their phenotypic differences.

The Palearctic group presents a different picture. It is also made up of six species in the process of isolation, which many authors consider to be subspecies: *A. maculipennis maculipennis, A.m. messeae, A.m. melanoon* or *subalpinus, A. sacharovi, A. labranchiae labranchiae, A.l. atroparvus.* The adult morphology is approximately the same. The best distinguishing features are in the eggs. These species demonstrate large biological differences that make them play different roles in the transmission of malaria. There are also differences in both their habitats and their reproductive behavior (see Chapter 1). In the wild these groups are quite consistently intersterile, but it is often possible to form hybrids under experimental conditions, although the ease of hybridization is not the same in each species and varies with the crossing attempted.

Thus we are dealing with sibling species that must have become differentiated quite recently along the pathway *atroparvus* → *labranchiae* → *maculipennis* → *subalpinus* → *sacharovi* → *messeae.* Their autonomy is much less marked than in the Nearctic species. In addition, the karyotypes of the Palearctic group are much more uniform and differ only in minor rearrangements, particularly inversions that provoke the appearance of loops in the hybrid but do not interfere with meiosis. At least in most cases, the hybrid offspring remain fertile. It is conceivable, as suggested by A. Grjebine *et al.* (1976), that the difference between the Nearctic and Palearctic complexes is based on the time of their splitting up. The first, Nearctic group is the older and was divided into autonomous species a long time ago. These took over different niches in a very wide area of distribution. Later, a Nearctic population must have migrated to Eurasia, but this migration was quite recent, thus explaining the less marked separation of the groups that now form the Palearctic complex.

Origin and Modalities of Chromosomal Reorganization

The recent work of Bernard Dutrillaux *et al.* (1981) on the cytogenetic evolution of primates has thrown new light on the role and probable modalities of chromosomal reorganization in speciation. Without going into detail concerning this very original research, we may note the following points.

1. Chromosomal reorganization during primate speciation is not random but regularly affects certain privileged zones on each chromosome. A typical chromosome, for instance, may show a dozen such points — always the same in all species — that are responsible for the modifications observed from one group to another. Thus we can find identical chromosomal mutations in different species, although phylogeny demonstrates that these mutations did not take place at the same time. Such convergence could not be due to chance but must involve a certain predisposition. Robertsonian translocations are frequently found in the lemurs, fissions are found in *Cercopithecus* monkeys, and pericentric inversions in the *Pongidae* and doubtless in man's direct ancestors.

2. The topography of these chromosomal rearrangements is often similar to what is found in human clinical practice, where we see on average one translocation per thousand births and one pericentric inversion per ten thousand births.

3. The same rearrangements appear in cultures of human lymphocytes subjected to X-rays. Thus, chromosomal accidents seen in the clinic or provoked in irradiated cultures are not the product of chance, as was once thought, but involve a series of zones that are always the same for a given chromosomal sequence, thereby reproducing in man an ancestral situation — a chromosome or chromosome segment with simian features. This may be considered a form of reverse chromosomal mutation.

Other experiments using irradiated fibroblasts from *Cercopithecus cephus* produced fissions leading to chromosomes identical in appearance to those of *C. aethiops.*

4. The appearance and nature of these chromosomal reorganizations may be determined or predetermined by a particular ancestral genetic constitution.

Dutrillaux's model is doubtless applicable to many other vertebrate groups if we consider that the relationship between karyotypes is demonstrated by the identification of homosequential zones, particularly in mammals.[1] It is based on solid arguments and probably is a generalized phenomenon. One of its beauties is that it reconciles the population theory and the chromosomal theory of speciation, the latter having long been considered by many people as the most recent avatar of typology.

Mutations that favor or provoke certain chromosomal recombinations would probably tend to be expressed preferentially in marginal populations undergoing a process of isolation that have already begun their genetic revolution — that is, groups where sexual isolation has a strong, positive selective value.

We may summarize as follows.

1. The process of speciation almost always begins before there is a visible modification in the karyotype. It seems to be most often linked with geographic isolation and its consequent ecological changes.

2. Karyotypic modifications appear later and make the sexual isolation of the species permanent. They are a sort of "lock," thanks to which speciation becomes irrevocable. These modifications are not random but involve a certain number of privileged zones. It is highly likely that genetic determinism plays a role. Their appearance in a group that is being isolated cannot remain selectively neutral insofar as it favors, reinforces, or even provokes this isolation and in any case makes it irreversible.

This explains why, since the beginning of the Neolithic revolution nine or ten thousand years ago, farmers have produced many new races but never new species. By applying rigid breeding criteria, using parents having the most marked form of the desired feature, and strict selection at birth through elimination of those offspring that do not correspond to the desired model, it is relatively easy to split sufficiently polymorphic species into a series of races that may be very different from each other. An example is the wolf, which has given rise to a wide variety of dogs differing considerably in their morphology, adaptive

---

1. We now know that the method of *banding,* which reveals the characteristic series of light and dark bands in each chromosome, reliably identifies the hereditary material in a given zone. Chromosome maps of different species made by using known markers have shown the close correlation between morphologically identical zones and the loci they bear. This correlation holds for related species, such as primates, but also for very distant species, such as primates, carnivores, rodents, and lagomorphs.

capacity, and behavior. However, all dogs, whatever their nature, can interbreed. In actual practice the creation of domestic races involves the selection of certain genetic combinations and the rejection of all others. The breeder begins with a concrete situation, a preexisting patrimony that he simply divides into races. Speciation is quite different. It needs a genetic revolution — that is, a remodeling of coadapted systems — and such a process takes a very long time to establish.

What is more, for speciation to be irreversible there must probably be a reorganization of the karyotype. This is a condition that the breeder cannot fulfill himself, in contrast to his ability to select desirable racial characteristics. Karyotypic changes need a long time and are difficult to accomplish. What is more, not all chromosomal anomalies lead to a new species. Rearrangements that lead to species must not involve phenotypic changes that are too drastic, as are found in some human conditions. They must be able to cross the barrier of meiosis in the first heterozygotes that carry them and ensure that the homozygotes are sufficiently sexually isolated compared with the ancestral group, as we have seen.

The probability that all these conditions will be fulfilled at the same time is low. Thus, a chromosomal rearrangement that is effective, in terms of isolation, and without danger, in terms of the phenotype, to appear and diffuse throughout the whole group requires much time — certainly much more than farmers have had since domestic breeding began.

For the same reason the animals and plants introduced in Europe by the Spaniards after the discovery of the New World have sometimes produced new races but never new species.

1. Most varieties brought back from America — such as corn, potatoes, tobacco, turkeys, and guinea pigs — were already domesticated and were thus genetically poor.

2. The environmental conditions in their new home prepared by man were not fundamentally different from their ancestral conditions.

3. In any case the discovery of America is so relatively recent that speciation has had no time to occur.

On the other hand we shall see how, during the whole history of life on earth, the introduction of plants or animals to new milieux has usually resulted in intensive speciation.

To conclude these remarks on chromosomal speciation, we should remember that in the majority of cases chromosomal reorganization does not create species. It consecrates them and makes speciation, which began as a populational phenomenon, an irreversible process. Karyotypic changes contribute essentially to the passage of a group from the stage of being a passive species, isolated for external reasons that do not depend on itself (extrinsic isolation), to the stage of being an active species with its own biological isolation mechanisms. Although chromosomal reorganization begins with a single individual, speciation is from the outset fundamentally a collective phenomenon. It is a populational adventure (Ruffié, 1978; Dutrillaux, 1979).

Prezygotic sexual isolation

Sexual isolation can manifest itself earlier and involve mechanisms that prevent fertilization or even mating. The block thus comes before the level of the couple. Almost always during speciation prezygotic isolation appears *after* postzygotic isolation. The selective advantage of this change is obvious.

First of all prezygotic isolation is more rigid. Since there can be no mating, there can be no fertilization. There is a total barrier. Then, it is more economical in terms both of biological material, such as gametes, eggs, and embryos, and of energy. An embryo that aborts or an offspring that does not reach reproductive age uses up a part of the resources of the ecological niche to no avail. It has occupied part of the species's space and time uselessly and thus represents a total loss for the group.

The mechanisms of prezygotic isolation involve several modalities.

1. Ethological isolation. We saw in Chapter 2 that behavioral characteristics mark the interface between individual and environment. They are thus the first to be tackled by natural selection; other characteristics, physiological and morphological, become involved in adaptation only when all behavioral features have been exhausted. This being the case, it is not surprising that behavior plays an essential role in sexual isolation. In many species sexual attraction depends on precise signals, whether auditory, olfactive, or visual, such as the song of the nightingale or cricket or the light display of fireflies. These stimuli can take the form of extremely complex, almost ritual, behavioral patterns. Many birds, fish, insects, and spiders engage in *nuptial displays* that have often been studied. (For a general review, see Leroy, 1978.)

Whatever signals are given, they can only be understood or decoded by individuals of the same species. For other individuals they are meaningless or even frightening. This behavior plays a triple role, according to Nikolaas Tinbergen (1954) as quoted by Ernst Mayr (1963): (1) "to advertise the presence of a potential mate"; (2) "to synchronize mating activities"; (3) "to suppress fleeing or attacking tendencies in the sex partner."

These signals are highly specific features. They often take the form of sounds, that is, the "songs" found in many species (Leroy, 1979). In spite of a few acquired modifications that form regional "dialects," most ornithologists can identify a bird by its song just as easily as by its morphological features. Female fruit flies will only mate when they have been stimulated by the sound of the vibration of the wings of conspecific males. They do not respond to sounds produced by males from other species. Many publications have described the acoustic analysis of these sound signals emitted during the nuptial parade. They reveal striking differences between sibling species that may well be the essential obstacle to their interbreeding, such as between *Drosophila pseudoobscura* and *D. persimilis,* and *D. melanogaster* and *D. simulans.* Thanks to an analysis of the male prenuptial sound, Dobzhansky promoted *D. pavloskiana* to the rank of a true species.

The song of the common cricket is another specific character. It varies very precisely from one species to another. The leopard frog of the United States was for long thought to be a single species, *Rana pipiens,* but is in fact a complex made up of four sibling species, strictly localized geographically in the north, south, east, and west. The males have a characteristic call that differs for each group in the duration of the notes, the number of impulses per second, and the time required for the amplitude of the sound to go from 10 to 90 percent of its maximum. All these call types are strictly allopatric and change suddenly at the geographic frontier of the species. There is no evolution in clines. The most one finds is a few intermediate song forms in the contact zones between two groups. This may be due to the presence of hybrids, although in small numbers (see Dubois, 1977).

Other research involves luminous signals. Fireflies are curious beetles; the male emits a light while flying, with species differences in duration, frequency, brilliance, and color, which may be white, blue, green, yellow, or orange. There are also chemical, olfactive signals found mainly in mammals, except for higher primates, which seem more sensitive to visual, auditive, and (in man) tactitle stimuli. Chemical signals have been described in insects: it has long been known that a newly hatched moth can be detected by a male several kilometers away.

There may be a quite complex exchange of signals between the two partners, a real "dialogue" that, if successful, will lead to mating. This exchange of stimuli, which are always produced in a rigorous sequence, is an extra guarantee of contact between individuals of the same species. The more complicated a system, the smaller is the chance of errors that might permit individuals from different species to mate. If this were not so, the independence of the two species would be threatened. In some cases it has been possible to follow the beginning of the ethological isolation between populations by the differentiation of their call while all their other anatomical, physiological, and cytogenetic features remained similar. An example is the Japanese newt *Triturus pyrrhogaster* (Kawamura and Sawada, 1959; Sawada, 1963). Such populations doubtless represent emerging species.

2. Physiological isolation. For reproduction to be possible the partners must have compatible sexual cycles. If, for instance, they have different hours or seasons of activity, contact is scarcely possible.

3. Anatomical isolation. Anatomical or "mechanical" isolation is due to the fact that for mating to take place the copulatory apparatus of males and females must be adapted to each other. The form of the genitalia is one of the most rigid specific criteria. Already in the last century, Léon Dufour (1844) believed that the sexual apparatus of females and males were the "lock and key" of the species. Samuel H. Scudder and Edward Burges (1870) attributed a great taxonomic value to the genitalia. It is certain that in many cases the mechanical mating of different species, even though closely related, involves the injury or even death of one of the partners. This phenomenon has a strongly negative selective value. Nevertheless, this rule is not always so strict as morphologists once believed, and a penis is often, anatomically speaking, more of a master key than a key designed for a single safe!

In fact, in most cases sexual barriers appear simultaneously at several levels. They involve a variety of mechanisms — ethological, physiological, and even anatomical — all of which act to guarantee the genetic autonomy of a species.

THE SPEED OF SPECIATION

Certain authors have claimed that there is a "standard" speed for speciation. This seems not to be true, for there are too many variables in the process of isolation of the group. Some groups acquires a rigid genetic autonomy quite quickly, while others continue to interbreed with neighboring populations at least in a limited fashion, for a long time. Generally speaking, the speed of speciation depends on the strictness of the geographic isolation, the intensity of selective forces acting on a group, the richness of its patrimony, and, most of all, on the number of its members. We saw in Chapter 3 that large populations with many members benefited from wide polymorphism, within which sexual exchanges were very free. There was a true pool of intercommunicating genes, producing a solid evolutionary inertia. On the other hand small, well-isolated populations are fragile and may prefer a genetic revolution. All these factors vary greatly between groups. This is why there can be no standard speed of speciation. In each case everything depends on the values of the different parameters that play a role.

natural selection

THE SELECTIVE ADVANTAGE OF SPECIATION

Speciation is a permanent phenomenon that has gone on since the beginning of life on earth, and it is continuing before our very eyes. No climatic changes or natural catastrophes or environmental modifications have ever threatened it. On the contrary, such perturbations favor it. Its consistency and variety show that speciation is a process oriented toward selection; but one might ask why, for by their very nature species are genetically stable groups. However, very few are satisfied with this stability, and many tend to escape from it at some time or other and create something new. What is the nature of the force that explains the paradox of speciation?

We said in the previous chapter that evolution is not related to a preestablished program or to any endogenous force or supreme will, but to a series of external constraints acting on material that has particular properties — living matter. In fact animals and plants are fundamentally pioneers. As soon as a habitat becomes available, a number of groups tend to invade it. To do this, preexisting types must be transformed to adapt to the new milieu. They can do this thanks to their genetic polymorphism, which offers them many alternatives and numerous "choices," as well as to their populational structure which, through the intermediary of sexual reproduction, permits them to effectuate the transformations necessary to respond to the new conditions. Living matter is endowed with an enormous faculty of adaptation. We saw in the previous chapter that when subject to different selective pressures, a population reacts first of all by modifying its gene frequencies and retaining only the best-adapted genotypes.

This phase represents a reversible racial differentiation. If it is not enough, the whole system of coadapted genes must be reexamined. The population then constructs another system, better adapted to the newly established constraints. It is now an emerging species. If this maneuver is successful—as is not always the case, for the great majority of attempts at speciation fail—an irrevocable evolutionary step forward is taken.

The newly formed species occupies a new niche and can in turn be the point of departure for an adaptive radiation. From being a shy and modest emerging species, it becomes an ancestral line that, if its offspring prosper and differentiate, will give rise to a whole genus, a family, or even a higher taxonomic level. Thus speciation is a means of survival for a group that has exhausted all the adaptive resources of its patrimony. It often constitutes the best response to environmental variations. It depends on many factors at the same time. For example:

1. The genetic variety of the population that is evolving, which is itself linked to a series of random factors such as the appearance of mutations, the size and direction of interpopulational gene flows, and perhaps introgression. As we saw in Chapter 4, we now know that the more a population is genetically polymorphic, the more it preserves its evolutionary potential. The experiments of Francisco Ayala described in Chapter 6 and those of many other workers prove this. Adaptive facilities and therefore evolutionary capacity are much greater in polymorphic populations than in monomorphic ones.

2. Speciation also depends on environmental variation. The constraints of the new ecological niches that evolving groups can exploit push evolution in a particular direction. They impose certain types of genetic combinations.

Nevertheless, when confronted with a given problem, living organisms almost always have several possible responses, and the choice between these responses depends on both selection and chance. This is why the most isolated, inhospitable locations such as caves, the depths of the ocean, and the polar regions, which demand adaptation at the limit of what is possible, can be occupied by members of very different lines, both invertebrate and vertebrate. Each group has its own solution to the problem of the constraints of a particular habitat and enters into numerous relationships with other groups living around it; trophic relationships grow up, food chains are organized, ecological niches are exploited.

We said earlier that stochastic elements were always factors of variation. Two lines originating from the same phylum and living in similar ecological and therefore selective conditions can give rise to very different groups if they are geographically isolated, unless there are phenomena of convergence to be considered, as is always possible.

Speciation is the price of conquest: the whole of natural history obeys this law. Species are more numerous when more or less isolated ecological niches become available. We saw earlier that the most important times of animal and plant evolution corresponded to the eras of geographical and climatic upheaval that the earth has known. In the end, speciation emerges as the construction of a system of coadapted genes giving rise to sexually isolated individuals capable of exploiting a new niche.

The Role of Ecological Factors in Speciation

It is impressive to realize that the whole of natural history is compatible with this ecological view of speciation. For the moment let us consider two pieces of evidence.

1. There is a relationship between the number of species and the size of animals. There are many more species of small animals than of large, for there are many more small habitats available than large ones. In the meadow beneath my window, where I often take a walk, there are probably no more than three or four mammalian species, including a herd of cows. The whole meadow constitutes a single ecological niche for the cow, which grazes without choosing anything in particular, or even for the field mouse, which eats whatever it finds. On the other hand I have observed dozens of insect species, and a good entomologist would certainly find more, for there are dozens of plant species growing in the meadow, each with its own particular features of shape, size, flower type, roots, leaves, and seasons, which make up as many different habitats for the insects. They do not see the world globally like a cow or an elephant, but in its minute details. The smaller a species, the more numerous are its ecological alternatives, so that it can become specialized in a particular habitat that no one will come to take away from it. In all phyla an increase in length of three times means ten times fewer species (May, 1978).

Take, for example, the main vertebrate lines. In general, the first to appear had undifferentiated shapes and occupied quite large habitats with indistinct limits. Then more specialized forms replaced them and took over more precisely defined habitats where competition was less acute, an obvious selective advantage. The initial line disappeared and specialized groups persisted, sheltering in their particular situations. This movement toward specialization and refuge-seeking affected the amphibians at the end of the Primary era, reptiles at the end of the Secondary, and mammals and birds during the Tertiary.

2. Speciation accelerated over the centuries each time there was a multiplication of ecological possibilities. We have spoken of the influence on speciation of the major changes in milieu such as ice ages and continental drift. They are not the only influences, for living organisms constantly modify their environment. The birth of new species creates new food chains and thus new, "virgin" niches.

During the course of the conquest of the landmasses, tetrapod vertebrates followed the plant cover that provided them with a hospitable milieu. They split up into an incalculable number of species that went on to populate every place that was populable.

The multiplication of insects followed that of the flowering plants, the angiosperms, which offered a multitude of ecological niches to terrestrial varieties. It is difficult to imagine how rigidly many insect species have become adapted to a specific host, and to it alone.

Birds developed from reptiles and colonized everywhere they could in the air.

All living beings constitute in themselves an ecological niche; thus the appearance of each new species prompted the differentiation of a host of parasites that also often demonstrate an astonishing specificity. And the parasites themselves bear "hyperparasites."

Man appeared on earth very recently and created a particular niche. His habitat quickly provoked the appearance of many species or subspecies of commensals (in particular the mouse *Mus musculus*, already described), but also, from Neolithic times, the birth of numerous domestic races, both animal and plant.

These data now permit us to reply to the question posed in Chapter 7 as to the true motive force for evolution and the precise nature of teleology. In other words, does life conquer for conquering's sake, driven by some pioneering spirit like cancer cells invading an organism, or does it rather obey more "objective" forces guiding it, quite simply, in the direction of natural selection? The second explanation imposes itself. Natural selection provokes the organization of new niches and thus speciation. But in order to understand this movement, we should fundamentally revise the idea of competition that was held for a long time.

THE TRUE NATURE OF COMPETITION

For many of Darwin's disciples, less subtle and not such good observers as he, competition represented a fierce struggle between wild forms and mutants. In this battle to the death one form had to disappear and the other take over, only to disappear in turn one day when a new, better adapted mutant came along. In reality, things are never quite like this. There certainly are frequent deadly struggles, but they are almost always *interspecific*.

We saw in Chapters 1 and 6 that predacity is a common phenomenon because it involves some of the most important steps in food cycles. This sort of combat mainly involves phylogenetically distant species such as birds and insects, carnivores and herbivores. It is very rare at the intraspecific level, at least under natural conditions. There is very little blood spilled or criminality in wild herds. Intraspecific killing is only seen in captivity. Such cruel animals as the fighting fish of Thailand or fighting cocks are not spontaneous products of nature; they are the fruit of long and patient selection by man, who finally succeeded in creating domestic monstrosities condemned to kill each other for the pleasure of spectators. Selective pressure never produces gladiators. This is hardly surprising, for any overagressive behavior at the intraspecific level would be a great disadvantage, particularly if the conflicts call for the death of the protagonists.

Natural selection prevents this by including in the genome of each species a number of ethological mechanisms that inhibit the killer instinct. It is only at the human level that competition can become fatal on a large scale. This modification — which is scarcely an advantage — is probably due to the reduction of our innate behavior and the extent of our acquired behavior. No species can destroy itself: it has neither the technical means nor the behavioral aptitude. The human race is the only one that is capable of massive self-destruction, of which we can find alarming examples in the world wars.

Nothing like this exists in nature. When two populations compete with each other, any modification that tends to allow them to exploit distinct niches is favorable and chosen by selection as a priority. This tendency might at least partly explain the advantage of the rare type studied in Chapter 6. Individuals that are clearly different from the average can colonize the frontier lands of the niche thanks to their genetic background and thus have an advantage over others. If they adapt sufficiently well, they find much less competition in this limited territory than in the crowded central areas. The same is true of populations. Selection tends to disperse, not to destroy. It encourages conquest, not murder. Rather than kill each other, two neighboring groups have a mutual interest in making peace and in separating.

Let us come back to what we said in Chapter 4 about introgression. Two geographically separated "species" whose members have no possibility of meeting can remain interfertile. We called these groups "passive" species, although for the geneticist the term "species" is not appropriate for two lines that can interbreed in the laboratory even if in nature this hybridization cannot occur. Now let us imagine that, following an upheaval, such as a migration or a climatic modification, these two groups meet. Two things can happen.

1. If their genetic divergence is weak and there is no rigid separation, the two passive species, which are in reality races, will be reabsorbed in a pool of common genes and form a single species.

2. If their divergence is sufficiently marked that sexual exchanges are inhibited, anything that favors their independence will be preserved. These two passive species will soon become active species. In such a confrontation, it is rare for one to eliminate the other; rather they tend to specialize, occupy two individual niches, and protect their specialization by reproductive isolation. Selection directs them toward a diversifying pathway, not a destructive one. This is probably why chromosomal reorganization, a disadvantage in a homogeneous group, may survive if it favors separation, as we saw earlier. This phenomenon explains the extreme richness of the living world.

An illustration borrowed from Diamond will help us understand the true nature of competition. In airports one frequently finds several car rental agents such as Avis and Hertz. It is rare that they open fire on each other with machine guns! Usually each tries to capture the clients that it can attract by what it has to offer.

The most favorable situation for each is to enlarge its own clientele in a sector where the other cannot compete. It tries to fill a slot that is not yet too competitive. Natural selection acts in the same way and rarely chooses war, which would be a biological disaster. If this had happened often, all species would have disappeared long ago.

Far from encouraging evolution, as the neo-Darwinians believed, elimination by competition would have stopped it. This is a fundamental law of biology that our contemporaries should not forget. By exploiting different milieux and engaging in complementary activities each group increases its own resources and diminishes the risks of confrontation. This tendency is generalized. It explains the breaking up of nonspecialized young lines that we have often discussed in the preceding chapters and explains the adaptive radiations observed in so many plants and animals.

Rather than impoverishing nature by elimination, selection enriches it by diversification. It favors noncompetition. Speciation is first of all an ecological conquest, and evolution is much more a constructive phenomenon than a destructive one.

THE LAW OF REINFORCEMENT AND CHARACTER DISPLACEMENT

The tendency for two diverging populations to "keep their distance" is observed in all groups. The main effect of intersterility between species is to protect what has been acquired by selection and thus make evolutionary progress. The intersterility that is established between emerging species is much stronger if these two species live side by side, and are therefore subject to a constant temptation to hybridize, than if they occupy more distant areas and thus have little chance of contact. This is the phenomenon of *reinforcement* described by Dobzhansky in the fruit flies *Drosophila paulistorum,* which are more strongly intersterile between neighboring populations than between distant ones.

Darwin had already noticed that species living side by side or slightly overlapping had clearer and more marked distinguishing characteristics in the sympatric zones than those in allopatric regions. But he attributed this divergence to the selective, and probably eliminatory, effect of competition. In fact, it is because selection causes divergence that cohabitation between groups that are no longer competitive is possible. It is obvious that the problem of noncompetition does not affect populations that live well apart, such as those of *Mus musculus,* described in Chapter 1.

Although there are some exceptions, this phenomenon is sufficiently generalized to have been called the rule of *character displacement.* All features — morphological, physiological, or behavioral — are involved in this character displacement and ensure reproductive isolation.

Very often the genitalia of sympatric populations forming emerging species are more different than the genitalia of allopatric species. This phenomenon has been observed in many insects. *Drosophila simulivora* and *D. cogani* live together in West Africa. Their degree of sympatry is such that one can find the two larval types living side by side on the same leaf, but the genitalia of the adults are very different.

Character displacement also, and perhaps especially, implies ethological factors. We have already seen the importance of the sound made by the vibrations of the male fruit fly's wings during the nuptial parade. It attracts females of the same species and only of the same species. Thus this sound is a strictly specific characteristic that plays an essential role in sexual isolation. Sibling species living sympatrically always make very different sounds, thus avoiding "illegitimate" breeding. On the other hand, phylogenetically or geographically remote species can have the same signals, but their situation is such that there is no risk of mating. This phenomenon has been described in several groups, and particularly in Central American and Brazilian fruit flies (Ayala, 1978). *D. willistoni* and *D. equinoxalis* can be divided into several subspecies. In each group it has been observed that flies belonging to two different subspecies but coming from widely separated localities are often interfertile, whereas they are intersterile when they come from the same locality or neighboring localities.

Crickets are found throughout the world. *Gryllus bimaculatus* from populations as far apart as Taiwan and South Africa have preserved the capacity to breed in the laboratory; whereas in nature they have no opportunity to meet, and thus their mutual independence is in no way threatened. *Gryllus assimilis,* living in Central America, is in the opposite situation. In the West Indies there are a series of isolated populations with individual morphological characteristics such that some authors have described them as independent species. But all these varieties, which have practically no contact, have remained interfertile and constitute a single polytypic species composed of isolated geographic races. On the mainland, on the other hand, where geographic barriers are less severe, there are species that have much less marked morphological differences but are strictly intersterile; otherwise the risks of contact and breeding would be much greater than in the island groups. This intersterility seems to be linked to a series of factors, particularly ecological ones (such as calls), biological ones (such as noncorresponding seasons), and so on. (See Dreux, 1977). The differences in the calls of anurans is always more marked between sympatric species than between allopatric species (Dubois, 1977). In fact sexual divergence between two species is only seen above a certain threshold, below which interpopulational gene flow is sufficient to maintain interfertility between neighboring groups. Such is the case with the herring gulls discussed previously.

We saw earlier that the first isolation factors to appear were usually ethological ones. The genetic mechanism of this segregation can be very simple. There may be a behavioral gene *A,* favoring breeding within the same population, or *stenogamy,* and its allele *a,* provoking breeding with distant populations, or *eurygamy.* Unless there exists sufficient interpopulational gene flow to keep the frequency of *a* high, *A* will eliminate *a,* which will fall to a very low level without actually disappearing, surviving in combinations where it has no phenotypic expression. The selective advantage of *A* is obvious. It encourages genetic isolation, thus favoring the establishment of the group in its own ecological niche. Such a movement separates it from its neighbors, thereby reducing the disadvantages and dangers of competition.

Competition is especially marked in frontier zones, where biological divergence is of most significance. It is of little importance to two very widely separated populations that have no physical possibility of meeting and therefore competing. In such a case sexual separation is of little interest. In the end the rule of character displacement encourages diversifying evolution between related individuals and favors cohabitation. It is a noncompetitive factor, an encouragement to peace.

species, races, and populations

RACES AND RACIATION

We saw in Chapter 1 that the concept of race was a direct consequence of a typological view of the living world. Like species, races can be defined in terms of a holotype featuring the ensemble of the characters that all members of the same race must manifest.

We also saw that species were made up of populations corresponding to genetically polymorphic groups. How can we interpret the racial groups that classifiers sometimes describe in the wild in the light of population genetics? We know that a species spread over a sufficiently wide area may be divided into varieties, each with its own particular geographic distribution. Almost always such varieties are defined as races. Their existence depends on three factors that may be present simultaneously.

1. Spontaneous mutations are not the same everywhere.

2. Selective pressure is not the same throughout the whole area of distribution, particularly if it is a wide and heterogeneous one.

3. The absence of sufficient sexual exchange between certain populations making up the species — as a result of natural obstacles or too great distances, for example — tends to maintain any differences that arise and even accentuate them.

On the other hand, if interpopulational gene flows are sufficient to ensure a suitable homogeneity from one place to another in the species's territory, no geographic variety will arise.

GENETIC DEFINITION OF RACE

Electrophoretic and immunological studies have consistently demonstrated that races are genetically polymorphic and that they are organized into populational structures. It is possible, but rare, for a race to be distinguished by the exclusive possession of certain alleles. Most often a race is formed of an ensemble of populations with more or less characteristic gene frequencies. If these gene frequencies are distributed in clines, as is frequently the case, racial frontiers become arbitrary. They are much less rigid than the frontiers of the species and allow a considerable margin for subjective decisions by a classifier. This is why Ernst Mayr has quite rightly stated that races are biological artifacts.

In fact races have always been described by typologists classifying for classification's sake on the basis of a few arbitrary morphological features. The existence in all groups of sibling species that are almost indistinguishable from each other, although genetically widely apart since they are intersterile, shows how artificial such a definition is.

In practice it is not always easy to classify a subject according to its racial group by using, for instance, blood-borne markers, which are the direct expression of a part of its genotype. At the most, they allow us to attribute a certain probability of racial origin to it. When we study the distribution of gene frequency in a species occupying a wide area, the results may lead us to split the species into a number of units, each containing populations with similar frequencies. These units correspond to races.

Thus a race is made up of populations that maintain greater gene exchanges between themselves than with populations of other racial groups. It is a sort of "superpopulation," within which there are preferential exchanges. This is why a race often coincides with a clearly circumscribed geographic zone. A race is partially isolated but is bigger than a population. In the case of man, natural obstacles are hardly a barrier. For us the most effective barriers are cultural, but they are also partially territorial, as a result of prehistoric or historic events. Because of his constant tendency to migrate, Homo sapiens has much less distinct racial frontiers than many other animal species.

Nevertheless the populational structure of races shows that this concept has only a relative value. As interpopulational flows are never completely absent, even between populations belonging to different racial groups, in practice the limits of a race are much less distinct than the limits of the species. Almost always one finds numerous intermediate populations forming a biological bridge between two neighboring races.

The Future of a Race

The future of a race is the same as that of the populations forming it. Like them, a racial group can have three destinies.

1. It can survive for a longer or shorter time: as long as the interpopulational gene flows within the race — the *intraracial* gene flows — are more intensive than the interpopulational gene flows between neighboring races — *interracial* gene flows.

2. It may begin to disappear if, for instance, interracial flows become as great as intraracial flows due to climatic or topographic changes.

3. On the other hand, the difference between races will become more marked if interracial flows diminish and intraracial flows persist. In this case races will finally become independent species after passing through their genetic revolution and achieving sufficient sexual isolation, at which stage we recognize an emerging species as defined in Chapter 8.

The race is, therefore, an unstable biological phenomenon, a temporary entity that may evolve toward speciation, or on the other hand become indistinguishable from the rest of the species. The border line between these two tendencies is represented by the critical phase discussed earlier that marks the beginning of the genetic revolution. Before reaching this limit, individuals in the two populations are perfectly interfertile. Frequently their offspring, which are true crossbreeds, exhibit a degree of luxuriance. This is heterosis, as defined in Chapter 6. Later, if the genomes diverge sufficiently for the two populations to be less interfertile, they form subspecies, almost always destined to diverge more and more to become semispecies, with even weaker interfertility, and then true species, with absolute intersterility, at least in theory. Chromosomal changes may intervene at any time to fix acquired features and make the movement irreversible.

This means that today the term "race" may cover very different situations, at least if we consider the future of the group in question. Races are transitional levels, often forming subspecies and representing the beginning of differentiation. At this moment raciation is simply the antechamber of speciation, and not an irreversible stage. Unlike the species which, once cut off from its original line by sexual isolation, can never go back, a race is not a permanent phenomenon. It may happen that a line evolves in the direction of racial diversification and then retreats to become once again mixed into the mother species. The human race is a good example of such to-and-fro movement. If the racial diversification that affected the dawn of humanity had persisted, we would today be divided into several subspecies, such as *Homo sapiens niger, H.s. albus,* and so on, and perhaps even into several species. Even more likely, we would have a species of mountain man, one of plains man, and another in dry regions, for skin color is a very secondary adaptive feature compared with the physiological constraints of such different milieux. After a certain period, interracial gene flows were so massive, especially at the frontiers of geographic races, that the forces of homogenization were more powerful than those of differentiation. And this movement is gaining in momentum even today. Humanity is destined to remain a single, genetically very polymorphic species.

In practice the very concept of race, derived directly from a typological view of the world, is extremely unstable, imprecise, and of little use. "Race as defined by the geneticist is merely a moment in the biological history of a group" (Ruffié, 1966).

RACES AND SPECIES

Raciation and speciation are only two successive stages of the same movement. The former is reversible, the latter is not. Only a populational concept of living groups allows us to explain the process that has for long been called microevolution. A still relatively undifferentiated group breaks away from an already overcrowded niche and occupies a new one. During its exile, which is also its victory, it first undergoes raciation; but it can only occupy its new milieu comfortably and permanently if it undergoes a more or less complete reorganization of its genome — that is, speciation. In this way the group achieves the sexual isolation that is necessary to preserve it. From this moment its new structure is protected and its conquests cannot be threatened. A species is simply an ecological adventure that has been a success and become genetically fixed.

Speciation and Living Fossils

Pioneer species may be satisfied with their new niche and remain there without particular modification for tens or hundreds of million of years. In almost all phyla one finds "living fossils" representing forms of which most disappeared long ago. They are the last witnesses of bygone eras.

Why have such archaic groups survived to our day? Probably because of the chance meeting of several factors that united to block the evolutionary process, such as genomes giving rise to very powerful coadapted systems that were therefore difficult to attack, and the absence or weakness of changes in selective pressure. At the end of the next chapter we shall describe a few examples of relict species, which sometimes mark important evolutionary crossroads.

The Death of Species

In the great majority of cases, a species is unable to adapt to environmental conditions that are very different from those in which it was born. If the milieu changes too much, there must be a genetic revolution for the species to be able to face its new ecological situation. If not, it is condemned. After a more or less prolonged death struggle, the species disappears irrevocably. This is the most common eventuality.

Of all groups that have ever existed, perhaps only a hundredth or a thousandth, or even fewer, are known to us. About a million species of the billion that must have existed since the Devonian period have been identified. We must then add to these successful species all the failures, probably even more numerous: groups that embarked on the way to speciation only to stop before reaching the stage of absolute sexual isolation. No species is immortal. Even the living fossils will disappear one day. Death is the normal destiny for all groups and will come at some time or another in their history, so that they are periodically replaced by newcomers, destined to repopulate spaces that are being left empty. We have no reason to believe that our fate will be any different. But all this takes a lot of time, at least if we behave rationally!

The Breaking Up of Species

The third possibility, the most glorious, has already been discussed. Resulting in the transformation of a species into one or more new species, it represents an exceptional success. It is the only one of interest to the evolutionist.

When a mother species has reached the limits of its genetic patrimony, it may send out a few "feelers" in the shape of peripheral populations that are becoming diversified. It attempts to occupy territory beyond its natural borders by means of specialized races.

Things may remain at this stage, in a state of equilibrium that may last a long time. In other cases, peripheral populations may become more than just races and cross the hurdle of speciation. They break up into daughter species, each occupying a new niche. In the long term the original, modest, restricted population will have built an empire through its offspring. This is branching evolution, long recognized by zoologists, whose mechanism was discussed in detail in Chapter 8. Its success depends on the richness and variety of the genetic background of the initial conquering species and also on the richness and variety of the available subniches. As we saw earlier, such a diversifying movement is beneficial because it avoids competition. Daughter branches are specialized and each exploits its own resources. Their victory reduces competition. We are here far removed from the idea of selection as an eliminating factor, as Darwin's successors conceived it.

CURRENT PERSPECTIVES ON SPECIATION

It has been said that "man arrived too late in a world that was too old" (Harant and Ruffié, 1956). This is true if we consider the major evolutionary events at the time of the massive invasion of new niches by species whose origin was the sea.

It is unlikely that there will be further profound modifications of our environment such as those the earth has already known. The times of the great crises are probably past, but speciation will not stop because of this. Even if all biotopes were saturated, a few openings would remain, without our being able to explain why. In spite of the wide variety of snakelike reptiles, no herbivorous snakes are known, and of all the multitude of birds that surround us not one is really a leaf-eater.

Many niches are in the process of modification every day before our very eyes, usually due to human activity. Man has created a new niche to suit himself that has become worldwide today because of the multiplication of the human race and the extension of our culture. The result was the appearance of a series of niches related to the anthropization of the environment, now occupied by a mass of vertebrate and invertebrate animals that have become commensals—flies, mosquitoes, pigeons, sparrows, rats, and mice—or domesticated animals. At the same time, man has destroyed—or so changed that they became unsuitable—a great number of habitats of wild species that have been forced to disappear.

Many others are endangered, and it has been estimated that before the end of the century 20 to 30 percent of species living today will have been exterminated. Biologically speaking nature is becoming dangerously impoverished, and it is our fault (Dorst, 1979). At the same time, new methods of *in vitro* genetic recombination allow us a glimpse of the future when man will be able to exploit all the genetic information created during evolution that can be of use to him. But until such time, how much of this information will be irrevocably lost?[1] Those groups that remain will perhaps become more and more closely associated with man, and in the not too distant future such survivors may be merely domestic or commensal animals. Hunting associations in industrialized nations are forced to repopulate their territory with artificially bred

1. Jean Dorst (1971) estimated that in the last hundred years one type of bird has disappeared each year on the average.

game every year. This is a forewarning of what nature may be like tomorrow, when wild life will be just a distant memory.

SPECIES AND TECHNOLOGICAL ADVANCE

Species and the Atom

A major evolutionary crisis could only be caused by a catastrophic change in ecological conditions. This has happened several times during the history of the world but is improbable in the foreseeable future unless man provokes a planetary cataclysm through a generalized atomic conflict. If such a disaster occurred, a few insular groups of animals and plants might nevertheless survive and form a food chain sufficient to repopulate the earth, but this might take tens or hundreds of millions of years. If the catastrophe caused all the warm-blooded vertebrates — the mammals and birds — to disappear (they being more fragile than the others), it is not at all certain that reptiles would once again give rise to birds and mammals as happened at the end of the Secondary era. For modern forms are much more differentiated than their ancestral lines and certainly have less evolutionary capacity. Even if a few primates miraculously survived, it is unlikely that they would once again follow the pathway of hominization. We should always remember that speciation depends on a series of random phenomena and particular local circumstances, such as the availability of a niche, competition, genetic richness, and the capacity of a given group to exploit its environment. The probability that the conditions that led to the birth of birds or man would once again exist is nil.

In the face of the threats that hang over us today, the myth of Noah's Ark is occasionally revived by certain sects. Man has become evil and God will punish him, not by drowning him in a universal deluge, but by burning him in an atomic holocaust, with the exception of a few righteous individuals who have chosen the path of salvation in time. In fact these promised events have little in common with the myth figure of Noah, who showed great wisdom by loading his ark with a pair of animals of every species. This diversity allowed him to replace each species in its own habitat and reestablish food chains as soon as the good weather returned. If he had been less prudent he might have only taken representatives of certain key groups as a measure of economy, hoping that they would give birth to the same lines again. Such a hope would have been illusory. The history of plants and animals is no more likely to be repeated than that of man. Lizards are unlikely to give rise to birds, just as mammals and monkeys will not produce man.

Species and the Cosmos

The adventure of the conquest of space is doubtless one of the greatest technological achievements of all time. People have wondered whether a new evolutionary surge could arise if different species were taken by man to other planets with new but acceptable ecological conditions: the conquest of another New World, not on a continental scale but on a cosmic one. It is hardly necessary to emphasize how unrealistic such a hypothesis is. When the earth was formed, environmental conditions were very different from those we know today; in particular, the atmosphere contained little oxygen. The appearance and spread of living groups, beginning with the chlorophyl-bearing plants, fundamentally modified the primitive conditions and prepared the way for our form of life. Our physiology is built on our history. The chances of such environmental modifications happening again are virtually nil, because they involved the appearance of a series of living beings from the first bacteria to man in a very precise sequence. Too many random phenomena were involved at each step in the development of life. But let us suppose—although it is absurd—that a planet with ecological conditions similar to our own exists with its own fauna and flora, and that it is discovered by astronauts who introduce plants and animals brought from earth. Insofar as the newcomers could find suitable niches —that is, become efficiently involved in preexisting food chains, their permanent establishment would be conceivable, for life is contagious; it tends to invade everything. Some species that were genetically rich enough might be the starting point of adaptive radiations if the ecological possibilities on hand were sufficient. But all this would need a lot of time.

On the other hand, no species could be "grafted" on an abiotic planet unless a minimum number of essential food chains were also introduced, beginning with bacteria, including plants, and ending with higher animals. Mammals would not be strictly necessary—for in general, and particularly since man appeared and began domestication, they take more than they give—but bacteria would be essential for their fertilizing value, as would plants for their synthetic value, humble earthworms to work the soil, and insects to pollinate the flowers.

As to the human race, it is of little use to biological evolution. It is in fact harmful, given the multitude of aggressions inflicted by modern man on his environment.

SPECIES AND EVOLUTION

The species now seems to be a privileged but temporary stage in the history of biological lines. It is part of a dynamic process that preceded it and is likely to continue long after. For the geneticist a species is merely an instant in the dynamics of a phylum, characterized by an equilibrium that has developed temporarily between a coadapted genetic system and an ecological niche. We have seen that, except in the rare cases of living fossils, equilibrium is not permanent. This instability is the foundation of evolutionary movements.

For the biologist today, the species has replaced the mutation that was once thought to be so important for the genesis of evolution. The species demarcates the stages of evolution, but not in a permanent fashion.

TYPOLOGICAL AND DYNAMIC SYSTEMATICS

When a naturalist studies a territory he collects individual animals or plants. Faithful to Linnean methodology, he classifies them according to a preestablished plan. He does this by giving each a double Latin name comprising the genus, with a capital letter and the species, with a small letter (*Mus musculus, Culex pipiens,* etc.). Every individual must be baptized with these two names. For traditional science this identification is the essential step in recognition. This is illustrated by the fact that recognition tests still play an important role in botanical or zoological examinations. The candidate who cannot give the correct double Latin name to the samples submitted to him is doomed to failure.

Nature has no respect for these contingencies: reality is quite different. Among the objects collected by the naturalist, some will represent true species and others different stages of evolution, such as groups that are beginning to split up after leaving the ancestral species. Although it is still widely utilized, Linnean methodology cannot encompass even such simple circumstances, for it makes use of a static form of systematics to study a perpetually moving nature. Indeed, the living individuals we observe all belong to lines that are being transformed and are made up of populations. Only some of these are true species. So what is the value of a taxonomic system that may be convenient but is incapable of a rigorous definition of what exists? Are we still justified in using a two-centuries-old method of classification that has hardly progressed since the time when no one doubted the myth of creation or the fixism that was its result?[2] Can we disregard more recent discoveries in the realm of fundamental genetics and population dynamics? Should we not now replace the binomial Linnean nomenclature with a non-Linnean system?

How can we define a group without knowing its precise degree of polymorphism or its geographic distribution, its populational structure or the way in which its populations are organized and the details of their exchanges? Linnean concepts lead, in anthropology, to descriptions of quite aberrant situations. They are scarcely more reliable for botanists or zoologists. In 1963 Albert Vandel emphasized the insufficiencies and limits of contemporary taxonomic methods, which are based on typological ideas that almost all workers recognize as erroneous.

We must agree with Jacques Daget and Marie-Louise Bauchot (1976) when they write: "It has become clear today that the typological concept of the species is not only arbitrary, but also gives rise to insurmountable difficulties. This is why taxonomists now consider it out of date, deserving to be abandoned for good."

There have been proposals for changes, but it is not easy to break old habits that are solidly anchored in our traditions. The establishment of a new system that would be satisfactory for everyone is fraught with obstacles, as François Vuilleumier (1976) explains in a review. For although we must respect biological reality, a naturalist still needs to identify and name what he is studying as precisely as possible.

Insofar as the Linnean method is still inevitable, it should remain a tool, with clearly defined limits, and under no circumstances become a concept. It is a method that is convenient for the mind of man, who finds it easier to work with classifications than with relational networks. In reality the basic systematic unit to which all living beings belong is not the species but the population, which may be more or less isolated or panmictic and can be defined as "the ensemble of individuals occupying a habitat sufficiently restricted — in relation to their ability to move about — that within this habitat cross-fertilization can lead to the creation of a common gene pool" (Daget and Bauchot, 1976). We may agree with Ernst Mayr (1963) that species are groups of sexually isolated populations that are mutually interfertile and that enjoy their own particular ecology whatever their typological characteristics.

It must be made clear that studying a single individual is of little use, for it is impossible to describe the whole group that it is supposed to represent from this single example. It is of great importance to understand the population to which it belongs, or at least a representative sample of this population. Such a sample, which can be used as a basic

---

2. According to an aphorism of Linnaeus dating from 1763, "There are as many species as the Supreme Being has created different forms since the beginning of time."

tool by the naturalist, was called the *operational taxonomic unit* by Jacques Daget and Marie-Louise Bauchot (1976). This unit must be based on the largest possible number of *variable* features — not *fixed* features, as in typological methods. On this basis it is easy to define the biological relationships between a series of populations making up one or more species, to establish a classification according to the biological distance that separates them, and to understand their probable interdependence.

Several methods can be used, of which the three most common include:

1. The projection method. If the number of variables n used to study a group is more than three, we obtain a multidimensional system that cannot be represented graphically because of its n dimensions. There are several ways to reduce this to a two-dimensional system that can be represented graphically. To do this one chooses techniques that lose as little information as possible, such as analysis into principal components and analysis of correspondence.

2. Dendrograms constructed from matrices of taxonomic "distance" and utilizing observed information. Their final form will depend in part on the algorithm used.

3. Cladograms beginning with the ancestral strain and giving a series of subdivisions by progressive irreversible evolution of the indentified characteristic.

These techniques, together or separately, allow each natural population to be allocated a place in the group. The results obtained are objective. When the information has been obtained correctly, we derive the same classification whichever method is used.

We must, however, recognize the limits of this dynamic approach. Although it indicates the position of one group compared with others, it does not indicate the taxonomic level of the group, and in particular tells us nothing of the degree of its sexual isolation. We must admit that in the present state of the natural sciences no analytic method suffices alone. A research worker must almost always utilize all available techniques, either typological or populational, Linnean or non-Linnean.

phylogenesis and transphyletic evolution

THE "CYCLES" OF MICROEVOLUTION

At the beginning of Chapter 8 we saw that naturalists have for a long time distinguished two sorts of evolution. Macroevolution creates new types of organization, such as the birth of arthropods or of vertebrates. This has been called progressive evolution, and it is the only form capable of major innovation and thus the establishment of evolutionary movements. Microevolution, leading to specialization and sometimes also called "regressive" evolution, creates species and is responsible for adaptive radiations, several examples of which we have mentioned.

A given line traditionally passes through three stages. First comes the birth of a generalized, unspecialized group exploiting a wide niche with unclear frontiers, but only superficially. Soon this group splits into several specialized forms that occupy a variety of niches. This is the "explosive" phase, during which adaptive radiations become apparent. Later these forms disappear one by one, as if the phylum were condemned to grow old after giving all it could. This stage of decline leaves only a small number of species—possibly even only one—to form relict groups. One of them may undergo a process of rejuvenation, often related to the phenomenon of neoteny, the acquisition of reproductive functions by embryonic organisms. This may be the point of departure for a new cycle.

For Albert Vandel, microevolution can be represented by a series of successive plateaux—the explosive phases and adaptive radiations—with a series of lines, most of which stop at a particular moment, a phase of decline. But these plateaux are linked by narrow stems that represent the birth of new groups, as seen in Figure 28.

These cycles always have three phases and are found in all invertebrate and vertebrate groups. But if we look closely at this model, evolution is far from constant. Some phyla do not split up but remain

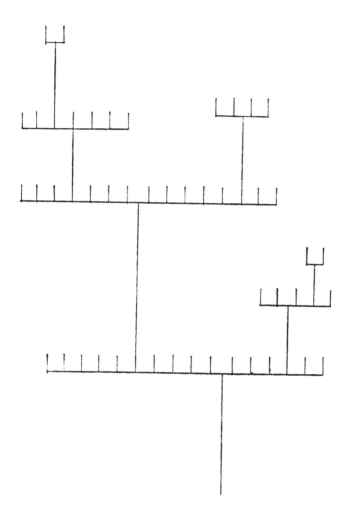

Figure 28. Each horizontal line represents the appearance of a new type of organization (macrogenesis) that will become differentiated into multiple specialized forms. The vertical lines linking them represent the transitional groups, almost always small in number and limited in time. (From Vandel, 1958.)

permanently at the initial stage, young and undifferentiated. These are the living fossils that we have already discussed. Others, on the other hand, flourish several times in succession, followed on each occasion by a phase of decline (Mayr, 1963).

P. P. Grassé (1978) quotes the very good example of the ammonites. Ammonites appeared in the Silurian era; they developed rapidly in the Devonian and declined, to disappear toward the end of the Primary, the terminal Permian period. But they were saved by the little family of the Xenodiscidae. This group had a great deal of success but was once again threatened by extinction at the end of the Triassic, when only two of the former sixty families survived. However, thanks to these two the phylum was maintained from the Triassic to the Jurassic, when there was a new explosion of the line into more than eighty families, with hundreds of genera and thousands of species. At the end of the Cretaceous period there was another crisis, this time a permanent one. No ammonites have survived.

In fact, the lines that demonstrate a typical microevolutionary cycle in three stages are not unique, but neither are they the most common. However, they remain the most spectacular, and naturalists always tend to concentrate on the cases that fit their theories best. They have sometimes formulated general laws from data they have chosen, neglecting other information that they have judged incomplete or imperfectly understood. This may be far from justified.

Today speciation seems to be an almost continuous phenomenon, involving all lines. Its success and diversity depend above all on the number of niches available at the time the new phylum is born. Some groups encounter favorable conditions and divide into multiple branches. Others, by contrast, appear in much less favorable circumstances and tend to maintain themselves either in their original form or after just a few secondary adaptations.

This model is reasonably satisfactory and is now accepted by most populationists, even if we must admit that we do not understand all the factors responsible for the rise, consolidation, or fall of certain lines.

It allows us to perceive a way of unifying micro- and macroevolution. Both depend on the same mechanism: continuous speciation that tends to endlessly exploit new niches. Microevolution, however, is typical of related groups that, having been separated for a relatively short time, have retained much of their genetic information in common and are characterized by only a few minor adaptive differences. It involves groups of species at the lower taxonomic levels — that is, belonging to the same genus, family, and even class. Macroevolution differs not in nature but in amplitude. It is typical of groups separated for a long time that have been subjected to waves of speciation taking them in different directions, such that their common pool of genetic information is much smaller. The fact that their divergence is so long-standing explains its magnitude and also the fact that many intermediate stages, which would be of value in helping us understand how their organization has changed, are lost forever. They existed too long ago.

Before going further, it might be useful to discuss some transitional forms, essential keys to phylogenesis.

TRANSITIONAL FORMS

Many of these forms are unknown. A few rare ones have been preserved as fossils found in geologic strata. In a few exceptional cases, very old organisms have survived in living form. They do not correspond exactly to the transitional groups that existed earlier, but represent ancestral lines that split off very early from the common trunk and invaded a particular isolated niche where there was little competition. They were able to survive quite easily on account of this isolation and absence of competition. They have lasted through the ages, protected in their ecological hiding place with almost no change. These living fossils give us a good idea of what their cousins, the real transitional forms, looked like as they passed the major milestones of evolution.

Why are these key groups so rare? First of all, because they occupied the scene for a relatively short time. Most disappeared quickly. These ancestral types did not have much of a chance compared with the better-adapted forms that they produced. Newly emerging daughter species took over specialized subniches with great efficiency, thus gradually eroding away a substantial part of the ancestral niche. So the less differentiated ancestors had little possibility of survival. What picture emerges of these transitional forms? The little that we know suggests that these rare, discrete, and precarious groups were never "harmonious" intermediaries, halfway between the different lines that they would engender. They seem rather to have had a mosaiclike evolution, with features that would later be distributed independently to the daughter lines juxtaposed side by side.

Good examples of these mosaics of characters are provided by several organisms that later split up in several directions. We may mention the Peripatidae, close to the common origin of worms and arthropods; the Pilina and Neopilina, close to that of molluscs and worms; the Stegocephalidae, key elements between fish and amphibians; those dinosaurs that became birds; the pelycosaurs and, later, the therapsids, from which sprang the mammals.

This is not surprising. When an ancestral species breaks up and evolves toward different niches, each new line retains and develops those features that are useful for its own adaptation. Selective pressure acts on these and these alone. There is thus a real dividing-up of factors, each group retaining only those that are of use to it and rejecting the others. The same rules applied to hominization, when the two lines of the anthropoid apes and the hominids separated. This is why we never find "absolute intermediaries" (Mayr, 1963).

Thus, generally speaking, ancestral groups that occupy a universal niche at the evolutionary turning point of several phyla are very likely to disappear first, in favor of more competitive forms that they have engendered. Only a small number of their contemporaries, usually of small size and protected in a marginal niche, have been able to survive to modern times. They are the living fossils.

This represents a general rule. At all major stages of evolution, when new groups appear the older groups that have the greatest chance of survival are those that have "kept their distance" by organizing more or less marginal niches where no other groups will follow them.

THE HARMONY OF THE EVOLUTIONARY MODEL

All this serves to support the proposal made at the beginning of this chapter that there is no real difference between microevolution, adaptive and diversifying, and macroevolution, which gives rise to new models of organization. Both are at the origin of speciation. Their only difference is one of duration. As paleontology has made more progress and our knowledge of refugelike habitats (such as the ocean depths, caves, and habitats occupied by parasites) has increased, the contrast between the two forms of evolution appears less marked, for transitional types have been found whose existence was not even suspected earlier. Figures 29 and 30 are highly simplified illustrations of the modalities and harmony found in evolution.

Groups C1, C2, C3 and C′1, C′2, C′3, C′4 are related species that

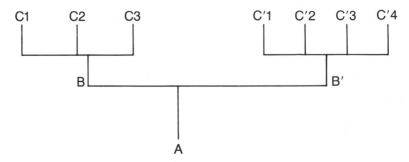

Figure 29. First example: Recent separation. The common ancestor (A) and the transitional forms (B and B′) are known.

differ in a few specialized details, which allows each to occupy a particular niche. The two groups can constitute two genera (B and B′), representing the first dichotomy from the ancestor A. The ensemble may represent a family. The phylum A belongs to a lower taxonomic level.

In Figure 30 the two groups F1 to F5 and F′1 to F′6 are the result of long periods of divergence. They are derived from the same phylum, but from two lines that adopted different directions very early. Thus, in the end, these two groups appear at first sight to belong to two independent organizational systems. But the surviving living fossil C′1 betrays their common origin by demonstrating that intermediate forms have existed.

THE MEANING OF TRANSPHYLETIC EVOLUTION

The meaning of what was once called microevolution is obvious. It represents the tendency toward a better adaptation to the ecological

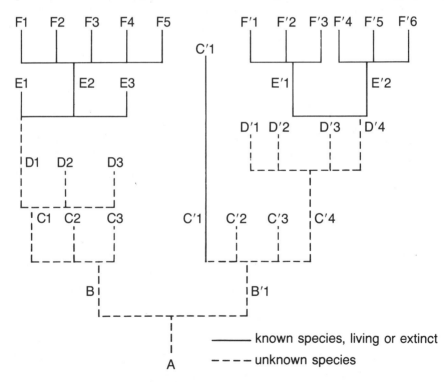

Figure 30. Second example: Much older separation. Neither the common ancestor nor the transitional forms are known, except for a "living fossil" (C′1) that has survived and allows us to reconstruct the probable genealogical tree.

niche and, if necessary, the organization of new niches. We have just seen that macroevolution is not fundamentally different, for it can be resolved into a series of microevolutions spread over long periods.

Nevertheless, if we consider the whole of the animal kingdom, evolution in the long term has a very precise aim: the regular development of the nervous system and the growth of intelligence. This phenomenon is most easily seen in vertebrates but can be found at all levels. There is little in common between the intelligence of an oyster or a snail and that of a cephalopod — an octopus, for example — which has a real brain enclosed in a cartilaginous skull that allows the animal to recognize geometric figures and show complicated behavioral patterns. Similar development can be found in the arthropods, the last surviving representatives of which are the insects (such as bees, termites, and ants), which organize themselves into complex societies. The phenomenon is even clearer in the vertebrates, reaching its climax in the hominids with their highly developed mental faculties.

But at the other end of the scale the unicellular organisms, the Protista, show similar tendencies. They do not have a true nervous system, but they have organelles in their cytoplasm that take its place and fulfill the function of sensory or motor organs. These take the form of small granules linked by communicating filaments of cytoplasmic thickenings. The best-developed of these organisms, the ciliates, have rows of cilia on their surface that beat synchronously, allowing the animal to move around quickly in an undulating fashion. These movements are not random but depend on stimuli such as heat, light, or vibration. It is possible to train ciliates to associate a light stimulus with an electric shock that will cause a flight reaction and ultimately create a conditioned reflex. The animal will then retreat when the light goes on, even without an electric shock. Protista have a form of associative memory.

If we consider the whole animal kingdom, the development of intelligence has two phases:

1. A quantitative phase. As it becomes perfected, the nervous system allows the animal to develop more numerous and more precise behavioral patterns. We saw in Chapter 2 that behavioral characteristics form the interface between the individual and the environment and are the first to be subject to natural selection. When it meets new constraints, an animal's first reaction is to change its behavior. Modifications of physiological or morphological features come much later, if the ecological response is insufficient. The first way in which the milieu can be exploited more efficiently is by adapting behavior. This is why it is so important.

2. A qualitative phase. As they develop, behavioral patterns change in nature. At the beginning, particularly in invertebrates, they depend on genetic programs. They are the fruit of natural selection and form part of the hereditary patrimony just as do organs or functions. At a certain threshold, which is reached most frequently by warm-blooded vertebrates, new behavioral patterns tend to be acquired that depend not on new genetic combinations but on experience. They are not dictated by a program but stored in memory. This replacement of the innate by the acquired is a vital part of evolution, and we should consider it in more detail.

Replacement of the Innate by the Acquired

This form of replacement, once believed to be restricted to higher vertebrates, can be found, at least in a primitive state, in many other groups, but it is most developed in birds and mammals and reaches its climax in man. It represents a considerable selective advantage, for innate behavior linked to a genetic program is rigid and difficult to modify since it implies changes in allele frequencies and even the creation of new combinations. Such changes depend on sexual reproduction and need many generations over long time periods. We can easily imagine how much they depend on chance.

Learned behavior is very different in nature and range. Once a certain level of intelligence has been attained, an animal can "invent" its own responses and modify its behavior in a given situation. It becomes able to adapt very rapidly to changes in circumstances without waiting for the long and difficult process of selection of new genetic systems. If the environment is modified further, an animal whose acquired behavior has already been modified can change its ethological features once again. It is not a prisoner of any organic constraints. Certainly acquired behavior is fragile. It can disappear as quickly as it was produced. But this fragility is part of its value. In a continuously fluctuating milieu, this lack of rigidity makes acquired behavior exceptionally efficient.

ACQUIRED BEHAVIOR AND SOCIETY

The development of the mind permits the appearance of communication, thanks to which individuals can leave behind their solitary status and organize themselves into social groups. We have discussed elsewhere the enormous advantage of social status (Ruffié, 1976). Whereas the solitary animal must provide for all the necessities of life — such as food, defense, reproduction, and the organization of its habitat — the social individual almost always chooses a "preferred" activity that it performs more efficiently than an isolated individual. The worker bee belonging to a swarm can collect infinitely more food than a solitary insect, for it does not restrict itself to its own needs, which it could fulfill very quickly, but rather helps satisfy those of the whole hive. On the other hand it does not have to deal with reproduction, for the queen takes care of that. Thus the social animal no longer depends only on its own resources; it benefits from the activity of the whole group. A baboon receives information about the outside world not only through its own eyes and ears but is kept informed by others. An alarm signal given by a single member of a troop is understood by all the others, which are immediately alerted. Although social advantages are always considerable, the nature of the society and extent of social interaction differ greatly in societies made up of individuals with innate behavior, such as the social insects, and individuals that possess acquired behavior, at least in part, such as birds and mammals.

The first type of society, found in insects such as bees, termites, and ants, is a rigid system held together by organic bonds within which individuals have no liberty and can make no choices. The hive, or the termite's nest, forms a sort of "superorganism" in which each member is just as rigidly programmed as the cells forming an animal. It is known that there can sometimes be a change of status during the life of a particular individual, as in the case of some wasps. But that is the exception and in any case is not "voluntary." All of this is no different when higher vertebrates begin to replace innate behavior with acquired behavior. The nature of the social bonds changes. Henceforth the animal is no longer condemned to perform the same gestures all its life, like the worker bee gathering honey; it is no longer a prisoner of its own behavior, for it can modify it according to circumstances. It does this as a function of its prior experience and also according to what it sees in its enviornment. It remembers and learns. It changes its methods, orientates and refines its strategies, and in turn communicates its discoveries to others, who adopt them by imitation.

Each member of the group has its own life history containing all its experiences, which gives it a "personality." Cultural polymorphism reinforces genetic polymorphism. At this stage the living being has become a real *individual.* Unlike a termite, a bee, or an ant, it is not interchangeable with its neighbors; its life is a unique adventure, its death an irreparable loss. What is more, in a society based on acquired behavior, no situation is permanent, no individual is trapped in the same status or the same caste for the whole of its existence. It can change according to circumstances and its own will. The worker bee will perform the same tasks from its birth to its death. It will never be promoted or criticized. Things are quite different in a troop of baboons where a dominant male that has enjoyed an almost total exclusivity of mating is not at all certain to preserve this privilege all its life. One day it will be replaced by a younger individual, more enterprising and better able to lead the troop. Then the old dethroned chief will ungrudgingly join the group of bachelor males living on the periphery, and continue to play an important role in the exploration of the territory and the surveillance and protection of the group.

We know how important this sociological mobility is in human societies. It allows supply and demand to be constantly adjusted, so that the potential corresponds to the needs and at all times each person's abilities are optimized.

In many socially advanced countries one readily changes one's job, or at least employer, just as one changes one's town, one's friends, or one's spouse! All university teachers or research workers think one day or another of leaving their laboratory to work elsewhere. This mobility is often an essential condition for progress. Those who become ensconced in comfortable habits are doomed. They refuse the advantages of the cultural society and are reduced to an organic form of life. They become sclerosed and lose all hope of making progress.

But there is another side to this coin. In our culture, mobility can lead to disastrous situations, particularly when promotion depends above all on seniority. We all know of cases of excellent juniors capable of supervising with great diligence programs conceived by others and who, if they achieve a position of responsibility, prove to be inconsistent. A useless person almost always reacts by fleeing or becoming aggressive, both of which represent failure. But in a cultural society failure is reparable. We need only nominate a new, competent chief.

The flexibility of societies based on acquired behavior affects not only the individual; it also affects structures. The organization of a hive or a termite's nest has probably changed little for tens of millions of years. By contrast, in less than half a century the French have known five constitutions, involving three republics (the Third, the Fourth, and the Fifth), a dictatorship under Marshal Pétain, and a provisional regime under General de Gaulle at the time of liberation. And this does not include constitutional modifications.

In the higher vertebrates the mass of acquired behavioral patterns, transmitted from generation to generation by example and modified and enriched over time, makes up the *protoculture*. This must already have existed in some of the last reptiles of the Secondary era, which probably lived more or less in social groups. Its volume increased with the appearance of man, with whom it has become *culture* and produced *civilization*.

PART THREE

sociobiology
or
biosociology?

the Darwinian society

SOCIAL DARWINISM

While elaborating his model of evolution, Darwin was subject to two influences, those of Lamarck and of Malthus. From Lamarck's *Physiologie zoologique,* published fifty years earlier in 1809,[1] Darwin adopted the idea that all groups tended to become more and more complex and adapt better to their environment. But he did not accept the French naturalist's explanation that these adaptive transformations were related to hypertrophy of organs through use or regression through nonuse. Although not excluding the direct influence of environmental conditions on the transformation of organisms, Darwin attributed the origin of evolution to the *spontaneous* appearance in natural populations of individuals showing variation. If the variant enjoyed an advantage compared with other "normals," it and its offspring would gain the upper hand. The others would be eliminated or at least swept into peripheral niches where they could survive, particularly if they were of some use to the dominant individuals in the form of resources or protection against predators. Variants that were unfavorable disappeared irrevocably. In his acceptance of the fact that the origin of adaptation was in variation, struggle, and elimination, Darwin simply took up the theory of Thomas Robert Malthus, the economist, demographer, and clergyman living at the turn of the nineteenth century. According to him, populations grew more quickly than their resources. Thus under the effect of demographic pressure, humanity tended to progress spontaneously toward a state of deprivation, enduring severe competition in each generation. This struggle ultimately kept the numbers of individuals within acceptable limits — that is, compatible with the natural resources. The winners were the best-armed, and the others disappeared. Malthus considered the poor to be "social misfits," elimi-

1. See the excellent analysis of Lamarckism by Jean-Paul Aron (1969).

nated because of their inability to gain the slightest wealth. But as a man of the church and great moralist, he proposed voluntary birth control as a way of escaping this cruel law; late marriage and limitations on sexual intercourse were the way to avoid wars and revolutions and ensure everyone a place in the sun.

There can be little doubt of the great influence exerted by Malthus on Darwin; he himself admitted it unequivocally in the *Origin of Species:* "A struggle for existence inevitably follows from the high rate at which all organic beings tend to increase. . . . It is the doctrine of Malthus applied with manifold force to the whole animal and vegetable kingdoms."

In fact, Charles Darwin projected the sociological situation of the mid-nineteenth century into the domain of biology. "The struggle for life in the organic world is nothing more than free enterprise — defended by the Manchester School at the beginning of the industrial era — adapted to biology. Biological utilitarianism was in conformity with the existing ideology" (von Bertalanffy, 1961).

THE SCANDAL OF DARWIN

The *Origin of Species* was the fruit of patient reflection. Darwin returned from his voyage around the world on October 2, 1836. During this long journey, lasting five years, he visited the archipelagos of the Atlantic, the South American coast from Brazil to Peru, Tierra del Fuego, the Galapagos Islands, New Zealand, the southern coast of Australia, Tasmania, the Indian Ocean, South Africa, and Brazil again before finally returning to Europe. After this he went into isolation, married, and worked almost nonstop exchanging material and information with contemporary naturalists. Little by little his model of evolution was developed. He wrote a sketch of his evolutionary theory in 1842 and expanded it into a long monograph in 1844, but never published them because he feared a scandal. In the first account of his voyage, a voluminous work published in 1839, filled with observations on both zoology and geology, Darwin gave a first glimpse of his ideas. He noted the efficiency with which the Europeans had invaded America, Australia, Tasmania, and New Zealand and had quickly replaced the indigenous population. He considered that "men seem to react to each other in the same way as other animal species, the strongest always destroying the weakest." But he did not stress the point. His prudence is quite understandable in the context of the time.

Darwin belonged to an upper-class family. His ancestors on his father's side were rich landowners, several of whom were interested in natural science and medicine. His grandfather, Erasmus, read the first works of Lamarck and published a medical treatise, *Zoonomia; or, The Laws of Organic Life,* that collected data about physiology, materia medica, pathology, and therapeutics. Erasmus already believed in the relatedness between the species making up the living world, including man, and seems to have foreseen the role of natural selection in their genesis. Erasmus was a Fellow of the Royal Society, a much sought-after title.[2] Indeed, there were eight Fellows in five generations of Darwins, including Robert, Charles's father.

On the maternal side Charles's grandparents, the Wedgwoods, owned the famous pottery and were liberals. His grandfather, Josiah Wedgwood, had even founded a model village for his workers at a time when few people could be bothered with the fate of the working classes. Darwin himself was philosophically and politically a liberal but profoundly marked by the spirit of his class, and in his innermost self respected the tranditional values that had made the empire great. So his discoveries disturbed him, for did they not threaten a whole way of thinking — particularly creationism, on which the ethics of Victorian society were based? He wrote to a friend who had expressed surprise at his hesitation: "It is like confessing a murder." So he waited, and only broke his silence in 1858. We have already seen the reason for this; that year, Alfred Russel Wallace, working in the Malay archipelago, sent Darwin a note, asking him to present it at the Linnean Society in London. In his text Wallace proposed a model of evolution comparable to Darwin's, and not yet published. In a spirit of fair play Darwin communicated Wallace's note on June 30, 1858, but added an abstract of his own work. Henceforth his mind was made up, and he began to write his book from the notes he had already produced. He simplified them, and they became *The Origin of Species,* which appeared sixteen months later on November 24, 1859. The first edition was sold out the first day of publication. Its success was enormous but was due to two contradictory reactions: scandal on the part of some people, and approbation on the part of others.

There was first the scandal, the fear of which had been responsible for Darwin's long silence. We must remember what Great Britain was like in the middle of the nineteenth century. The country was enjoying

2. We tend to forget the influence that Erasmus's writings had on the young Darwin. His grandfather, who died in 1802, seven years before Charles's birth, was considered one of the greatest physicians of his century. He was a scientific naturalist, especially a botanist, but was also interested in mechanics and geology. Believing the earth to be much older than commonly supposed, he dated it at several hundreds of millions of years, a very audacious idea at the time. He was a fervent admirer of the French and American revolutions and championed the abolition of slavery and a humanization of the treatment of the mentally ill. A complete atheist, Erasmus Darwin had formulated the hypothesis of the evolution of living beings and believed that man was descended from monkeys. It seems that "Darwinism" originally applied to his ideas, long before his grandson Charles published the *Origin of Species.* See King-Hele (1977).

unprecedented technological, economic, and commercial success (about which we shall speak later). But from the moral point of view, the English refused to sacrifice tradition for modernism. On the contrary, the church was highly respected. It was one of the fundamental pillars of society. The Great Exhibition, inaugurated in London by Queen Victoria on May 1, 1851, was a triumph for the country's young industries: out of 14,000 exhibitors, 7,500 were British. There were 6 million visitors, an enormous figure for the time. The phrase "with steam and the Bible, the English travel the Universe" was coined.[3]

In this traditional, deeply religious society, proud of its success and certain of its rights, a scientific naturalist from a good family, (and who had studied theology) dared to declare and attempt to prove that God had not created species but that they descended from each other. According to this new way of seeing things, men and monkeys stemmed from common ancestors.[4] The myth of the Creation was questioned.

Queen Victoria incarnated the power and glory of the nation. By the end of the century this sovereign lady would be the grandmother of all the crowned heads in Europe, from St. Petersburg to Madrid. Could she also be a cousin of the chimpanzees in London Zoo? One can easily understand the reactions that Darwin's book provoked. Some considered it the Gospel of the Devil. Among its opponents, Samuel Wilberforce, the Bishop of Oxford, was one of the most virulent, multiplying his attacks against an "immoral and antichristian doctrine." He even wrote in an anonymous article in the *Quarterly Review* for July 1860 that Darwin's manner of treating nature entirely dishonored the natural sciences. "Is it credible that all favourable varieties of turnips are tending to become men?"[5]

Although Darwin had been very religious when he embarked on the *Beagle,* he soon lost his faith. At the end of his life he admitted to being not an atheist but rather an agnostic. Nevertheless, as a rich member

3. Quote from *Tallis's History and Description of the Crystal Palace,* London, 1852. The largest part of the exhibition was housed in an enormous glass building, the Crystal Palace, a sort of monumental railway station 520 meters long and 125 wide. Its area was 9,000 square meters, three times that of St. Paul's Cathedral (which is, together with St. Peter's at Rome, the biggest church in Christendom). Almost half a century before the Eiffel Tower, the Crystal Palace marked the end of the era of the cathedrals and man's entry into the age of technology and steel.

4. In fact Darwin, whose caution is well known, only touches on the origin of man in his 1859 work. It was left to Haeckel, a convinced Darwinian, to announce in 1868: "The human race is a tiny branch of the group of catarrhine monkeys; it developed in an ancient world and springs from the long-extinct monkeys." See also Haeckel (1874). But these concepts are already implicit in Darwin's work and were developed in works of 1871 and 1872.

5. Darwin avoided quarrels and publicity. Although he quickly became famous, he never sought fame. Some of his admirers, especially T. H. Huxley, took much firmer stands and were more vociferous about Darwinism. Huxley — an agnostic, to use his own term — sang Darwin's praises, leading the battle against the forces of the church, especially Bishop Wilberforce.

of the upper classes, he always supported his local parish. The religious controversies that surrounded his work upset him. On May 22, 1860, he wrote to Asa Gray, "With respect to the theological view of the question: this is always painful to me. I am bewildered. I had no intention to write atheistically."

The quarrels between creationists and transformists did not cease with Darwin's death. They continued long after, particularly in the United States. After the First World War, several states passed laws banning the teaching of evolution as contrary to the faith in God that was in large part the basis of the American social system. Not until 1968 did the U.S. Supreme Court declare these antievolutionist bans unconstitutional under the First Amendment which separates church and state. Nevertheless the traditionalists did not accept defeat. They still present creationism as a true science, founded on the Bible but compatible with scientific observations. They have organized several groups such as the Creation Science Research Center in San Diego and the Institute for Creation Research, whose aim is to prove biblical theories scientifically by basing their arguments mainly on the insufficiencies of Darwin's theories.[6] In 1981, one of the leaders of these groups, Kelly Segraves, took the state of California to court to force it to ban the teaching of evolutionary theory in schools, on the grounds that it was an immoral doctrine. The court ruled the action unjustified and rejected Segraves's suit, but directed that the Department of Education should present evolution as a theory and not as a proven scientific fact. During his first presidential campaign, people pressed Ronald Reagan to state his position. He declared that he was against the removal of creationism from curricula and pledged himself to maintain it at the same level as evolution.[7]

DARWIN'S SUCCESS

Despite this violent criticism, Darwin enjoyed immense success very quickly, for his theory, inspired directly by the ideology dominating nineteenth-century Western civilization, provided scientific justification for the situation prevailing both within the country and abroad.

Industrial expansion in England and in most European nations, and later in North America, was overwhelming. Between the beginning and

6. Among the projects undertaken by these institutions was a search for the remains of Noah's Ark on the slopes of Mount Ararat in Turkey.
7. Reagan's position caused a degree of agitation in England. Andrew Huxley, the grandson of T. H. Huxley, Darwin's admirer and companion, devoted his inaugural speech on becoming President of the Royal Society in 1981 to refuting creationism and defending transformism.

the end of the century national income had increased eightfold, while the currency had remained stable. In 1806 Great Britain had a single steamship; in 1836 it boasted five hundred, thus reinforcing its dominance of the oceans and permitting it to exploit its colonies in grand style. In the middle of the century Britain furnished 60 percent of the coal and steel for the whole world and almost 50 percent of cotton materials. It was responsible for a third of the manufactured goods — and an even higher proportion of high-technology goods. At the time, the nation's gross national product was the highest on earth. It occupied a position that today would be equivalent to that of the U.S.A., Japan, and Saudi Arabia combined.

Proletarianization

This success was not without cost. It was founded on inhumanity. The rural population began to desert the countryside, where unemployment was soon felt. The number of manual workers needed fell rapidly as a result of two factors. The first was the development of machines that, appearing in agriculture for the first time, tended to reduce the need for manual labor.[8] But even more important, arable land was turned into pasture. The new, rapidly expanding textile industry was highly profitable, and it needed large flocks of sheep.

Former peasants out of work moved to the towns, looking for employment in the factories and workshops being set up everywhere. This massive migration was accentuated by a hitherto unknown demographic explosion. At the beginning of the nineteenth-century there was only one agglomeration with more than 100,000 inhabitants, London. At the turn of the century, there were thirty-three. Leeds, based on the wool industry, increased from 53,000 in 1801 to 123,000 in 1831 and 430,000 in 1900. In the same period Birmingham's population increased from 73,000 to 200,000, and then to 760,000. Liverpool and Manchester underwent similar transformations. The migrants formed a poorly paid subproletariat, working in very hard, almost intolerable conditions. Sixteen-hour days were not rare, but more generally the working day was from six in the morning to six in the evening. Many companies insisted on sixty-four hours per week; only the textile industry permitted sixty hours after 1850. Workers went to their factory six days a week, five and a half after 1850. They had no right of appeal and no social protection. When they arrived from the country, most were crowded into overpopulated slums without light or air, sometimes with

8. McCormick patented the first "automatic" harvester in 1834. Further stages in the mechanization of agriculture followed quickly.

twenty-six people in a lodging. The enormous increase in the number of people looking for work caused a drop in the already precariously low wages. In 1821 a worker could earn sixteen shillings per week. This had fallen to six by 1831. In this social group life expectancy was less than forty years. Faced with such poverty, all members of a family tried to obtain work, including children of nine, seven, and even four, many of whom died at work (Engels, 1845). Whenever demonstrations broke out, they were severely suppressed. A few isolated acts of terrorism occurred, such as arson, murder, or vitriol attacks against owners. Workers began to form secret societies. Not until 1835 was a Working Men's Association officially formed to demand universal suffrage, but in vain.

In a country rapidly becoming richer, the poverty of the working classes worsened. In 1863 an official inquiry revealed that a large part of the population was undernourished. Thus the success of Victorian England was built on the terrible misery of the workers.

Although less advanced, the other industrial nations could not boast of a better situation. In France, for instance, the law of March 28, 1841, tried halfheartedly to stop the inhuman practices that were rife. Its first two articles read: "In factories and workshops with mechanical motors or permanent fires and in their dependent buildings and in all factories occupying more than twenty workers in workshops, children must be at least eight years old to be admitted. From eight to twelve years they cannot be employed to work more than eight hours in twenty-four, with a break for a meal. From twelve to sixteen years they cannot be employed to work more than twelve hours, with a break for a meal. This work can only be between five in the morning and nine in the evening." The new bourgeoisie thought that through this measure it was achieving a great work of charity. On January 11 of the same year, M. Cunin-Gridaine, minister of commerce under Louis-Philippe, declared: "The admission of children into factories after the age of eight years is for the parents a means of surveillance, for the children the beginning of an apprenticeship, and for their families a resource. The habit of order, discipline, and labor should be acquired early, and most industrial work requires a dexterity and alertness that can only be acquired by long practice, which can never begin too early. A child that starts in the workshop at eight years old becomes used to working and, having acquired the habit of obedience and the basic elements of primary education will be, at ten, more able to tolerate fatigue, cleverer, and more educated than a child of the same age brought up in idleness and only then beginning work for the first time."[9] In other words, poor

9. Quoted by François Mitterrand (1978).

children should be broken in like circus animals and given a series of conditioned reflexes that they would keep all their lives. Yet the law of March 28, 1841, that Cunin-Gridaine was defending represented a considerable progress over the previous, practically uncontrolled regime, under which children of the poorest families often worked fourteen or sixteen hours per day. In fact, the first social laws were established in Germany by Bismarck, and industry there was not weakened — to the contrary, as France learned to its cost.

The Colonial Adventure

The situation abroad was just as shocking. Developing industries encouraged the extension of colonial empires, from which European nations obtained the cheap raw materials needed for their young industries. The deportation of black populations was condemned, at least in principle,[10] but slavery was still common throughout the Americas. In the United States it was officially abolished in 1865 at the end of the Civil War. It lasted longer in Brazil, in the West Indies, and elsewhere. And even liberated slaves led a very precarious existence.

Darwin's book appeared seventeen years after the Opium War was ended by the Treaty of Nanking, which subjugated China economically, if not politically, to the Western powers, England in particular. This situation troubled many consciences, for it was far removed from the Christian ideal subscribed to by some, or the rationalist precepts proclaimed by others. Europe was enjoying unimaginable prosperity, but not without a degree of worry and remorse.

By proposing a model of evolution based on variation, struggle, elimination, and subjection, Darwinism legitimized the inequality of classes within a nation and of races outside it. Struggle came to be considered the motive force of progress. Thus the Darwinian model had an enormous impact on the sociology and the politics of its time. We shall see later how this influence is still felt by our contemporaries, especially the *sociobiologists*. [11]

10. This condemnation was based not only on moral grounds, as has often been proclaimed, but also on infinitely more down-to-earth considerations, for the rapid development of industry imposed a need for a revalorization of the European powers' African colonies, which were to furnish them with growing quantities of raw materials. Hard-working, underpaid native labor was needed on the spot. Thus the deportation of blacks to American nations now independent of their former European masters was no longer of any value but was, on the contrary, poor economics.
11. In Britain Darwinism still has a basically biological meaning: the application of the selective model to organic evolution. Social Darwinism gets a poor press! But the term is used throughout the world. Indeed, social Darwinism was born in England in the aftermath of biological Darwinism, with Francis Galton (Darwin's cousin), Karl Pearson (Galton's successor), and many others, as we shall see in the following pages. One of Darwin's grandsons, Charles, also believed that natural selection could apply to human societies. I have therefore adopted the expression "social Darwinism," which seems to be accepted usage, and apologize to any readers whom it might shock.

GALTON AND "MENTAL" FACTORS IN COMPETITION

We said earlier that Darwin was very reserved about applying his theory to human society — not because he lacked conviction, but because he was prudent. Throughout his life he took care not to revive the violent polemic that had accompanied the appearance of his book on the origin of species. In the end it was his disciples, particularly his cousin Francis Galton, who used the Darwinian model to explain the existence of a hierarchy of races. In a work published in 1869, *The Hereditary Genius,* Galton asserted that human races were defined not only by their morphological features — size, shape of the head, color of the skin and eyes, color and aspect of the hair, and so on — but also by cultural or (as was said at that time) "mental" characteristics.[12] For Galton, it was much more intellectual features that determined a scale of values between races than physical aptitudes like muscular strength or running speed. According to him, the white race

12. This confusion between hereditary physical characters (often, in fact, under environmental influence) and acquired cultural characters had already been made a hundred years earlier by Linnaeus, who in *Systema Naturae* recognized four types of *Homo sapiens,* in addition to a worldwide race of "monsters" into which he put deformed individuals. They were:

    1. The white race: *Europaeus albus* — ingenious, inventive, and governed by laws.

    2. The red race: *Americanus rubescens* — happy in his fate, liking liberty, swarthy, irascible, and governed by habit.

    3. The yellow race: *Asiaticus luridus* — arrogant, greedy, jaundiced, melancholic, and governed by opinion.

    4. The black race: *Afer niger* — cunning, lazy, negligent, black, phlegmatic, and governed by the will of his masters (cited by Salmon, 1980).

This amalgamation lasted a long time. At the end of the Second Empire (1865–70), the *Atlas of the 89 French Departments and Colonies* by A. H. Dufour and T. Durotenay (edited by A. Logerot, Paris) described a particular "temperament" for the inhabitants of each department. This is taking the idea of the psychological holotype a long way! More recently, in a French primary-school textbook widely distributed between the two world wars (*Geography Through Pictures and Maps,* written by a group of teachers in 1934) and known to many French children, one could read in Lesson 20: "The Races of Man": "All men are descended from Adam and Eve. But they do not all look alike. Their skin does not have the same color, and their faces are different shapes. It is according to the color of their skin that we name the three main races of man: the white race, the yellow race, and the black race. We belong to the white race and have white skin, with healthy beards and hair, often wavy or curly. It is the most intelligent race. It has given us many scientists who have produced great inventions: printing, telegraphy, railways, steamboats, airplanes. White people live in Europe, America, western Asia, and North Africa. The yellow race has yellowish skin. Their faces are flat, with eyes slanted towards their temples and very little beard. They are hard-working and patient. Their dress and houses are not like ours. Yellow people live mainly in Central and East Asia. The black race is so called because their skin is black. Their cheekbones are prominent, their lips thick, their beard sparse, and their hair woolly and fuzzy. They are usually less intelligent and less hard-working than other races. Since they live in warm lands, they wear few clothes. They live in huts. Black people live in Central and South Africa, and that is why we speak of Black Africa. They are also found in parts of the South Seas and America."

This mode of thinking is scientifically absurd, but is found today in typical manifestations of racist thought.

was superior, *biologically* created to invent and direct, while the yellow, and particularly the black races were conceived to obey and work. Galton divided the white world into a series of hierarchies, at whose summit he placed the English. He estimated that in England there existed one genius in four thousand, and in a genius's family a particularly high number of other geniuses would be found. They represented the summit of the pyramid, and consisted of company directors, politicians, writers, musicians, judges, and, obviously, Galton's own relatives.

There is no doubt that this was the beginning of a form of pseudo-scientific racism, whose development we shall follow. At the time of the International Exhibition of 1884 in London, Galton set up a little laboratory where for threepence visitors could learn their "anthropological profile," based on a certain number of morphological measurements, such as weight, height, and shape of head, and physiological criteria, such as visual and auditory acuity, color sense, respiratory capacity, reaction time, and general strength—all of which, he believed, were associated with intelligence.

Galton and Social Classes

Under Galton's influence it was assumed that poverty or misery corresponded to a biological given and that the poor remained poor because they were incapable of anything else. Everyone occupied a place in the society that he earned by his capacity for struggle. The proletariat was relegated to the bottom of the ladder by its constitutional mediocrity. We saw earlier that this idea had already been proposed by Malthus. But Galton went further. He condemned relentlessly any form of solidarity or even charity in favor of the poor. It seemed to him useless and dangerous to help these people, for such a transfer of wealth would damage superior members of society for the benefit of the inferior, who were incapable of using it profitably. In this light the condition of the proletariat was as "normal" and as irrevocable as slavery had been for the blacks. It was the fruit of selection and thus a motive force of progress. Therefore, all social reforms were to be condemned because they acted against normal selective forces and slowed the progress of the people.

In his autobiography in 1908, Francis Galton divided the English population into a small class of "desirables"—the elite; a large class of "acceptables"—the middle class; and a smaller, undesirable class of the poorest people. He judged that these last were being helped too much, whereas it would be in the general interest to help the most favored people. "It is obvious that it would be an advantage for the country for a social and moral assistance and an opportune material aid to be given to these desirables and not monopolized, as is the tendency at the moment, by the undesirables."[13]

However surprising it may seem, this form of reverse social assistance had its partisans, like Georges Vacher de Lapouge in France, the father of modern racism and anti-Semitism, of whom we shall speak later. Just like Galton, Vacher de Lapouge (1896) stigmatized charity and social assistance, which hindered the beneficial action of selection: "Assisted people are in general heirs to laziness and debauchery and sometimes crime. They are antisocial, living on the margins of society, generally inept for sustained work, incapable of providence; primitive people who through parasitism have escaped from selection, which has brought whole populations out of the gulf that preceded civilization. Help from the hands of charity leads to an increase in vice and crime."

One could hardly be more explicit. His views brought him to question the very foundations of democracy. "The conflict of races is beginning openly, and one wonders if ideas of fraternity, of equality among man, are not contrary to the laws of nature. Scientific politics prefers the reality of force, of law, of races, of evolution to the fiction of justice, equality, and fraternity. In the footsteps of economy, anthropology is tightening the noose around so-called human liberty" (1899).

As to Galton, he pushed his social Darwinism to its logical conclusions, ending by denying the validity of democracy, which was responsible for drowning superior beings in a mass of "average" or mediocre people. "The great majority of people of our race proclaim the *vox populi* even when they know that it is the expression of a pack of non-entities and make it the *vox dei*" (Galton, 1883). On the same subject Vacher de Lapouge (1899) even questioned law, which he considered a hindrance to the forces of selection as factors for progress. "The very idea of law is a fiction. There is only force. Laws are pure convention, transactions between equal and unequal powers. As soon as one of these ceases to be strong enough for the transaction to be valid for the other, law ceases. For members of a society law is what is sanctioned by the force of the collectivity." Half a century later Charles Maurras was still contrasting the "legal country" — the result of universal suffrage — with the "real country" — representing the profound values of an elite, trampled underfoot by mediocre, sheeplike masses.

The companion of social Darwinism, which ensured its scientific legitimacy, was liberalism, which adopted as its motto "Let do [*laisser*

---

13. In fact, this idea also occurs in Darwin's work. In *The Descent of Man* (1871), he writes: "Man, like every other animal, has no doubt advanced to his present high condition through a struggle for existence consequent on his rapid multiplication; and if he is to advance still higher, it is to be feared that he must remain subject to a severe struggle. . . . There should be open competition for all men; and the most able should not be prevented by laws or customs from succeeding best and rearing the largest number of offspring."

faire], let pass, and may the best man win." This idea guided the industrial development of the nations of Europe and America and influenced their political relations. It still inspires the whole of Western industry. Nothing has changed since the beginning of the century, when John D. Rockefeller asserted that "The growth of a large business is merely the survival of the fittest."[14] James J. Hill, one of the American railway magnates, added a few years later in 1910, "The fortunes of railroad companies are determined by the law of the survival of the fittest."

These phrases are just as current today, in spite of the advances of modern biology. We still live with the same principle: bitter struggle and selection, which allows only the best to survive. Certainly, after many years of struggle, workers in the older industrial countries have obtained better living conditions and an enormous improvement in their social welfare compared with the last century. But the state of mind has not changed.[15] Liberal industrial societies have never been moved by the collective interest, which is usually beyond them. They have a series of particular interests, more or less contradictory, that can easily be hidden behind the myth of universal suffrage. Too often liberty, indispensable for the progress of human society, and unlimited liberalism, which in the end will threaten it, are put on the same footing.

Galton and Scientific Racism

As we have seen, Galton established, on the basis of Darwin's model, a scale of classes and races derived from variations in intellectual qualities. In the middle of the nineteenth century the preeminence of the English was taken for granted. "The Englishman lives for movement and for struggle. He is there to conquer and to build, to cross the seas, to spread the genius of his nature among nations. Industry, Protestantism, liberty are products of the Anglo-Saxon race, a race to which God has entrusted the care and diffusion of truth and on whom, in the first place, civilization and the progress of the world depend."[16]

During his long voyage in the 1830s, Darwin was very impressed by

14. Cited by Ghent (1902).
15. This social-Darwinist mystique is deeply rooted in our contemporary consciousness, or sub-consciousness. Bitter rivalry, in spite of its obvious disadvantages, remains the golden rule of modern society. In 1972 Christian Chavanon, then vice president of the French Conseil d'Etat, summed up this concept admirably in an interview in *Le Figaro*. "The accentuation of competition is a source of mutations. They may be painful, and social measures must be used to alleviate their most distressing effects. But progress must not be hindered. . . . A new economic society is being built thanks to competition. This trend will increase. . . ." Jean-Jacques Servan-Schreiber in his very successful book *Le Défi américain* took American society as an example and stated: "Progress is a battle, as life is a combat."
16. From Hood (1850).

the rapid development of Australia and New Zealand. He wrote: "Progress in the Southern Hemisphere has been achieved through the philanthropic spirit of the English nation. No Englishman could visit these distant colonies without feeling pride and deep satisfaction. Wherever the English flag is raised prosperity, wealth and civilization are ensured in that place." We might ask ourselves what the Tasmanians might have thought of that, for they were exterminated in less than a century and, unlike the South American Indians invaded by the Spanish and the Portuguese, never even had the right to crossbreed.

Later, in *The Descent of Man,* Darwin stated: "The astonishment which I felt on first seeing a party of Fuegians on a wild and broken shore will never be forgotten by me. . . . He who has seen a savage in his native land will not feel much shame, if forced to acknowledge that the blood of some more humble creature flows in his veins. For my own part I would as soon be descended from that heroic little monkey, who braved his dreaded enemy in order to save the life of his keeper, . . . as from a savage who delights to torture his enemies, offers up bloody sacrifices, practices infanticide without remorse, treats his wives like slaves, knows no decency, and is haunted by the grossest superstitions."

At the time, the ideology of the superiority of the white race and civilization was so widespread that even the best minds did not escape it. Ernest Renan, whose thoughts illuminated the second half of the nineteenth century in more than one respect, wrote in 1871 in *Réforme intellectuelle et morale:* "Widespread colonization is a political necessity of the utmost importance. The conquest of a country belonging to a inferior race by a superior race that establishes itself there is in no way shocking. Although conquests between equal races should be condemned, the regeneration of inferior or bastard races by superior races is part of the providential order of humanity." He goes on to settle an account with the people of the south of France. "Our foolishness comes from the Midi, and if France had not involved Languedoc and Provence in its circle of activity we would be happy, active, Protestant, and parliamentary." He went further in 1876, in his *Dialogue philosophique:* "The black race is made to serve the greatness sought after by the white."

Theories became facts, and this epoch saw an unprecedented development in the industrial countries, founded largely on the colonies, in accordance with the newly acquired vision of evolution.

A few more farsighted minds expressed a certain reserve or even alarm. Their merit was all the greater because almost all of public opinion, whatever its tendency, accepted the Darwinian movement. Jules Ferry, who instituted compulsory, nonreligious public education in France, was influenced by the racist ideology that marked his era, even though he and some others showed evidence of a degree of generosity. In 1885 he proclaimed in the Chamber of Deputies, in the face of his opponents, who criticized his colonial policies: "I defy you to support to the extreme your thesis based on the equality, liberty, and independence of inferior races. Gentlemen, we must speak more forcibly and more truthfully. We must state openly that superior races have a right over inferior races."[17] But Georges Clemenceau contested the widespread argument that the superiority of white people over people of color justified conquests that resulted in "waves of civilization." The "Tiger," who thirty years later led France to victory in 1918, believed that this form of reasoning could boomerang. "Superior races, inferior races! Easy enough to say. For myself, I discount the idea, especially since I saw German scientists demonstrate that France would be beaten because the French are a race inferior to the Germans."[18]

In fact the social-Darwinist ideology impregnated the whole of French and European political thinking at the end of the nineteenth and the beginning of the twentieth century. At that time racism and anti-Semitism were neither right-wing nor left-wing; they were everywhere. Vacher de Lapouge himself claimed to be a socialist, calling the Jews usurers, and Semitism the cult of money. The Dreyfus Affair in 1899, in which a Jewish officer was unjustly accused of betraying France to Germany, brought the first breach. From then on racism tended to be the prerogative of the right wing. In 1933 and 1934 many Jews fled Central Europe, and particularly Germany, as they felt the danger of persecution menacing them.

These *métèques,* as they were called at that time, called forth anti-Semitic sentiments from a number of "patriots" and "nationalists." These feelings were exacerbated in 1936 when Léon Blum, also a Jew, became president of the French government. Charles Maurras, Pierre Gaxotte, Xavier Vallat, Louis-Ferdinand Céline, and hordes of right-thinking literary personalities took up the banner and slogan of Edouard Drumont—"France for the French"—proposing to slaughter Blum with a kitchen knife, Blum whose "ancestors picked their fleas in the shadow of the Carpathians." Incitements to murder the socialist chief increased (Shirer, 1969). On May 15, 1936, just after the victory of the Popular Front, Maurras wrote in *Action Française:* "It is as a Jew that we must see, conceive, hear, fight, and kill Léon Blum" (cited by Lacouture, 1977).

17. Debate in the French Parliament, July 28, 1885. Cited by Girardet (1966) and Salmon (1980).
18. From Erlanger (1979). The victory of 1870, which marked the sudden and unexpected collapse of the France of Napoleon III, was interpreted in Germany as a confirmation of racist theories and a proof of the inevitable superiority of the Germans, descendants of the Aryans. Later, the defeat of 1918 was considered a transient accident to be rapidly repaired, as Hitler seemed to do in 1940.

Gustave Hervé, a former socialist gone over to the right wing and director of the newspaper *La Victoire,* the organ of "national socialism," wrote: "Anything rather than this repugnant anarchy! . . . How many people are now muttering between their teeth: 'Long live Mussolini, long live Hitler.' " Others took up the refrain: "Rather Hitler than Blum" (cited by Shirer, 1969). The state of mind of many of the French explains the mass indifference to the purges by the German invader of elements considered foreign and responsible for the country's woes. Not until the occupying forces became extortionate, particularly after 1942, did public opinion begin to change; but resistance to Hitler and aid to the persecuted were long restricted to a small, lucid, and courageous minority (see Amouroux, 1977).

This ideological explosion was accompanied and supported by a so-called scientific movement that, from the end of the nineteenth century, gave a preponderant position to physical anthropology based on morphology. It became a fashionable science in Europe in a way that we can scarcely imagine today. There was a passion for defining races, and people seriously believed that the anatomical characteristics of a race were related to its sociological, economic, cultural, and political characteristics, and that these features were a factor in the role the race had played or would play in history.

The industrial powers were invaded by a frenzy for measurement. "Europe is full of people measuring skulls, tibias, and nasal indices; impecunious charlatans, neurotic statisticians whose favorite instrument is an improved version of the folding ruler. Two measurements are the key to history. The future belongs to the dolichocephalics, whose skull is longer than it is wide" (Le Bras and Todd, 1981).

This hardly represents a caricature of the state of mind that reigned almost everywhere at the end of the nineteenth century and appeared in most writings of the period. The physicians, and particularly the surgeons, of the colonial armies cared for native populations efficiently and devotedly, introducing them to Western medicine, particularly Pasteur's discoveries about vaccination; but some did not escape the anthropological myth.

As André Langaney wrote in 1977: "The worthy descendants of Molière's physicians began abandoning the 'human physic' of their predecessors to concentrate their energies on a single problem, the racial classification of human beings. Often military doctors, they adopted the shortcomings of both medicine and the army — a passion for verbiage and an obsession with order. This brilliant synthesis, usually occurring in a colonial context, finally led to the diversion of a fundamental science to the service of an unjust order, maintained with difficulty. The scientific basis was lost; only its form survived in a caricatural proliferation of vocabulary and the Homeric quarrels of specialists."

In the 1860s Paul Broca, the eminent French anatomist, who had described human races by their morphological features, created the Anthropological Society of Paris and the School of Anthropology, both aimed at diffusing the new science. But Broca's mind was particularly clear and critical, and he was one of the rare scientists to express reservations about Darwin's model (Conry, 1974) and to criticize the use being made of it. He wrote in 1876: "The races of Europe are capable of such expansion and dispose of means of aggression that are so irresistible that they will doubtless have the time to exterminate several native races before philosophy and science are strong enough to stop these systematic attacks on humankind. It is said that this is the law of progress, and there is no lack of orators who claim that the substitution of superior races for inferior races is the means employed by Providence to spread civilization to the whole earth. What is now happening to civilized people happened previously to the barbarians, and this so-called law of progress is nothing other than the law of the strongest." He concluded on this timid note of hope: "As long as there are regions that spell death to Europeans, some native races will remain in possession of the territories they have occupied since time immemorial" (cited in Monod-Broca, 1980).

But Broca was an exception. The greater part of Western thought at the end of the nineteenth and the beginning of the twentieth century was guided by social Darwinism, which legitimized wars and aggression.

In 1912 the Prussian general Frederik von Bernhardi (1849–1930), wrote in a book translated into English as *Germany at War*, "War is a biological necessity of the first importance. But it is not only a biological law, but a moral obligation, and as such an indispensable factor of civilization."[19]

It would be wrong to think that only German officers believed in the ideology of the inevitable and useful war. When the famous colonialist General Faidherbe arrived in St. Louis on June 12, 1856, he declared, "From nothing we wish to become everything. Only through war have we been able to achieve this." The same sentiment is still present in many of our contemporaries. For example, Alexandre Sanguinetti in 1980 claimed that war was one of the essential motive forces of progress and refused to adopt the idea of universal peace: "It would probably

19. This inscription can be seen at the entrance to an enormous underground hospital constructed by the Germans on the island of Jersey during the second World War as they prepared the invasion of England. In order to finish the work as rapidly as possible they used prisoners and deportees, many of whom died as a result of this forced labor.

be the end of the human race because it would be no more than a living body animated by its contradictions."

THE NEW IDOL: HUMAN EUGENICS

Throughout his life Galton wished to improve the human race by judicious selection, just as breeders had done for ages in creating domestic races. Why not apply to man the methods that had produced such good results with English racehorses, for instance? Imagine the impact on the future of humanity of the appearance of a new race of "domestic man" as powerful in his intellectual abilities as are thoroughbreds on the race course.

On this basis Galton drew up the outlines of a eugenics (which in fact was not entirely absent from Darwin's works).[20] He wrote that when he had understood that the heredity of mental qualities on which he had undertaken his researches was real and that heredity was a means of developing human qualities that were more powerful than the average, he wanted to develop the scale of qualities, in order to establish to what extent, at least theoretically, birth could modify the human race. A new race could be created, he thought, with a mean quality equal to that thus far found only in exceptional cases.

In 1865 he declared, "The improvement of the breed of mankind is no insuperable difficulty."

Galton made a new religion of what he considered eugenics to be, sincerely believing that this science had solid reasons for becoming an orthodox religious dogma of the future (1909). Together with one of Charles Darwin's sons, Leonard, he founded the Eugenics Society, and Leonard presided over the first international eugenics congress, held in London in 1912. At the same time a French Eugenics Society was formed. Galton had died in 1911, leaving an endowment for the creation of a Chair of Eugenics at University College, London, a chair that was occupied by his disciple, Karl Pearson. Pearson went even further

20. In *The Descent of Man* Darwin wrote: "Man scans with scrupulous care the character and pedigree of his horses, cattle, and dogs before he matches them; but when he comes to his own marriage he rarely, or never, takes any such care. . . . Yet he might by selection do something not only for the bodily constitution and frame of his offspring, but for their intellectual and moral qualities. Both sexes ought to refrain from marriage if they are in any marked degree inferior in body or mind. . . . All ought to refrain from marriage who cannot avoid abject poverty for their children; for poverty is not only a great evil, but tends to its own increase by leading to recklessness in marriage. On the other hand, as Mr. Galton has remarked, if the prudent avoid marriage, whilst the reckless marry, the inferior members tend to supplant the better members of society."

Galton may have constructed his theories on Darwin's principles, but Darwin was, for his part, certainly a Galtonian.

234 THE POPULATION ALTERNATIVE

than his master, believing that the simple coexistence of different races in the same place was undesirable, even in a system marked by imperial domination. The stronger race had to dislodge the weaker if it was not to degenerate. In support of his thesis he cited the example of America, where the replacement of the Indians by Europeans "gave us benefit that largely compensated the initial hardships." He went on to say: "We shall never enjoy a healthy social situation in South Africa until the Bantu have been pushed towards the Equator" (1905).[21]

The Two Poles of Selection: Aryans and Semites

By making permanent competition and the replacement of some races by others a general rule of evolution and a necessity for progress, Darwinism provided a pseudoscientific justification for racism. Actually such theories were not absolutely new, and racism existed before Darwin. Even in the time of Malthus, Friedrich Blumenback, a professor of medicine who died in Göttingen in 1840, claimed, in a work called *De Generis Humani Varietate Nativa Liber* and published at the end of the eighteenth century, that the noblest race of man was the white race, originating in temperate climates, the most hospitable, where it dominated. The other, inferior races lived on the periphery in harsher climates, where their function was to serve their white masters. Blumenback related the degeneration of the colored races in large part to the influence of the tropics. He accepted Lamarckism. In its essentials his theory was shared by Georges-Louis Leclerc de Buffon.

Nevertheless, the person who played perhaps the most essential role in the development of Western racist thought was a contemporary of Darwin: Joseph Arthur, comte de Gobineau, poet, diplomat, and phi-

21. We might recall here the utter contempt for indigenous natives manifested by the British colonizers, who either systematically destroyed them or isolated them in reservations. In 1744 settlers in Massachussetts offered fifty pounds to anyone bringing in the scalp of an Indian woman or child and a hundred pounds for a man or boy over twelve years old! Thus, when Darwin visited Tierra del Fuego, Tasmania, and New Zealand and saw such massacres he was not very surprised, although he deplored these brutal measures on humanitarian grounds. Nevertheless he recognized, as we saw earlier, the enormous benefits of British colonization, which could transform a country and even a continent in a very short time. These benefits were not usually reaped by the natives, who tended to disappear. By contrast the Latin colonizers, especially the Portuguese and the Spanish, preferred to mix the races and make slaves. This is the difference between Brazil and North America.

As an anecdote, we might mention that R. A. Fisher succeeded Karl Pearson in the chair at University College and introduced a quite different orientation. He established the first research laboratory concerned with human blood groups, in which Taylor and Race came to work immediately. The laboratory was transferred to Cambridge during the second World War and welcomed A. E. Mourant. By a curious reversal, the first center consecrated to the study of blood groups thus was born in an establishment originally created by Galton, the hardened typologist and racist. Blood groups quickly revealed the consistency and extent of human polymorphism, a discovery that was to deliver the first blow to the typological concept of the living world and became a major argument for the concept of populations.

losopher, who had pretentions of being an anthropologist. In the middle of the last century, Gobineau took it into his head to look for the superior race that, during the history of man, had produced civilization, as others invented pottery or the casting of bronze. Using linguistic and cultural criteria that we now know to be valueless or even completely erroneous, Gobineau "invented" an Aryan race that supposedly lived in the north of India about two thousand years before Christ. From this privileged epicenter, the roots of civilizations — Chinese, Hindu, the Mediterranean, and European — spread out. Gobineau was a great admirer of Germanic culture, and the Germans represented for him the purest descendents of the Aryan race. The value of a people was directly related to the quanity of Aryan blood in its veins. Gobineau set out his ideas in his *Essai sur l'inégalité des races humaines,* published in 1853. The success enjoyed by the Aryan myth in the first half of our century is well known.

Although Gobineau was a racist and put the Aryan race at the vanguard of humanity, he was not anti-Semitic. "Scientific" anti-Semitism was formulated a few years later by another Frenchman, Georges Vacher de Lapouge, born at Neuville, near Vienne, on December 12, 1854. After studying law and beginning medicine, he became a magistrate, then abandoned his career to go to Paris to study philology, linguistics, zoology, and anthropology. In the end he became university librarian at Montpellier from 1886 to 1893, at Rennes from 1893 to 1900, and at Poitiers from 1900 to 1909. At the same time he taught in various places, mostly unofficially, and produced a certain number of written works. His philosophy was simple. There was no doubt about the superiority of the white Aryan race of Nordic origin, large in stature with blond hair and blue eyes, and dolichocephalic. The only race to produce real geniuses, it was well represented by the Germans. All other races were inferior but were no threat to the Aryans as long as they remained in peripheral, geographically isolated areas. They could even be useful, for example by working for the Aryans. The Arabs, pushed beyond the Pyrenees after the battle of Poitiers won by Charles Martel in 732 and finally forced out of Moorish Spain at the time of the *reconquista* by Ferdinand and Isabella in 1492, were no threat to the Nordic races. Black races were concentrated in Africa and yellow races in the Far East. There was little danger of crossbreeding between these inferior groups and Aryans; thus colored people were no threat to the superior race but could even be useful working for them in the colonies. But this was not true for the Jews. They came from the Near East, with the same Semitic racial origin as the Arabs, with whom they shared many features, and had undergone a certain negroid crossbreeding throughout the ages (as was probably true of all Mediterranean races). But unlike the Arab, black, and yellow people, who had remained in remote countries, the Jews had invaded the whole of Europe. They were to be found in the biggest cities, in Germany, England, Holland, Belgium, and France, and often occupied important positions in the press, banking, and commerce. "Vehicles" of inferior characteristics, they permanently threatened the superior race by bastardization. Vacher de Lapouge's theories are set out in a series of publications, especially in 1896, 1899, and 1909.

Vacher de Lapouge, closely associated with Woltmann, theoretician of social aristocracy and director of the *Politisch-Anthropologisch-Revue,* enjoyed considerable success in Germany, where the Aryan myth fell on fertile ground (M. Thuillier, 1977). Wilhelm II made the well-known remark: "You have only one great man in France, Vacher de Lapouge, and you ignore him" (cited by Thuillier). The national pride of the German intelligentsia was flattered by de Lapouge's ravings, and they were easily seduced into racism and anti-Semitism. In 1881 Richard Wagner wrote to Ludwig II of Bavaria: "I consider the Jewish race to be the born enemy of humanity and all that is noble. It is certain that the Germans in particular will perish because of them, and I am perhaps the last German to resist Judaism, which already dominates everything" (Poliakov, 1968, cited by Salmon, 1980).

Even Marx, himself of Jewish origin but of German nationality, did not escape this tendency. In *The Jewish Question,* he wrote about the secular roots of Judaism. "We should not look for the secret of the Jew in his religion but rather the secret of religion in the real Jew. If our age could emancipate itself from trading and money, and this means from real, practical Judaism, we should be really emancipated. We must recognize in Judaism a modern, universal antisocial element that, through a historic development in which the Jews have actively collaborated in a deplorable way, has reached particular heights in the contemporary epoch" (Poliakov, 1968, cited by Salmon, 1980).

Vacher de Lapouge's concepts were adopted in their entirety by the theoreticians of National Socialist Germany, in particular by Hitler himself. In *Mein Kampf,* conceived between 1924 and 1926 while he was in prison in Bergam Lecht, Hitler defined the respective positions of the Aryans and the Jews in history: "The Aryan is the Prometheus of mankind; the divine spark of genius has always shone from his illumined forehead. He has always fanned the fire of knowledge that brightened the nights. He conquered and subjugated men of inferior races, placing their practical activity under his command, according to his wishes and his aims. By imposing useful, although painful, activity on these people, he not only spared their lives but perhaps reserved for them a better fate than if they had enjoyed what they formerly called liberty. As long as he rigorously maintained his moral situation of master, he remained not only master but also guardian of the civilization that he continued to develop. If he disappeared, a profound obscurity would descend on earth. In a few centuries human civilization would disappear and the world would become a desert." As to the Jew, Hitler expressed his position just as clearly in *Mein Kampf:* "The Jews do not satisfy one of the essential preconditions of a civilized people. They have no idealism. They have remained parasites, scroungers, like harmful bacteria, spreading ever further as soon as they find a fertile soil. Wherever they establish themselves their hosts disappear after a time. They poison the blood of others. They attempt by all possible means to destroy the basis of the races they wish to subjugate."

When he took power in 1933, Hitler knew that war was inevitable and even desirable. According to Darwinian and selectionist ideals, war was the only means of establishing the German people in its proper place, of leaders of the world. From that time the National Socialist government undertook two actions, fighting on two fronts. First of all, there was intensive military preparation, which resulted after many sacrifices in a largely mechanized army based on high technology and essentially designed for offensive warfare. When the time came, Germany had to be able to destroy France and England to the west, Poland and Soviet Russia to the east. But at the same time Hitler waged war internally. He had to eliminate the Jews from the machinery of state and give all posts of responsibility to Germans of pure race, who were by definition the only people fit to command. Beginning April 7, 1933, a series of laws removed Jews from the army, the administration, and, with a few exceptions, from the liberal professions. Later, through the laws of citizenship of the Reich and for the protection of the blood and honor of Germany, it was decided that Jews would no longer enjoy certain civil and political rights and could no longer marry or even have sexual relationships with German citizens. From 1939 onward offenders risked the death penalty. Finally, on January 20, 1942, the so-called Wannsee conference in Berlin decided to apply to the Jews the "final solution." Whatever their age, sex, or country of origin, all Jews must be physically eliminated.

The racists of the Reich were convinced that if the German people regained their original Aryan superiority they would inevitably triumph over other peoples, for their domination would become a biological destiny that nothing could oppose. An extraordinary system of arrest and destruction was set up that would in a period of a few years bring six million innocent people to the gas chambers and the incinerators. Looking at some of the documents used at the Nuremberg trials, one is amazed at the logic and perfection of a system designed to destroy a whole human community after having first exploited it to the maximum in terms of work and raw materials. For example, fat was recovered from the bodies of deported persons to make soap and their hair used to make cloth and blankets. All these operations were carefully studied to achieve maximum productivity. There were rules that in the incinerators thin bodies, called "Moslems," should alternate with fat bodies — "fat" not because they had eaten too much but because they were edematous due to lack of vitamins — apparently in order to economize on coal. This represented industrial crime on a continental scale.

The system went to absurd extremes. While wanting to destroy bad genes that might contaminate German blood, the Nazis also claimed that they would save the good genes that had left the Fatherland in the course of history. Thus in 1940 an enormous project began to recover Aryan-type children in Poland, first in orphanages and then in the street. During the winter of 1941 secret order number 6711 of the SS Gruppenführer Ulrich Greifelt, chief of the SD in Poland, stated: "In the old Polish orphanages and in Polish families, there exists a large number of children who must be considered as of Nordic parentage because of their external racial aspect. These children recognized as having valuable blood for the German community must be integrated in the German nation." Himmler told his generals on October 14, 1943: "I believe that we must remove these children from their surroundings and bring them here, even if we must take them by force and abduct them. We cannot take the responsibility of leaving this blood outside our frontiers so that our enemies may have great leaders" (Hillel, 1975). This hunt for good genes involved the capture of large numbers of young Poles, perhaps up to two million. Of these, only 100,000 were recognized as "racially useful" by the Nazi anthropologist doctors. Others disappeared without trace, mostly exterminated in concentration camps.

The racist mystique was such that Hitler himself and the leaders of his party believed to the end with a blind faith that the superior race would finally triumph. In the last days of the war, as the eastern front was crumbling and Stalin's tanks advanced toward a Berlin consumed by flames, the Führer, sealed in his bunker, continued to give orders: essentially to use to the maximum the new weapons, particularly the V-1 and V-2 being launched on London, and to complete the extermination of the Jews, pitiful living cadavers forced by their captures to drag themselves along the roads from one concentration camp to another, as the Russians advanced.

On April 29, 1945, the day before his suicide, with the Soviet army only a few hundred meters from his underground refuge, Hitler drew up his political testament, which he ended with a plea to the Germans to maintain racial laws in all their rigor and to pursue implacably "those who poisoned all nations, the Jews" (cited by Cartier, 1966).

Nazism collapsed, and Aryan Germany was occupied by the victors, who included Jewish, Arab, and black soldiers. After all these horrors and absurdities, one might have expected that the typological concept of race, shown by modern genetics to be scientifically false, would be abandoned forever. Unfortunately, this is not the case and this concept is still accepted by many, doubtless more out of habit than reason.

One could quote many examples. The Breslau school of anthropology was from the 1930s one of the most brilliant and most appreciated in Germany. It was led, among others, by Egon von Eickstedt, who devoted himself to the measurement of a large number of somatic racial characteristics. He even went so far as to define social races: races of cleaning women, of dockers, and so on, all based on morphological, mental, and sociocultural features that are all, to a great extent, dependent on environment.

In 1961 Ilse Schwidtzky, von Eickstedt's assistant, wrote in the *Handbuch der Biologie,* scarcely sixteen years after the end of the period of torture: "The role of racial differences in character is even more probable now, although less well studied. Within the same cultural environment and assuming a similar body structure, a character test showed German students belonging to the Nordic racial type in bodily appearance to be more introverted than Mediterranean individuals, who were much more extroverted."

This nonsense would be of purely anecdotal interest if it did not provide a basis for racist ideas that have reappeared here and there in the last few years, deceiving many honest people who know little of anthropology. Thus in June 1980, Ilse Schwidtzky received the Broca Medal in Paris during an international colloquium organized by the CNRS (Centre National de la Recherche Scientifique) in commemoration of the hundredth anniversary of the death of the illustrious anatomist. Broca, who never accepted Darwinian ideals and was a confirmed antiracist, must have turned over in his grave.

The Genetic and Environmental Basis of Intelligence

Intelligence is an extraordinarily complex function, difficult to define and even more difficult to measure. In 1896 a French psychologist, Alfred Binet, and his student Henri Victor developed a series of tests designed to quantify intelligence on the basis of eleven apparently fundamental qualities—in particular imagination, comprehension, esthetic sense, memory, concentration, moral sense, and muscular effort. These tests were improved in 1905 by Théodore Simon to produce the Binet-Simon scale, used to help in selecting backward children for special classes that had just been created by the French Ministry of Education. At the beginning of the First World War, the Binet-Simon test was further modified by Lewis M. Terman of Stanford University and became the Stanford-Binet test for determining intelligence quotient (IQ). We do not intend to discuss the problem of heredity of intelligence here, for several excellent works have already been published on this subject, such as those by Dobzhansky (1973) and Larmat (1979). The hereditary theory that makes intelligence a purely genetic and therefore racial characteristic cannot be defended today, although it has recently been taken up by a few authors such as A. R. Jensen, professor of psychology at the University of California, and Hans J. Eysenck, director of the Institute of Psychiatry at the Maudsley Hospital in London. These hereditarists include psychologists and sociologists, but very few geneticists. Intelligence depends on a number of very complicated factors that have never been defined precisely. An anecdote relates the joking remark made by Binet when asked what intelligence was, in his opinion. He replied: "It is what my test measures." In fact, just like body height and weight and physiological function, intellectual faculties depend on both extraordinarily complicated gene combinations and educational factors, which seem to play an essential role.

There must be hereditary factors involved in mental ability, but there must also be as much genetic polymorphism as is observed in relation to other functions that depend, at least partly, on the genome. Nevertheless, differences in intellectual ability cannot be considered to have a quantitative basis but are, rather, qualitative. We are all born with brains of a certain capacity that are never the same from one individual to another except in identical twins. An individual's performance depends, first of all, on the amount and meaning of the information received, and second, on the degree of mobilization of the individual's mental functions. These can be developed by training, like swimming or running, but they depend much more on education than other physiological functions because they are based on what we have learned, the education we have received. In man no other feature is as much influenced by milieu as his psychological makeup and activities. Society — whether the family, the school, or the state — bears a considerable responsibility for the education and development of youth. We shall discuss the problems of teaching a little later.

In the last century an Italian criminologist, Cesare Lombroso, claimed to be able to define a criminal "type" from his somatic features. He would have a congenital tendency toward delinquency whatever his social environment. This theory did not survive a close examination of the facts. Morality or perversion is not hereditary. One is not born a saint or a scoundrel, a physicist or a farmer, an organist or a cook; one learns to become these things. "It has never so far been demonstrated that there exists a difference in intellectual level of genetic origin between different social classes, nor even between human races" (Larmat, 1979).

The IQ can be modified by training and education much more easily than we can change stature or weight by our life-style or diet. Nevertheless, diet is not totally remote from the development of intellectual faculties. Today we know very well that protein deficiency in childhood can affect intellectual development, whatever the nature of the genome. Underdevelopment through deficiency can be seen in all countries where there are very poor populations, particularly after several years of famine. Once these changes are physiologically fixed, recovery is almost impossible. From birth to age seven or eight, children are particularly sensitive to protein deficiency. This period represents the time when they are most educable and after which individuals learn much less easily.[22]

DARWIN AND MARX

To say that Darwinism was merely a basis for liberal ideology would be unjust and would reduce its field of influence in an exaggerated way. Indeed Marx and Engels, the creators of scientific materialism, were much inspired by it, as we can judge from the letters they exchanged or those sent by Marx to Darwin. Unfortunately, much of this correspondence has been lost (but see particularly Prenant, 1938; Marx and Engels, 1973; and Naccache, 1980). The nineteenth century was marked by transformist ideas. In all branches of knowledge fixism was abandoned for more dynamic models. In astronomy Immanuel Kant and Pierre-Simon de Laplace had shown that the solar system could not be reduced to a series of never-ending revolutions around the same orbits but that the system had a history. James Hutton, and particularly Charles Lyell (1797–1875), influenced geology in a similar way. Defying traditional fixism, they proposed that the earth had evolved under the physicochemical influence of water, wind, and sun in particular, and that this evolution was continuing in modern times. In 1833 Lyell published his *Principles of Geology,* the first scientific history of the earth. Darwin was very impressed. Later, Lyell became one of the most ardent defenders of the naturalist and applied his evolutionary scheme to humanity (1863). Herbert Spencer (1820–1903) was one of the first to try to explain the evolution of organized life by the laws of mechanics. He was initially influenced by Lamarck but later accepted Darwin's concepts, which he tried to apply to human society. His theory was published in *Principles of Biology* between 1864 and 1867. The zoologi-

22. For instance, children who have been lost in the wild very young and remained without human contact tend, when discovered, to be very difficult to educate—even to speak, read, or write in an elementary way—if they are more than seven or eight years old. See Ruffié (1976).

cal aspects of the text were reviewed by Thomas Huxley and the botanical aspects by Joseph Hooker.

This transformist movement, which influenced the whole of the nineteenth century, inevitably affected sociology and politics. Marx and Engels postulated that societies evolved. They looked for the motive force behind "sociological dynamics" and tried to define its laws. Man was a living being, a mammal. So they anticipated great things from the natural sciences, which according to Marx in 1844, would "become the basis of human science. History itself is a real part of the history of nature." At the same time he emphasized "the difficulty of ridding the conscience of the people of the idea of creation." Even before the publication of the *Origin of Species,* Marx had an "intuition" about transformism and put forward in 1857 the idea of "a common origin of man and the other animal species, the idea of an interrelationship of living beings" (Naccache, 1980).

How could the authors of dialectical materialism resist the first proposals of an objective explanation of natural history in which the appearance of *Homo sapiens* was the latest chapter? This idea was expressed by Henri Lefebvre (1966) when he wrote: "We cannot understand the development of dialectical materialism in the decisive years without considering the publication by Darwin of the *Origin of Species.*"

It seems that Engels was the first to discover Darwin. On December 11 or 12, 1859, a few days after the appearance of the famous book, Engels wrote an enthusiastic letter to Marx: "I am reading Darwin and it is quite sensational. No one has ever attempted in such a broad manner to demonstrate that there is a historical development in nature, at least never with such joy."

This enthusiasm, quickly shared by Marx, was due above all to the materialistic content of Darwin's ideas. Marx and Engels were profoundly atheistic. They considered religion as an epiphenomenon of feudal or bourgeois societies, a convenient instrument for the governing classes to control the workers and peasants, based essentially on fear. For to contest the established order set up by God was a sin and would result in damnation. This threat had a consolation: those who had not known happiness in this world would know it in the afterlife. By suffering on earth, one entered heaven. Faith led to resignation and a respect for established structures. The Portuguese and the Spanish, who converted and baptized the black people they captured as slaves, had applied this form of reasoning. For the first time, Darwin introduced a scientific justification for atheism. He destroyed the myth of Adam and Eve, our first parents, created by God in his own image, who, seduced by the temptations of the Devil disguised as a snake, took the apple of the tree of knowledge of good and evil and were evicted from paradise on earth. Fallen into original sin, our first ancestors lost those attributes that made them like God, the first of which was immortality. They began to experience suffering, disease, and death. The humanity descended from them had to accept misery, which would allow it to expiate their fault and find paradise — not on earth but in heaven.

According to Darwin, reality had fewer consolations. Man had descended from ancient vanished primates that had also produced the great apes of today. We were not the result of a deliberate, divine action. We had never committed any particular sin, and paradise on earth had never existed. Like the other species around us, we had developed from variations through the intervention of natural selection. Humanity was the fruit of experiments that had been successful and battles that had been won.

The Darwinian revolution had a much greater impact on religious belief than did Galileo's. To say that the earth revolved around the sun may have been contrary to something written in the Bible, but it in no way challenged the existence of God. However, Darwin implicitly denied creation. If God existed, of what use was he? We are really confronted here with the "gospel of the Devil."

The similarity between the philosophies of Marx and Darwin goes still further. When Marx and Engels published their first works, they introduced the revolutionary notion of "social transformism" to a civilization that believed blindly in revelation and fixism and — as far as its ruling classes were concerned — was persuaded of the superiority of its organization. The fathers of scientific socialism found a convincing argument in Darwin's concept of the transformation of species. If animal species evolve in a search for a better equilibrium, why should human society not do the same? Socialism would thus be derived from capitalism as a superior species was derived from an earlier inferior species. At the social level, revolution was the same as speciation in the biological domain. The dawning of a new humanity was near.

Just like Darwin, Marx saw the prime mover of social change in struggle, hierarchy, and elimination. When the *Origin of Species* was published, Marx and Engels were refining their theories of class struggle, considered inescapable and necessary for the progress of humanity. In his manuscript of 1844, Marx quotes Buret (*De la misère des classes laborieuses en Angleterre et en France,* 1840): "In the present economic regime, industry has become a war, and commerce a game. Left to themselves, these economic interests must enter into conflict. They have no other arbiter but war, and the decisions of war mean defeat and death for the one and victory for the other. This perpetual war is called competition" (cited by Naccache, 1980).

The same idea can be found in the *The German Ideology,* published by Marx in 1845. "Industry made competition universal. All nations are dragged into the struggle of competition."

Later, in 1877, an identical theme was developed by Engels when he applied Darwin's model of competition in life to economic competition. "Industry and the establishment of world markets have made struggle universal and have given it at the same time an unequaled violence. The existence of individual capitalists, as well as of whole industries or countries, is decided by natural or artificial conditions of production, according to whether they are more or less unfavorable. The loser is eliminated mercilessly. The condition of an animal in the wild seems to be the apogee of human development. It is the Darwinian struggle for the existence of the individual transposed from nature to society with tenfold fury." Further on he adds: "Struggle and frantic competition mean an ever-increasing elimination of workers" (cited by Naccache, 1980).

Nevertheless, Marx and Engels differed from the Darwinians, and particularly from Galton, by the way in which they defined the groups involved in the struggle. According to Galton, the competition was between biologically different individuals, superiors and inferiors. This organic difference gave each his social situation. Marx and Engels, on the other hand, remained egalitarian. In this they stayed faithful to one of the fundamental ideas of the Enlightenment, but they were nonetheless aware of the limits of the bourgeois revolution of 1789 that had brought man political equality but not economic equality, the latter being the only way of giving full significance to the former.

The French Revolution spelled the end of the aristocracy, that is to say the power given by birth. But it replaced it with the power given by fortune, which was accentuated with the development of industrial societies. According to Marx and Engels the inequality of man was due not to biological factors but to socioeconomic structures. These, not the hereditary patrimony, defined class.[23] Each class was characterized by the relative importance of the profit made by each individual according to his work. At the bottom of the ladder, the proletarians worked hard but only received a minute part of the fortune they produced, most of which went to the directors. Between these two extremes there were intermediate classes whose number and size varied, according to Marx, from one country to another according to the level of industrialization. In all he counted eight in Germany and seven in France, all of which

23. In the name of egalitarianism, Marx remained a *de facto* anticolonialist, even if colonial problems have little place in his works. He wrote: "A nation that oppresses another is not a free nation," an elegant and true statement. But if he were to be reborn now, what would Marx think of the attitude of the Soviet Union toward its satellites?

could be grouped into three principal classes — the workers, the capitalists, and the proprietors (see R. Aron, 1964). There was no doubt in Marx's mind that the whole of human history depended on class struggle. In the *Manifesto of the Communist Party,* published in 1848 with Engels, Marx wrote, "The history of all hitherto existing society is the history of class struggles. Freeman and slave, patrician and plebeian, lord and serf, guild-master and journeyman, in a word, oppressor and oppressed, stood in constant opposition to one another, carried on an uninterrupted, now hidden, now open fight, a fight that each time ended, either in a revolutionary re-constitution of society at large, or in the common ruin of the contending classes."

This model is often described in Lenin's works. "The dictatorship of the proletariat is indispensable, and it is impossible to conquer the bourgeoisie without a long, stubborn, relentless war" (from *The Immediate Tasks of the Soviet Government*). In *Left-Wing Communism: An Infantile Disorder,* Lenin evokes "the stubborn struggle, bloody and bloodless, violent and pacific, military and economic, pedagogic and administrative, against the forces and traditions of the old society" (cited by Fontaine, 1978).

This class struggle would first of all provoke a simplification of the system. At the beginning, the number of classes would be reduced, for the rich would be richer but less numerous, and the poor poorer and more numerous. Finally, the only opponents left would be the bourgeoisie, reduced in number but possessing all the means of production and capital, and the enormous mass of the proletariat, providing the work but deprived of everything else. Since the latter group was condemned to grow in size as the middle classes and the peasants became poorer, the balance would finally swing in its favor. It would overcome the bourgeoisie and impose its dictatorship. Then, since classes would be abolished, the causes of conflict would also disappear; there would be no more war or revolution. The world would finally enjoy an era of peace and prosperity.

This forecast by Marx did not come true any more than did his others. Far from joining the proletariat, the middle classes of industrial societies became richer and richer, and at the same time their political influence continued to increase in most developed countries, where they often took over the leadership. This form of evolution was precisely the opposite of what the father of scientific socialism expected.

Since their whole argument was oriented around class struggle, it is hardly surprising that Marx and Engels were so seduced by Darwinism. In 1850, nine years before the *Origin of Species,* Marx published a series of articles that later appeared as *The Class Struggles in France,* in which he hoped to find a "naturalistic" justification for his own theories.

Curiously, historians and philosophers have for long ignored or under-estimated the place of Darwinism in the ideas of Marx and Engels, in spite of the fact that neither of them made any secret of it. On December 19, 1860, Marx wrote to Engels that he saw in Darwin's work "a foundation given by natural history for our way of seeing things." He wrote to Ferdinand Lassalle on January 16, 1861, that he found in Darwin "a basis given by natural science for class struggle" (cited by Naccache, 1980). Nevertheless, in spite of repeated overtures by Marx, his relationship with Darwin always remained remote, due to the lat-ter's reserve.

In 1872 Marx sent the English naturalist a copy of the second Ger-man edition of *Das Kapital,* dedicated flatteringly "To Charles Darwin. By his sincere admirer." On October 1, 1873, Darwin wrote to Marx to thank him, still keeping his distance: "I believe sincerely that I would deserve this homage from you more if I understood better the profound and important subject of political economy."

This volume is still to be seen today in the Library of Down House, Darwin's former home, now a museum. Only the first few pages are cut, proving how little interest the navigator of the *Beagle* had in Marx's theories. Eight years later, in 1880, Marx tried again. He wrote to Darwin to ask him to review Chapters 12 and 13 of the English edition of *Das Kapital,* which were based on the *Origin of Species,* in order to be sure that his ideas had not been misquoted. At the same time he proposed to dedicate the second volume of *Das Kapital,* which was being prepared, to Darwin. This letter has unfortunately been lost, but we do have Darwin's reply. "The publication, in whatever form, of your observations on my writing does not need my agreement and it would not be serious of me to agree when there is no need. I prefer that the book or the volume is not dedicated to me, for that would imply in a way my approval of the whole work with which I am not familiar. I regret to have to decline your offer but I am old and weak and reading the Editor's proof tires me very much" (cited by Naccache, 1980). A polite but firm refusal.

In truth, Darwin's influence on Marx and Engels was profound. (Of the two, Engels was an accomplished naturalist and chemist, while Marx was particularly expert in sociology and mathematics.) But there was no reciprocity; to the end of his days Darwin ignored Marxism and probably mistrusted it. His attitude is not surprising in someone from a solid upper-class family who had already been responsible for much disturbance in his milieu when the *Origin of Species* appeared. The current that flowed from Darwin to Marx was never seen to flow in the other direction.

THE MARXIST MODEL AND SELECTION

Marx transposed Galton's scale from the domain of biology to that of economics and turned it upside down. The bourgeoisie had usurped its position at the summit, a form of historic but temporary theft. It had gathered the fruits of work that did not belong to it. In the end the working class would win. The world could only evolve toward the triumph of the proletariat. Thus Marx and Engels fixed the limits of Darwin's model as applied to man. For at the human level, the factors that intervened in forming groups and giving them their value were not only biological, as in animals; they were also sociological and cultural. But Marx and his disciples gave these last aspects a rather restricted meaning, for they considered work as the essential factor determining the place that each must occupy in society. And even then it was work such as was conceived in the industrial civilization of mid-nineteenth-century Europe. It is easy to understand why, due to its very ethnocentricity, Marxism was never able to become universal, in spite of what its founders claimed. It was much more of a socioeconomic model, limited in space and time, than a sociocultural analysis with more generalized applications. As to Marx's predictions, we have only to consider the present situation of the industrialized nations, in the East as well as the West, and even more so that of the poor nations that have never undergone an industrial revolution and may never do so, to see to what extent the father of scientific socialism strayed from reality. This is a fundamental point, about which we shall speak later.

The essential criticism of Darwin by Marx was that he did not sufficiently take into account the factor of work in the evolution of mankind. Darwin sketched a pathway, but as a good bourgeois zoologist he stopped halfway. "In order to find in Darwin's work its full content of revolutionary truth and at the same time to denounce its scientific dross Marx and Engels were needed, men of the same epoch but linked to the proletariat," wrote Marcel Prenant in 1938. But he added later: "Marxism developed and condensed the revolutionary germs contained in the heritage of the great Darwin."

In fact, Darwin's theories represented for scientific materialism, which was looking for an identity, a "thunderbolt that provided the occasion for a lightning love affair" (Naccache, 1980).[24]

We may ask what the successors of Marx and Engels thought of Darwinism. For the founding fathers themselves, class struggle was only a transitional stage, and a communist society should lead to a

---

24. In fact it was more of a nonreciprocal love affair, according to Jean-Pierre Gase in *Humanité*, February 26, 1982.

classless, stateless structure living in liberty and peace. "We are now advancing rapidly toward a development of production, so that the existence of classes not only ceases to be a necessity but even becomes an obstacle to production. Classes will disappear as inevitably as they are formed. As the classes disappear, so the state will inevitably disappear too" (Engels, in a letter to Bobel, March 18, 1875).

Stalin altered this optimistic vision. "Certain comrades have interpreted the thesis of the abolition of classes, the creation of a classless society, and the disappearance of the state as a justification for idleness and placidity, a justification for the counterrevolutionary theory of the extinction of class struggle and the weakening of the power of the state. The disappearance of the state will result not from the weakening of its power, but from its maximum strengthening, indispensible for removing the remains of the disappearing classes and organizing our defense against the capitalism that encircles us and that is far from being destroyed and will not be destroyed for some time" (Stalin, report on the first Five-Year Plan, 1930). Half a century after this proclamation, the "remains of the disappearing classes" are in good shape in the Soviet Union (see Bettelheim, 1974–83). Stalin was a Darwinian without realizing it.

DARWIN, MARX, AND JESUS

The relationship between Darwinism and Marxism may be fairly clear, but it is more difficult to establish the relations there might be between these two ideologies and Christianity, which preceded them by nearly two thousand years. At first sight they seem in opposition. Marxist doctrine is basically atheistic, just like Darwinism, which essentially denies creation and does not reserve a privileged place for man in the zoological hierarchy. What is more, the Gospels almost always defend the weak against the strong, the oppressed against the oppressor.[25] It

25. One apparent exception is the parable of the talents. "So it was with a man who went on his travels; he called his trusted servants to him and committed his money to their charge. He gave five talents to one, two to another, and one to another. . . . The man who had received five talents went and traded with them, until he had made a profit of five talents more; and in the same way he who had received two made a profit of two. Whereas he who had received but one went off and made a hole in the ground, and there hid his master's money. Long afterward, the master of those servants came back, and entered into a reckoning with them. . . . But when he who had received but one talent came forward in his turn, he said, Lord, knowing thee for a hard man, that reaps where he did not sow, and gathers in from fields he never planted, I took fright, and so went off and hid thy talent in the earth; see now, thou hast received what is thine. And his lord answered him, Base and slothful servant, thou knewest well that I reap where I did not sow, and gather in from fields I never planted; all the more was it thy part to lodge my money with the bankers, so that I might have recovered it with interest when I came. Take the talent away from him, and give it to him who has ten talents already. Whenever a man is rich, gifts will be

suffices to recall the Beatitudes from the Sermon on the Mount, which summarize this doctrine admirably: "Blessed are the poor in spirit; the kingdom of heaven is theirs. Blessed are the patient; they shall inherit the land. Blessed are those who mourn; they shall be comforted. Blessed are those who hunger and thirst for holiness; they shall have their fill. Blessed are the merciful; they shall obtain mercy. Blessed are the clean of heart; they shall see God. Blessed are the peace-makers; they shall be counted the children of God. Blessed are those who suffer persecution in the cause of right; the Kingdom of heaven is theirs" (Matthew 5:3–11). The same text is to be found more briefly in Luke 6:20–23.

Here Jesus simply develops an old Hebrew tradition referred to in many passages of the Old Testament. "For our sakes a child is born. . . . Father of the world to come, the Prince of peace. Ever wider shall his dominion spread, endlessly at peace; he will sit on David's kingly throne, to give it lasting foundations of justice and right" (Isaiah 9:6–7). Later the prophet adds: "Out upon you, that enact ill decrees, and draw up instruments of wrong; suppress the claims of the poor, and refuse redress to humble folk; the widow your spoil, the orphan your prey!" (10:1–2).

And verses 12 and 13 of Psalm 71, attributed to Solomon, state: "He will give the poor redress when they cry to him, destitute folk, with none to befriend them; in their need and helplessness, they shall have his compassion."[26]

As we said earlier, Darwin, son of a respectable upper-class family,

made to him, and his riches will abound; if he is poor, even what he accounts his own will be taken from him. And now, cast the unprofitable servant into the darkness without; where there shall be weeping, and gnashing of teeth" (Matthew 25:14–30).

This parable, which contrasts with the spirit of the rest of the New Testament, has given rise to many explanations. According to Jean Delumeau, Jesus takes the opportunity of repudiating the Pharisees for their scrupulous observation of the Mosaic law calling for the hoarding of talents. It is an "invitation to risk and action." One must act and take risks to enter the Kingdom of Heaven.

26. Altruism, the Golden Rule of religions, is a sort of cultural "invariance" in direct contradiction to social Darwinism. René Dubos (1974) mentions the following quotations:

Christianity: "Do to other men all that you would have them do to you; that is the law and the prophets" (Matthew 7:12).

Judaism: "What is hateful to you do not do to your neighbor. That is all the law; the rest is merely commentary" (Telmud Sahhar 31a).

Brahmanism: "This is the sum of duty: do not do to others what would be hurtful to you" (Mahabharata 5-1517).

Buddhism: "Do not harm others with what would harm you" (Udana-Varga 5–18).

Confucianism: "Here is the maxim of love: not to do to others what we do not want others to do to us" (Analects, 15–23).

Islam: "None of you is faithful unless he desires for his brother what he desires for himself" (Sunnah).

Taoism: "Consider that your neighbor earns your bread and that your neighbor loses what you lose" (T'ai Shang Kan Ying Pien).

Zorastrianism: "The only good nature is that which controls itself so as not to do to others what would not be good for itself" (Dadistan-i-dinik 94-5).

had been a theology student and was therefore presumably a Christian in his youth. Later his scientific work made him agnostic, and he did not change thereafter. Darwinism and its sociological applications may seem fundamentally anti-Christian. In reality, the situation is rather different. Although basically Christianity preaches a profoundly altruistic moral code, it is deeply influenced by temporal powers exploiting its Manichean concept of the world, based on the struggle between good and evil, angels and devils, and the forces of light and darkness. The collective conscience of Western Christianity remains clearly marked by this Manicheism. Darwinism found a hospitable terrain. Its struggle was between variants and normals, superiors and inferiors. Progress depended on victory, which for the Christian is also a condition for salvation. Darwinism would probably never have been born outside an established Christian country with such a divided view of the world.[27] We have described the close ties between the philosophies of Darwin and of Marx. A century has passed, and passions that divided earlier generations have now subsided, so that we can ask ourselves in all serenity about the role played by Christianity, or rather by the church, in the genesis and development of Marxism.

For many years historians and sociologists have been struck by the fact that Communism, the concrete expression of Marx's visions, has developed particularly in those countries with long-standing ecclesiastical structures, either Orthodox or Catholic. This does not apply to the "people's republics" that were made satellites of their great eastern neighbor, often against their wishes. The uprisings in Hungary, Czechoslovakia, East Germany, and Poland in the last thirty years leaves this in little doubt. Let us rather consider what happens in Western countries with pluralistic democracies, where all political parties can organize and express themselves as they wish and even take part in the government if they are part of the parliamentary majority.

The Communist party is powerful in nations with strong Catholic influences, such as France, Italy, Spain, and Portugal. It is weak or nonexistent in areas of Protestant culture. Great Britain was the first European country to undergo an industrial revolution that generated a large, poor proletariat. Marx and Engels predicted an imminent popular revolution there and also in Germany, which was not far

---

27. We might recall that the aggressive urge of Christian nations throughout history to go and convert others was in no way less marked than that of revolutionary states. It is found in the distant past and can still be seen in national emblems, often chosen long before the Reformation and representing fierce predators: "The eagles of Germany, Austria, Russia, Poland, Spain, or America; the lions and leopards springing curiously from the foam of the North Sea or even the Baltic. The most Christian nations have voluntarily pictured themselves throughout the centuries in the form of beasts of prey. Less impressive, but no less aggressive, is the Gallic cock, rather a poor relation in this particular zoo!" (Fontaine, 1978).

behind. They never took place. England has always remained a model of parliamentary democracy, evolving under the monarchy toward a liberal socialism without the dictatorship of the proletariat that had been foreseen. As to the crisis in Germany after the First World War (which was principally a capitalist crisis), far from bringing revolution, it brought Hitler to power by legal means and banished the Communists to concentration camps. With the reestablishment of peace the Federal Republic of Germany, the only part of the former Reich that could decide its own political destiny freely, joined the pluralistic nations, among which it is today a model. This state of affairs is far removed from the prophecies of the fathers of scientific socialism. We may wonder how it is that the Latin nations, neither poorer nor less educated than Nordic peoples, are able to assimilate political Communism so readily. The reasons are certainly complex, but their ancient Catholic culture certainly has a role to play. Through the view of the world it implies, and the mental structures it encourages, Catholicism favors a monolithic and strongly hierarchical system. When they become party members the faithful rediscover their church with its Manichean dogma, its articles of faith, and its infallibility. The background is familiar. No system, no salvation. There can be neither criticism nor discussion. The heretic, now called revisionist or deviationist, is eliminated irrevocably.[28] Exclusion from the party, with its consequences, is the same as excommunication. The same public confessions are required. In modern times, infallibility only exists in the Vatican and the Kremlin. Pope Pius XII and Stalin were contemporaries, and both reigned over masses to whom protest was impossible. Over the years, dogma has become less strict but discipline has persisted, although the church has become more open since the latest Vatican council, and the Western European Communist parties have mostly refuted the dictatorship of the proletariat. Another striking similarity is the way in which the two systems take care of an individual throughout his life — from birth (or even from conception) to death, and even after. They make free thought unnecessary or prohibit it. Decisions come from the outside. This "permanent providence" ends by sterilizing a human being and depriving him of all initiative. He is secure, but stifled. So he is promised paradise on earth by the Marxists, and in heaven by the Christians. Both ideologies exhibit the same proselytism and the same intolerance. They both demand the same purity and the same lack of self-interest from their members, and sometimes create the same mar-

---

28. François Mitterrand (1978) writes: "My mother constantly repeated that all wars were due to religion. She had not read Marx, and was perhaps mistaken. But if she had been aware of the Marxist-Leninist adventure she would doubtless have discovered more reasons to confirm her convictions."

tyrs. The most sincere and devoted militants tend to be found among Communists and people with deep religious conviction, for they have a "faith that can move mountains" *(perinde ac cadaver).* [29]

But both systems, or rather the people who run them, have progressively lost from view the highly praiseworthy aim of their founders: the betterment of mankind's condition by the introduction of justice and liberty. Very early in its history the Catholic church became involved in politics and ceased to fulfill this role, just as Soviet Russia forgot its Marxist ideals and abandoned the proletariat it was supposed to represent in favor of the party — a new class holding the reins of power and later the wealth of the nation.[30] Both are supported by a heavy, meddlesome bureaucracy, remarkably efficient as an organ of control and even of repression. Similarities also exist in the way in which others are seen — for, as we saw earlier, Marxism is no less ethocentric than Christianity. In all his works Marx dealt mainly with the industrialized countries of Europe, particularly England, Germany, and, to a lesser extent, France. He hardly touched on Russia and hardly at all on America. He did not deal with purely agricultural nations nor with people of color, who in the middle of the nineteenth century must have already formed about three-quarters of humanity. Traditional local culture was considered suspect and backward (R. Aron, 1964; Bettelheim, 1974–1983). It is not surprising that many Asiatics and Africans see in Marxism an internal quarrel of the white man and look upon it as the most recent form of European imperialism. Léopold Sédar Senghor stated in 1971: "Marxism must be not revised but rethought by black people and according to their black values." But we might wonder what would be left of it then. Marxism has already been rethought by the "yellow" people, and we know very well what the result was.

At the present some nonaligned nations look toward Soviet Russia, long considered hostile to colonialization (at least by other nations). But more and more others see little difference between Gorbachev and Reagan, each imperialistic in his way and both seen as descendants of Julius Caesar. It is easy to understand how the present situation can appear monstrous to an orthodox Marxist. Marx and Engels had foreseen a war between industrial nations — particularly between England and Germany, perhaps with France caught in between — and consid-

29. "Like a cadaver." Colorful formula used by St. Ignatius of Loyola, founder of the Jesuit order, in his *Constitutions* to exhort all members of his society to absolute obedience to their superiors, as if under military discipline.
30. These striking parallels may explain the very similar images of the two competing sides. Just like the clergy of former times, the party holds the power, the fortune, and the education, in the name of truth. It represents the father figure, which explains the mistrust of both Catholicism and Marxism of a Freudian approach to human relations.

ered such conflicts as historically inevitable. Capitalism secreted war as a liver secretes biles and socialism peace.[31]

Recently Soviet Russia's worst enemy has been China, the new ally of America. Lesser countries faithful to the same ideology have waged, or are still waging, bitter wars, like Ethiopia against Somaliland, Vietnam against Cambodia, and so on. One must not forget little Albania, pure and hardheaded, isolated between the sea and the mountains, spreading its anathema passionately and courageously on all the great socialist powers, accusing all of them of deviationism in various directions. If we study the evolution of Western Philosophy we are struck by the relationship between Christianity — or rather the use that great powers have made of Christianity — Darwinism, and Marxism.

Darwin would certainly not to have become what he was had he been born in Japan or in India and if he had not thought of becoming a minister at a crucial moment of his adolescence. It is difficult to imagine Karl Marx imposing his system other than in Christian countries, particularly those in which the church is an important feature.[32] One cannot ignore all that Manichean reasoning has achieved in Europe. The assurance that it alone possessed truth legitimized its territorial expansion as well as its accumulation of material wealth and its generalized rise in living standards. It is not pure chance that since the Renaissance the richest and most powerful countries have been those with a Christian cultural background, with the exception of Japan, which has enjoyed such success only relatively recently. This situation gives them an enormous historical responsibility, as was rightly emphasized by the International Eucharistic Congress held in Lourdes in July 1981. But we must also remember what this attitude is costing us. By prohibiting or limiting freedom of thought and expression, by imposing a single "mental" holotype, ideological or religious dictatorships have reduced the cultural polymorphism of individuals who make up these societies. Such a dictatorship not only impoverishes them dangerously but often

31. We should say that at the end of the nineteenth century England and Germany, the two greatest industrial powers, were in constant conflict — in international markets, overseas, technologically, economically, and militarily — in spite of both populations being of "Aryan" origin. The emperor Wilhelm II had even married a daughter of Queen Victoria. Such a situation was inexplicable to the racists, but confirmed the Marxist vision of the world.

32. It is interesting to realize the ideological conservatism of the Soviet Union, the homeland and cradle of Communism, where Marxism has hardly evolved for sixty years in spite of the social upheavals in the world that have almost always invalidated Marx's predictions. From this point of view things seem to have stopped with Lenin's death. The only Marxist philosophers who have tried, sometimes successfully, to improve the doctrine or simply to modernize it are from Western nations. In the East, theoreticians either align themselves with the ruling powers and quickly become "sterilized" by the party apparatus and bureaucracy, or they become dissidents and are condemned to silence, exile, or the psychiatric asylum. Marxism has been able to develop historically only in liberal democracies. The only Communist regime to have innovated, Tito's Yugoslavia, paid the price in the form of a violent rupture with Stalin that almost led to armed conflict.

leads them toward a dead end. The result has often been the suppression of the very originality that is the hallmark of man. We shall analyze this problem in more detail in Chapter 14.

THE LATEST AVATAR: SOCIOBIOLOGY

The selective theory is periodically reborn from its ashes in spite of the discredit that recent history has thrown on it. Its latest avatar is *sociobiology,* a theory proposed by E. O. Wilson, professor of zoology at Harvard University, whose work has mostly concerned social insects.

In the summer of 1975 Wilson announced the birth of sociobiology with great publicity in a voluminous seven-hundred-page work of twenty-seven chapters. He is an entomologist, and his whole theory is to a large extent inspired by his observations of insects. They can construct complex societies, but within them all individuals are united by innate behavior. Wilson generalized this scheme. According to him, all social structures in all zoological groups depend on genetic determinism. If these structures are accepted by natural selection, it must be that their features bestow a certain advantage. All the phenomena related to them are biological necessities controlled by heredity. They include a constant hierarchy dominated by certain individuals, the existence of caste, tendencies toward altruism or aggressivity, conflicts and struggles. They represent the fruit of selection in just the same way as do the color of our eyes or the way in which we metabolize sugar, and were preserved by evolution because they were favorable for the conservation of the species, or rather for the spread of the corresponding genes. On the human level, class struggles and wars between nations thus are as inevitable as digestion or walking. By reducing everything to the insect model Wilson ignored the principle, referred to in Chapter 11, that in warm-blooded vertebrates learned behavior tends to replace innate behavior. This replacement reaches its maximum in man, in whom behavior is to a great extent conditioned by apprenticeship, example, and education. Wilson's system is rigidly deterministic, leaving little place for experience and freedom. By reducing humanity to a community of termites, he overlooked those features that are responsible for all its value and its originality.

According to Wilson the origin of all behavior is in each individual's tendency to spread his own genes in two ways. The first is by reproducing as often as possible and as for as long as possible. Human beings have always given great importance to love, which has even been deified in many civilizations. Sexual taboos imposed by Judeo-Christian culture are a recent phenomenon, both limited in its extent and avoided in many ways by the people concerned (see Duby, 1981). In sexual "strategy" it is often the male who plays an active role, while the female, so to speak, executes the orders of her partner. This would explain the differences, not only physiological but also psychological, that exist between man and woman. The former is the conqueror who must be obeyed; the latter must follow his orders. It should be noted that sociobiology ignores matriarchal societies, whose form of tyranny is often just as extreme as that of patriarchal societies.

The second pathway for this "genetic conquest" involves an individual helping to diffuse the genes of his relatives, using "relatives" in the widest sense of the word — particularly children but also father, mother, brothers, sisters, and even cousins, for all these have one or several common ancestors and thus share a fraction of the same genetic patrimony. Thus by favoring the spread of the genes of related individuals one is in fact spreading one's own genes. This activity would explain many forms of behavior, such as altruism toward our entourage and aggressivity toward strangers. In this way, Wilson explains the "paradox" of altruism, which can be disastrous in terms of individual selection — when, for example, it incites an individual to sacrifice himself for others — but can be very advantageous for the spread of a particular patrimony. Wilson goes further. He extends this system to what one might call "friendly relations," meaning that individuals who are not biologically related are led to help each other. Indeed, the protection given us by our friends, which we return whenever we can, can be a selective advantage because it helps us to raise our children and increase their chances of survival, and thus the probability of the spread of our genes. On the same basis Wilson explains celibacy and homosexuality, which are by definition not directly transmissible because they result in no offspring. However, the genes that are responsible for these states have a certain selective advantage. The unmarried and homosexuals are available to help rear the children of their neighbors. Thus they can be tolerated in the population.

In Wilson's view of the world the individual has little importance in his own right. He is simply the vehicle of genes that can be expressed through him. In spite of its apparent novelty, this reasoning takes up the neo-Darwinian model developed at the beginning of this century, according to which evolution can be reduced to competition between alleles. The "best" are those that ensure egoistic and aggressive behavior, with altruism being only the individual's most efficient form of egoism when the problem is to ensure the survival not of himself but of the patrimony that he bears. In such a case the most "active" alleles are retained to make up the heritage and will inevitably eliminate less active alleles — or, if one wishes, less aggressive ones. We are back once again to the holotype, this time defined in terms not of traditional morphological features but of behavioral characteristics.

Man arrived very late in a world made long before him. He became involved in this genetic "war," the fundamental force of evolution, which he can in no way change. Thus the social ascent of certain people or certain races is part of biological fate, like being born with blue eyes. Wilson prophesies in concluding: "The principal goal of a general theory of sociobiology should be an ability to predict features of social organization from a knowledge of these population parameters combined with information on the behavioral constraints imposed by the genetic constitution of the species." This is Galton arising from the ashes. Sociobiology is just as elitist as Galton's Victorian theories. Wilson considers egalitarian democrats as ignorant or simple-minded. Sociobiology claims that "capitalism, like competition and interest, is inscribed in our genes" (cited by Rouzé, 1980). Pearson, Galton's successor, and Vacher de Lapouge preached the same message. Wilson is a zoologist. His theory uses new ethological terms to propose a model that has been out of date for a long time. It is incompatible with modern genetic data. Dressing up a false theory with neologism does not make it more credible. Wilson considers the hereditary patrimony as an assembly of genes side by side and the gene as an autonomous entity acting selfishly and in isolation. This is the "genetics in beanbags" of the years from 1910 to 1930 that we discussed in detail in the first chapter of this book. What is more, the naïve hypothesis of competition between alleles terminating in the triumph of the most "egoistic" ones is totally opposed to the idea of that "colossal multipolymorphism," which as we have seen, is a characteristic of all natural populations.

One cannot judge the biological value of a group by its ability to retrieve and spread the "best" genes and eliminate the others. On the contrary, its wealth lies in its variety, which, faced with the constraints of the environment, ensures for it the greatest number of possible responses and thus maximum efficiency. At the present time a typological view of the world based on genetic monomorphism of species cannot be defended even using behavioral arguments.

Faced with such excesses, due more to his admirers than to Wilson himself, the father of sociobiology retreated, declaring recently to a journalist: "Man is not so dependent on his genetic heritage as we might have thought. His mind and his culture confer on him a liberty that may make him capable of overcoming his hereditary predispositions. Indeed liberty is not a vain word" (Dorozynski, 1981).

In a new work, *Genes, Mind and Culture* (1981), written in collaboration with the the physicist Charles Lumsden, Wilson admits that by the very nature of his brain and its possibilities, man cannot be compared with an insect. He has "softened" his model without rejecting it completely, because he still considers that there is a relationship between genetic patrimony and culture, but that this relationship is certainly less strict than he thought originally. According to Wilson, man does not have the same instinctive, rigidly programmed behavior as an insect but always has the possibility of a certain degree of choice. Nevertheless his choice is conditioned by genetic structures. According to these structures, an individual, and through him a population, will select certain customs, practices, and values that are best adapted to his genotype. These cultures in turn favor the diffusion of the corresponding genotypes. These pairs of genes and cultures, called *culturgens* by Wilson, are supposed to evolve simultaneously by permanent interaction. This process is what Wilson and Lumsden call *co-evolution.* According to them this co-evolution explains the development and divergence of civilizations. Nevertheless, this ensemble imposes less constraints than a true genetic program, and at any time an individual can change from one culture to another. History is full of such conversions.

Even restructured and softened, this neo-sociobiology cannot escape from the criticisms just outlined and presents the same dangers. For instance, it allows one to suppose that the poorest populations live in an economic, social, and even political environment conditioned by their culturgen complex. Thus a sound knowledge of the relationships between genes, mental faculties, and cultural factors would make possible efficient social control and avoid the dangers inherent in the rapid introduction of new forms of development, like industrial expansion, in traditional societies that are genetically unprepared. So generalized misadaptation with the appearance of a subproletariat and all sorts of unbalance would be avoided. It is easy to understand why such an obviously erroneous theory, unanimously condemned by geneticists, caused such a stir (see Lewontin, 1976; Ruffié, 1979; Jacquard, 1980; Sahlins, 1976). Recent history provides the explanation.

In publishing the *Origin of Species* in 1859 at the apogee of the Victorian empire, Charles Darwin was influenced by the ideology dominating nineteenth-century Western society. We have seen how struggle and competition emphasized by this naturalist were merely a projection into the biological world of the wild capitalism, unlimited competition, exploitation, and colonial adventurism that marked his time.

On the other hand the typological way of thinking, creating hierarchies of races and human classes of different values, made everything sound legitimate. Even Marx did not escape the Darwinian illusion. We can see where this has led us. Today the world is sinking into a new crisis whose nature we should try to understand. The most favored social groups and nations wish to preserve their privileges, even if these are unjustified and in part responsible for our difficulties. For want of anything better, everyone is trying to make his point of view legitimate by looking for arguments in data accumulated during 150 years of biological history, whether valid or out of date. As Albert Jacquard (1980) wrote: "References to science have replaced in many speeches references to the philosophers of antiquity or the fathers of the church." The myth of selection, which many believed dead, reappears on the scene whenever we have need of it.

crisis in the
Darwinian society

THE DEFEAT OF SELECTIVE IDEOLOGIES

The crisis in which humanity finds itself today is generalized. Nothing and nobody is spared; all nations are involved, whatever their political system. Its widespread nature illustrates the defeat of all ideologies, liberal or socialist, built on the selective utopia of the last century.

Unbridled liberalism born of the industrial revolution has in general raised the living standards of the majority but also led to inflation, unemployment, and war. Scientific socialism born of a romantic wave of generosity and the naïveté of scientific faith has scarcely succeeded better. In spite of undeniable social progress it has brought neither the well-being, the development, nor the freedom it had promised. Marx and Engels conceived a radiant, opulent, pacific society without constraints and without states. A century later what remains of their dream, and what would Lenin, the venerated prophet, say if he came back today? In their practices his successors turned their backs on his teachings.

Let us imagine a visitor from another planet. Before leaving he has acquired a general knowledge of the history of modern humanity. He will see that our world is divided into two intransigent blocs, each with a superpower at its head. Their ideas seem irreconcilable to him, so he wonders who is right and who is wrong. He reads the American Constitution. It is attractive, and he decides to go the United States. When he arrives in New York he expects to find a sort of fertile, liberal, peaceful paradise. Like the young Buddha leaving his family palace, he immediately encounters a very different and cruel reality.

In the midst of overwhelming and disturbing luxury he meets the unemployed and the poor, the robbers and the drug addicts. He sees the violence and injustice that are our daily experience. In going from Fifth Avenue to Harlem he has the impression of entering a different world and cannot understand how such different people can live under the same laws. Such a system, he thinks, cannot be the right one and is a form of imposture. So he turns toward the other side of the world and reads the democratic and generous Constitution of the Soviet Union. Believing he has found the truth, he goes to Moscow, finally hoping to have found a society conceived by reasonable men who make their speeches work practically. His disappointment will be just as great. At best he will find long lines of people in front of the almost empty shops. At worst he will encounter psychiatric hospitals or the Gulag Archipelago. He would be highly shocked by the difference in status, not *de jure* but *de facto,* that separates high dignitaries from the common people, although all call themselves "comrade." Disappointed, he will return to his own planet thinking that the people of the earth are very peculiar. Whether liberal or Marxist, they have pretenions to worthy principles that are hardly applied in reality. They are powerless, for they are up against dead ends.

THE MODALITIES OF THE CRISIS

Today we are suffering from a triple crisis, whose different aspects are related. They are linked to the enormous size of the human race, to the imbalance between populations and their environment, and to the growing disparity in the distribution of wealth between nations and, within an individual nation, between different classes. People have very unequal chances of escaping from their misery. Some become rich and others survive with difficulty, but many have no other perspective than absolute poverty and famine in the short or long term.

Before going further, we might examine the three aspects of the present crisis—demographic, ecological, and socioeconomic—all of which are interdependent.

The Demographic Crisis

The demographic crisis is related to the rapid increase in the size of the human race. Three million years ago, when our first ancestors, *Australopithecus* and *Homo habilis,* established themselves around the Great Lakes of East Africa, there were, according to paleontologists, no more than a few thousand individuals.

Then, 500,000 years or more ago, a new link, *Pithecanthropus* or *Homo erectus,* made refined stone tools and domesticated fire. He invaded the whole of the Old World. *Pithecanthropus* formed small, isolated groups of relatively few people. There were perhaps a million in all. At the end of the Würm glaciation 15,000 years ago, *Homo sapiens* developed a new culture and more advanced techniques. He was better able to exploit his ecological niche than his predecessors. There were perhaps 10 million of him. During the millennia that followed there was a new population explosion thanks to the Neolithic revolution, which made agriculture and husbandry widespread with the introduction of domesticated plants and animals. Society thus possessed much more abundant resources, and prehistoric man was liberated from the risks of hunting and food-gathering. In 5,000 B.C. there must have been a hundred million humans. Thanks to Neolithic development the population had increased tenfold in a few tens of centuries. Later it increased more slowly, for the very concentration of humanity provoked wars, famines, and epidemics, which until recently were often devastating. For example, the Great Plague that ravished Europe in the fourteenth century probably destroyed between twenty-five and thirty percent of the population.

From the eighteenth century onward progress in nutrition and hygiene led to another increase in the population, a tendency that has accelerated in the recent past with the advent of mass vaccination, the discoveries of modern medicine, and, above all, the rise in the standard of living.

In 1830 the whole of the earth's population reached its first billion. Reaching this stage had taken 4 or 5 million years, but only a century later the second billion was reached in 1930. The third was reached 30 years after that, in 1960, and the fourth 15 years later, in 1975. At the present rate of growth the human race increases every 3 years by as many people as were alive at the time of Christ, that is, 250 to 300 million. Today, mankind is increasing in number by the population of France every 6 months, or of a town like Bordeaux every day, according to Jean Dorst (1979). We shall be 7 billion at the end of the century, 14 billion in the year 2030. If nothing stops it, the world population will double every 32 years. Supposing that we can occupy all the landmasses with the same density as found in Japan—that is, 200 inhabitants per square kilometer—the world could support a maximum of 45 billion inhabitants, ten times the present population. If the overall demographic curve remains what it is today, we shall reach this limit around the end of the twenty-first century, according to Albert Jacquard (1978). At this rate humanity would multiply by a factor of 2,000 in 3.5 centuries and a factor of a million in 7 centuries, an absurd hypothesis! It is probable that regulatory phenomena, voluntary or spontaneous, will intervene long before these astronomical figures are reached. Already fertility is diminishing in numerous areas. This phenomenon was seen first of all in the rich industrialized regions. More recently it is beginning to involve underdeveloped countries, particularly the largest, such as China, India, and Brazil.

Nevertheless, demographic stability, or even reduction, is still on the faraway horizon. Taking into account the men and women already born, a ceiling will not be reached before the middle of the twenty-first century when the earth will have to support between 8 and 12 billion inhabitants—that is, two or three times today's figure, according to Jean-Claude Chesnais (1980; see also Leridon, 1982). This is a minimum forecast that is difficult to escape.

One thing remains certain: in order to provide for its needs, mankind must double or triple its agricultural resources in the next century. Will it be possible? The present situation leaves one in doubt. At the beginning of 1981 the world population was increasing by 200 people every minute—that is, there were 200 more births than deaths—while at least 30 people, including 10 children, died of hunger. For the situation not to become worse the rate of agricultural food production must catch up with the birthrate very quickly. This seems improbable when one considers the often irreversible degeneration of the natural environment and the generalized depression of production in the poorest countries due to lack of fertilizers, fuel, and machines, which have become too costly compared with the money available for buying them.

The Ecological Crisis

The demographic crisis represented by the generalized overpopulation of our planet is accompanied by an ecological crisis involving the degradation of our natural environment. This problem is due above all to the vast number of human beings; there will be 7 billion before the end of this century, a record among mammals with the possible exception of our commensal companions, the rats. In addition, modern man has become an enormous consumer, at least in some countries. But the ecological crisis is also due to man's technological progress. We have described elsewhere the important role of the fabrication and use of tools in the genesis and development of man (Ruffié, 1976). Armed with a spade and a fork, man is capable of imitating the mole, for we use tools to replace the organs nature has not given us or to greatly increase the power of those we possess. They allow us to intervene in the natural world every efficiently. the variety of our tools makes us polyvalent animals capable of the most diverse activities. They allow us to become specialists in many fields, according to our needs, without the necessity of organic modification.

We have described how the first men appeared in a hostile world without great physical strength and would certainly not have survived the onslaught of predators had they not been armed with stone tools. These were the first arms, used very effectively instead of the fangs and claws our ancestors lacked. The industrial use of stone and then metal allowed man to survive and even to spread. But these had little effect on the environment as long as they were wielded by simple muscular force (although already man's efficient hunting techniques led to the extinction of some species, particularly some big mammals in America). The appearance of domestic animals six or seven thousand years ago provided Neolithic man with new energy that, however, remained "physiological" in origin and within fairly modest limits. Man's ecological niche was little modified even if his frontiers became wider.

All this changed with the technological explosion that accompanied the industrial revolution 150 years ago. It was characterized by two trends:

1. An increasing sophistication of tools, which became complex, high-performance machines, built for a particular purpose and capable of accomplishing the most sophisticated tasks.
2. The use by these machines not only of muscular energy (perhaps aided by that derived from water or wind by means of mills) but of thermal energy, virtually unlimited in supply and derived initially from steam engines and then internal combustion motors. By transforming it into electricity this energy can be distributed over great distances at little cost.

The new era of mankind has little in common with times gone by, in which work depended on the muscular energy of man and animals. Now machines that can undertake the most complex and difficult work, precisely and without the slightest rest, constitute overwhelming extrasomatic extensions of our capacity. They have enabled us to enlarge our territory in a way that no one would have imagined two centuries ago. Today no part of the earth or the sea is protected from man.

Where once several years were necessary to clear a forest with axes, even by mobilizing armies of peasants, now machines fell trees and raze the forest in an instant. We can flatten mountains and deviate rivers because we possess the necessary energy (Dorst, 1979). Modern man utilizes 100 times more energy than prehistoric man — 250,000 calories per day in the United States, compared with 2,500 in Paleolithic societies, some of which still exist today. What is more, the earth is a thousand times more populated. Between 1950 and 1970, energy consumption doubled in North America and tripled in the European Economic Community, which started rather later than America. Since the beginning of the century the production of electricity has increased a hundredfold (Dorst, 1979). All countries, whatever their regime or ideology, share the same desire to model their standard of living on the American example and even to exceed it, as Khrushchev promised the citizens of the Soviet Union.

Even if it is often qualified as imperialist or diabolical, the American dream has its attractions. But taking into consideration the resources of the biosphere and the present or future capabilities of technology, it is impossible for our planet to accomodate seven or eight billion people living like Americans at the beginning of the twenty-first century.

Nowadays it is not particularly the magnitude of the changes the world is undergoing that poses the problem: it has seen many others. The trouble is their rapidity. No species has ever transformed its natural environment as brutally as our contemporaries are doing. Jean Dorst uses a simple comparison to illustrate this acceleration. We need not take into account prehistory which, as we have seen, had little effect on the ecological conditions in which our distant ancestors lived for a period of three or four million years. Let us simply consider the historical period, which begins with writing — that is, around the time of the Bronze Age — and has lasted a few thousand years. If we reduce the whole of this period of history to a single year, Louis XVI was guillotined on December 29, and at that time the energy utilized to exploit the environment had scarcely changed since the beginning of history. It was still derived from the muscular force of men and domesticated animals, from mills and from wood fires in the forges. On the same scale, the industrial revolution that is exhausting our reserves of fossil fuel has only lasted two days. On this scale, today is December 31!

Thus, seen from another planet, the industrial revolution becomes a geological phenomenon that by its speed takes on catastrophic proportions.

The State of Want

In 1972, while industrialized nations were growing in an unprecedented way, a group of experts who had worked for four years under Aurelio Peccei raised the alarm by stating that the principal raw materials of the world would soon be exhausted. The oil crisis that happened soon after made the pessimism of this report by the Club of Rome more credible. Ten years later their opinion must be taken with a certain reserve. If our resources and the way in which we utilize them remain the same, it is likely that many of the products that are indispensable for our industries would be in short supply in a century or two. But things are a little different in reality. Claude Guillemin of the mining and geological research bureau, Wassily Leontieff, the American economist, and many other people estimate that there still exist enormous quantities of raw materials, particularly in the depths of the oceans that cover the greatest part of the earth's surface. But access to them will become more and more difficult and their extraction more costly. Whatever the case, we should not fear a lack of basic resources if we exploit reserves that were hitherto inaccessible or unknown and if we invent new techniques and find means of substituting certain products. In addition, the deceleration in population growth that has been noticed in many places has produced a cooling off of markets and a reduction in demand, in spite of a continuing high supply. Ten years after the beginning of the oil crisis, which one might have imagined to indicate an exhaustion of natural supplies, the world is in a state of overproduction that is forcing the Organization of Petroleum Exporting Countries (OPEC) to freeze its prices to avoid stagnation of the market.

At present the crisis humanity is facing is primarily economic and political, related not so much to the exhaustion of raw materials as to the increase in prices and the necessity of developing ever more sophisticated and costly techniques. This irreversible trend is increasing the separation between the rich and the poor.

This situation has serious consequences for renewable resources, particularly in the domains of agriculture and food, which depend more and more on industrial techniques. Having first expanded very rapidly due to the speed of technological progress, which allowed exploitation of all the "useful" corners of our planet, our exaggeratedly broad niche is finally approaching its limits. Arable land is disappearing regularly, due to the combined effect of urbanization, highways, and drought, either natural or provoked by man, particularly by the destruction of forests. In the Sahel the desert is advancing by 8 kilometers per year. Those cultivated areas that have survived are overexploited and are beginning to show signs of exhaustion. Little by little our forests are disappearing. According to a report produced a few years ago by an American committee of environmental experts for President Carter, the world's forests are disappearing at a rate of 12 million hectares a year. The Ivory Coast had 11.8 million hectares of forest in 1956; in 1981 it had 1.5 million less. Every minute 5 hectares of tropical rain forest disappear, and with them the wild animals that live there. We are very rapidly losing our biological "capital," which was built up and diversified by evolutionary processes that needed a great deal of time.

In addition to this crisis in our basic resources we must now add the problem of pollution, which is changing the environment in industrial societies, often irreversibly. It is commonplace to talk of the dangers of atomic power plants, (which we shall discuss later), but it is easy to forget the equally serious problems caused over the last few decades by the use of coal and oil and the vast quantities of industrial waste that poison the air of our large cities or regularly destroy the fauna of our waters. We tend to forget about the aggression represented by certain products that we manufacture, such as the plastics that are so widespread now. These macromolecules are not the result of a long biological process but are artificially synthesized. They were produced too recently and too quickly to have allowed the co-evolution of bacteria capable of degrading them and reintroducing their constitutents into a natural cycle. They represent a form of inert material, indestructible and increasingly voluminous, that even if reduced to tiny particles will finally occupy an exaggeratedly large part of our biological environment.[1]

Unless there is a radical change in policy at the international level, it is difficult to conceive how our natural environment can resist for very long the brutal aggression of the 14 billion industrialized *Homo sapiens* that we expect to be on the earth after the year 2030.

The Disparities

In fact, the problem that will arise is even more complex. Although the crisis affects all countries in one way or another, it is not uniform in all. First of all, its demographic character is different. We have already seen that the present rate of growth will probably continue to the middle of the next century. This phenomenon is not geographically homogeneous. Certain parts of the globe are overpopulated, while others are depopulating. Until the nineteenth century, Europe, and to a certain extent Asia, was growing the most rapidly. This led Europeans to occupy America where, together with their black slaves, they practically replaced the native Indians, except in the most isolated areas such as the Amazonian forests and the high Andes, where the natives are relatively protected and have survived. The white man has also invaded Australia, New Zealand, South Africa, certain parts of North Africa, Siberia, Alaska, and so on.

Today this movement is reversed. European population is diminishing, as in West Germany and the Soviet Union, while Southeast Asia and certain countries in tropical and North Africa and in Latin America are experiencing a marked increase. This has led in turn to a mass of immigrant workers, often people of color, immigrating to Europe to search for work. Many established themselves permanently with their families.

1. The injection of "new" organic molecules into a preexisting ecosystem sometimes leads to the accumulation of these molecules in the form of waste when the ecosystem fails to reintroduce them into a new biological cycle. The consequences of the recent introduction of cows in Australia are well known. There are no indigenous coprophagous insects capable of destroying their dung as in other countries, so it persists indefinitely until the meadows are suffocated. The accumulation of organic matter that had not been entirely degraded has given us our reserves of oil and gas.

This movement has serious consequences. At the biological level we can certainly foresee a total change in the near future, not of the overall genetic patrimony of mankind, but of its distribution. In other words the genetic polymorphism of man will not increase or decrease but will be spread differently. For example, in 1970 the populations of France and Mexico were about the same, about 50 million each. At the beginning of the twenty-first century Mexico will have 200 million inhabitants, France only 60 million. Thus at that time the world will have nearly four times more "Mexican genes" than "French genes."

Of course, we do not know of any specific Mexican or French genes, but the two populations do not have the same frequencies of alleles. Thus man is certainly headed toward a different distribution of gene frequencies (Jacquard, 1978). Similar phenomena, perhaps less abrupt and spread over a longer time, must have happened on many other occasions — for example, when Neolithic societies began to replace Paleolithic man and when the Indo-Europeans migrated to the west. More recently the barbarian invasions overwhelmed the Roman Empire. Very recently, the conquest of America by the Europeans — Spanish and Portuguese in the south, English, Dutch, and French in the north — together with the import of black Africans revolutionized the original distribution of populations in the New World in less than two centuries.

This disparity between different regions can be observed at the economic level. Natural wild populations generally have fairly constant numbers as a consequence of severe regulatory mechanisms that adapt the number of individuals to the available resources. Thanks to his intelligence and his technology, man has succeeded in breaking away from these regulatory constraints, an obvious success on the moral level. Nevertheless, it poses a peculiar problem, that of establishing new voluntary limits based on criteria established according to ethical views.

At the present time the countries that have the highest birthrates are almost always the poorest. In poor populations that have no form of social protection, a child is the only guarantee for the future of a family. It is the family's working capital, its old-age insurance, and its health insurance. On the other hand, in rich countries with well-developed social systems the child is no longer profitable but more of a liability. In general, birth control develops in parallel with living standards. Overpopulation drives countries into a vicious circle. Poverty causes high birthrates that absorb a large part of the productivity of the nation, thus paralyzing productive investment and blocking expansion.

In 1960 the ratio of average income in the poorest nations and that in the richest was 1:10; it is now 1:14 or even higher.[2] The gross national product (GNP) *per capita* is about $17,270 per year in Kuwait; $16,950 in Qatar; $15,590 for the Gulf States; $14,240 in Switzerland; $12,280 in Luxembourg. In Sweden, Denmark, and West Germany it is over $11,000. On the other hand, fourteen of the most densely populated nations have an annual *per capita* income of less than $200!

The poorest countries scarcely manage to survive and are constantly on the brink of famine. Even if one could hope for increased productivity through improved agricultural techniques, it is hardly likely that these countries could overcome the needs implied by the foreseeable growth in their population. These critical areas pursue an unequal and endless fight to balance the evolution of the available resources and that of the needs of their inhabitants, for the needs are constantly ahead of the resources. Under present conditions these nations cannot even be sure of being able to buy the foodstuffs indispensable for their survival from abroad, for all imports require hard currency, and their balance of payments — the difference between what underdeveloped countries export and what they must buy from the industrialized nations — is worsening continuously. It is now the worst it has been for thirty years, so that the debt of Third World countries has reached astronomical levels, far above what they are capable of repaying. Since 1982, just paying interest has absorbed 11 percent of the money they have earned from exports, compared with 7 percent in 1979. What will become of these enormous insolvent areas? The margin that protected them from absolute catastrophe is becoming narrower and narrower while the gulf between them and the rich countries is widening (Dadzie, 1980).

At present the United States, the Soviet Union, Japan, and the European Economic Community, which together represent about a billion people, or rather less than a quarter of the population of the world, consume 90 percent of its electricity. The United States alone uses half the fuel extracted from the land or the sea. Overall the North Americans, who represent 5 percent of the world's population, use 60 percent of its raw materials and 35 percent of its foodstuffs.

The city of New York utilizes each day as much fuel as the whole of India. The comparison is even more shocking if we consider the poorest countries. For instance, the food consumed in America would allow from ten to fifteen billion people to be fed at a standard familiar to the people of Upper Volta, which is very low.[3]

---

2. From the report *Reshaping the International Order* (1976), drawn up by a group of twenty experts from Third World and industrialized countries, under the direction of Jan Tinbergen, Nobel laureate in economics. The report concludes that it is necessary to create an international solidarity fund, a sort of worldwide Marshall Plan.

3. One might add that the industrialized nations spend more calories each year to feed their cats

The Four Worlds

At present one can, schematically, divide the countries of the world into four groups, which are growing further and further apart.[4]

1. First of all there are the superpowers, possessers of technology and raw materials: the United States and the Soviet Union. Tomorrow there will be China, and perhaps even Brazil. Their chances of survival are good in spite of the crisis they are undergoing just like the others. For the moment the Soviet Union and the United States are in the vanguard; each has its customers and its satellites, often created by the scars of the second World War. We shall discuss the consequences of this division of humanity later.

2. The second group consists of the other technological nations, which have fewer raw materials. These traditionally include the countries of Western Europe, particularly those of the European Common Market, and a few people's republics, such as East Germany, Czechoslovakia, and Hungary. Outside Europe, Japan was for a long time the only nonwhite country in this category.[5] But very recently other young industrial nations have followed Japan's example. Almost all of them are in Southeast Asia: South Korea, Taiwan, Hong Kong, Macao, the Philippines, Malaysia, Singapore, Thailand, and Indonesia. They can provide goods of high quality and low cost because of the technical development of their production (with a high degree of automatization), but also because of the low salaries paid and the almost total absence of social welfare. The difference in the cost of labor between the European Common Market and the new industrial countries of the Far East is 1:20, and between Europe and China 1:50.[6]

Thus these newcomers constitute formidable competitors for European industry, of which whole sectors have already collapsed — such as textiles and shipbuilding — or are seriously threatened. The survival of the second group of nations depends on the supply of raw materials, and they are obliged to establish, by force or reason, privileged relationships

and dogs than are available to human populations in the poorest countries. See particularly Fourastie and Bazil (1980).

4. This division is obviously schematic. There are some intermediate cases, such as India, where the situation could evolve in either direction according to circumstances. But this reservation in no way invalidates the basic classification, which corresponds well with the contemporary world situation.

5. Israel is a special case, a high-technology nation without raw materials, allocating a large part of its budget to defense, whose young industries are not a serious threat to European or Asian countries. But here, once again, things could change with the international situation.

6. According to the French review *50 Millions de Consommateurs, 125,* May, 1981.

with their suppliers, often former colonies that now enjoy independence, although of a purely formal nature.

In addition, medium-sized industrial nations have often been forced, through economic or political necessity, to follow in the wake of one of the superpowers, which becomes a sort of "protector." For them, and in spite of official statements, neutrality has no place. Declarations of dignity or sovereignty are merely historical dreams or electoral arguments.

3. Next are countries that possess raw materials but little technology. At the moment, the most representative are the oil-producing nations belonging to OPEC. Not only do they possess indispensable products for the industrial nations but also, in the last few years, a substantial part of world finance, which allows them to modify the equilibrium of whole regions according to where and how their floating capital is invested.

These *nouveaux riches* nevertheless do not have a guaranteed future. Most of their oil reserves will be exhausted within a few decades. Let us take the example of Algeria, which today balances its commercial budget to the tune of some 95 percent by selling oil and gas. In this way it has been able since independence to make considerable progress in sectors such as schooling, agriculture, housing, and public health. But what would become of Algeria if it did not succeed in its technological development and if in thirty or forty years it was exporting at less than 10 percent of its present level? The same problem arises for all nations that live substantially by the sale of nonrenewable raw materials. Some of them have taken measures for their future prosperity, while others have paid little attention to this. La Fontaine's fable of the grasshopper and the ant is still pertinent.

4. Finally there are countries that have neither raw materials nor technology and are now termed the "Fourth World." They are the most numerous. Their only resources are agricultural and are almost never sufficient for subsistence. They are forced to export half their meager production in order to import oil and indispensable manufactured goods such as chemicals and machines. They become ever poorer due to their worsening balance of trade, as explained earlier, and their inability to control their demographic explosion. There are degrees of misery just as there are degrees of wealth, and not all the poor are housed in the same conditions. At the bottom of the scale are the thirty-one least advanced countries, two-thirds of which are in Africa and one-third in Asia, the South Pacific, and the West Indies. Twelve of them are French-speaking countries. It is difficult to see how in the light of the current international context we can hope to save these castoffs of our planet, which many people prefer to ignore, without an immense "survival operation" (Drouin, 1981b). The nations that come just above the least advanced countries on the scale face a situation that is hardly more encouraging in the medium term. The same danger threatens them. Their progressive debts will probably force them, in their turn, toward absolute poverty, after a certain latency due to the existence of meager reserves.[7] But their "emergence"—a term now often used—encounters another, still more formidable obstacle. Since the industrial revolution and its natural offspring, colonialism, many native populations have been forced, often against their will, to replace their traditional agriculture, which at least ensured a minimum food production, by monocultures such as sugar cane, coffee, or cotton. Because of their high market value compared with basic food production for local consumption, they ensure larger profits for the rich land-owners.

The division of the world into four very different groups from the point of view of wealth, living standards, potential, and future prospects has nothing to do with the traditional divisions into bourgeois and socialists. It is a new phenomenon that Marx had not foreseen.

In 1975 the world had 4 billion inhabitants, with 1 billion living in security while 3 billion were poor; of these, 1 billion were very poor. Of the 7 or 8 billion inhabitants of the earth at the end of the century, there will be 2 billion in security, 4 billion poor, and 1 or 2 billion starving. This situation could rapidly become explosive if no remedy is found.

THE PERVERSION OF THE SYSTEM

Today it is possible to analyze precisely the causes of the crisis we are facing. They are complex, but the direct responsibility of the richest nations cannot be denied in the present situation. The crisis was not historically inevitable but is rather due to the perversion of the system in which we live. Let us see how. During the whole of prehistory and the greatest part of history, man's activity was essentially aimed at covering his legitimate needs and raising populations out of the physical and physiological misery in which they were struggling. It was a question of life or death. Our ancestors had few resources and only modest reserves. Their existence was precarious. Since the Renaissance, but particularly since the industrial revolution, the standard of living in developed countries has grown considerably. Production was soon deflected from its essential aim, that of fulfilling natural necessities, in order to become a profit-making machine (Robin, 1980). Human activity is now programmed as a function of this profit by those who enjoy

---

7. Some simple figures give an idea of the tragedy of the situation. In 1930 the Third World nations *exported* 10 million tons of cereals. In 1978 this same Third World had to *import* 60 million tons, and by now this figure has been far exceeded.

the power and the fortune. The end is becoming confused with the means. Nowadays the value of an enterprise is judged not by its social utility but by its profitability. The same is true of everything concerning our environment. While an increasing proportion of humanity goes hungry, we continue to base the price of land not on what it can produce but on its real-estate value. Land is no longer an instrument to satisfy our needs but a target for speculation. To ensure the functioning of the system and constantly stimulate production, consumption has to be increased, almost always in an arbitrary fashion. In the rich nations overproduction has only been maintained by a growing overconsumption that represents an enormous waste.

Let us take, for example, one of the commonest and most costly of our aberrations, our unbalanced nutrition. In Western Europe we generally consume too much fat and sugar at the expense of meat, fish, green vegetables, and fruits. We now know that a diet rich in fat and starch favors the development of cardiovascular and metabolic disorders. In spite of this, most television advertising concerns foodstuffs based on fat and sugar. These two categories are easy to stock and preserve and have been widely commercialized, providing high profits, but to the detriment of the consumers who, becoming ill, then become consumers of drugs. Thus an economic cycle has developed, taking away part of the national income and transferring it, with the aid of advertising, first to the food industry, through the intermediary of healthy people, and then to the pharmaceutical industry, through the same consumers when they have become sick and dependent (see de Rosnay and de Rosnay, 1979). This is no small problem if one remembers that almost one-third of hospitalizations concern cardiovascular diseases, which represent more than a third of the causes of invalidism or death in people over forty.

This situation seems even more paradoxical when one considers that we possess today the means necessary for the effective prevention of many of the diseases of civilization, as we begin to understand the factors that precipitate these diseases.

In other domains artificial needs have been created, such as those related to changes in fashion, which is considered an external sign of social status. Our society of wastage is a society based on disposable goods (Pelt, 1977). A few years ago everything was recycled. The village grocer charged a deposit on bottles, which were therefore brought back to him scrupulously. Nowadays we throw away not only the wrappings but also the objects, even before using them. We live in a world of debris and wrecks. For the biologist such a concept is absurd and suicidal. The living world tolerates waste very badly. Natural food chains are based on constant recycling (Dorst, 1979). If this were not the case, life would have disappeared from the surface of the earth long ago; but today, because of the requirements of profitmaking, it is better to plunder the environment than to recover those substances we already have on hand in enormous quantities in the waste of our industrial society, and that could represent an important source of raw materials. Given the technological power of modern nations and their financial resources, the logic of profit has become disastrous. Nature can no longer defend itself. It is being destroyed on a massive scale.

The New Powers: The Multinationals

In recent decades real power has passed from the hands of politically constituted states to those of the multinational corporations, which have spread out from the capitalist countries, particularly the United States, invaded the Old World, and are now even penetrating the people's republics, so desirous of technology and consumer goods. Their resources largely exceed many national budgets.[8] The integrating power of these new conglomerates is considerable. They cross frontiers, have little respect for the real needs of people, manipulate prices, control markets, and speculate in foreign exchange, flouting servile or bypassed governments.

The multinationals decentralize their production lines, building factories where labor costs the least by virtue of low salaries and absence of social benefits, while selling their products where the market is most favorable. They are responsible for the young industrial countries mentioned earlier, which now threaten European industry. Goods are manufactured in Macao, Taiwan, and even Hungary for reselling in New York, Paris, or Geneva. But the natives of Macao, Taiwan, and Budapest will have little chance to profit from the objects they have manufactured. They are destined for others, and the standard of living of the workers is scarcely bettered. Only the native bourgeois who have assumed the function of the compradors of nineteenth-century China will benefit.[9] Similar dangers threaten the agricultural scene since multinationals have begun to control the markets. They dispose of means that are infinitely more powerful than those of the old latifundia.

Today three multinationals control 75 percent of the market for bananas, and five others 75 percent of the market for cocoa;[10] but a new threat is appearing that could have even more serious consequences. As

8. According to Jacques Robin (1980), the 1980 budget of the fifty most important multinationals exceeded the GNP of Holland or Belgium. Since then the difference has increased.
9. As a typical example, certain multinationals have installed some of the most advanced factories in the world in Indonesia, especially in Java. Completely automated, they have created very little employment, and no improvement in the living standards of the population, more than 80 percent of whom continue to live a life of absolute poverty (Lesourne, 1981).
10. Editorial, *Le Monde*, August 22, 1981.

the price of oil increases, people are beginning to envisage the manufacture of fuel from biological products, with alcohol as an intermediary. This requires large-scale cultivation of rapidly growing plants. Faced with the pressure of a particularly open market, the temptation will be great to valorize enormous territories by devoting them to this new "energy" agriculture, designed to produce "green" oil. These practices could constitute a veritable genocide for the poorest populations, which are least protected.[11] In terms of international integration, the multinationals have realized in less than a generation what Christianity took two thousand years and Marxism a century to achieve. At present the power of the multinationals is enormous. On account of them, European nations enjoy less liberty and guarantees in relation to America than do the fifty states of the Union, protected as they are by the Constitution and federal law. Washington must pay more attention to Mississippi or Alabama than to Belgium or West Germany. Having only one *raison d'être* — profit — the multinationals ignore the human context. One should not be surprised at this, for such problems do not concern them. They are preoccupied with other business, involving technology and profitability.

This new form of creeping colonialism, which began in the second half of the twentieth century, is probably more to be feared than traditional colonialization with military and administrative occupation, for it is more efficient and less obvious. And it is also more profitable because it does not have the same obligations that the mother country used to have toward its dominions — such as public health, schools, transport, and telecommunications.

This economic and technological imperialism is always accompanied by cultural imperialism. As far as concepts, innovations, and decisions are concerned only one way of thinking is admitted, that of mid-nineteenth-century England, where a form of Darwinism developed that exceeded the boundaries of biology to influence social and political life. In its wake and in a spirit of imitation, well known throughout history, we have built an absurd world where overconsumption rubs shoulders with famine and overproduction with stagnation, and where for the first time economics come before culture.[12] We should be careful not to ignore the danger of a system that in the end could shake even our most solid structures.

11. When Volkswagen, for example, invests in Brazilian agriculture, as it is doing today, it is more concerned about the profitability of the investment than the nutrition of the native population. The logic of the system requires it. Mgr. Helder Camara, archbishop of Olinga and Recife, in a conference at the Collège de France on March 16, 1971, emphasized the threat of such a policy to the local population.
12. The multinational phenomenon is now seen in Eastern countries, which are using these companies to try to penetrate Western trade. See the report by McMillan (1979) and Grapin (1981).

THE DANGER TO NATIONS OF SOCIAL DISINTEGRATION

The first signs of social change have been visible for a long time, particularly in the industrialized nations, which are engaged in a merciless commercial battle that involves a systematic lowering of salaries (or, in the same spirit, a reduction in paid working hours) and the curtailment of protective measures. Every business must try to remain competitive toward its peers at all costs. In order to avoid the demands of the workers, labor is imported from poor countries and forced to live under the permanent threat of being sent back home as the result of a simple, administrative decision. Such workers are much more docile and their demands more moderate. Almost everywhere temporary-employment bureaus have sprung up to provide qualified labor without giving any real form of protection; they represent the slave traders of modern time. The continual search for more advanced technology that economizes labor also worsens unemployment unless working hours are reduced in parallel. Unemployment, real or camouflaged, is a most serious phenomenon, which we should consider for a moment.

Unemployment

We often forget that man is above all a social being. His fundamental requirement is to occupy a specific place in the group to which he belongs and to be able to identify himself in relation to his neighbors, to whom he gives and from whom he receives. This is called social integration; it allows our biologically unfavored species to achieve exceptional success. Man's intelligence confers on him an extraordinary facility for sociability and communication. He has thus been able to negotiate the most perilous barriers in his history successfully in spite of his physical weakness.[13]

The first men faced attacks by the big carnivores with expertly manufactured stone weapons, developing group strategies that enabled them to become formidable predators. Later, thanks to the domestic use of fire and the construction of clothes and suitable lodgings, they were able to survive the ice ages, although they were better adapted to the tropical climate in which our species was born. Later still, social development made agriculture and the rearing of animals possible, for these are typically social activities. This Neolithic revolution was responsible for a considerable increase in food resources, which allowed reserves to be

13. Olympic running champions reach a speed of twenty miles an hour. A moderately frightened chimpanzee can do twenty-five. If they could participate in sports events, many primates would beat us, except in sports requiring intelligence and strategy, or that have complex rules, like fencing or team sports. We shall never see a baboon rugby team.

stored in the form of both harvested crops and herds of cattle. Famine became rare. What is more, Neolithic life encouraged man to settle and change from nomads into sedentary groups. Thus the first towns grew up, then the first empires. Since what was produced was not always the same everywhere, trading grew up, which, whether peaceful or warlike, resulted in cultural exchanges.

Still later, the social order helped humanity overcome another danger — that of the epidemics, such as the plague that devastated fourteenth-century Europe, killing as much as a third of its population. Cholera in the nineteenth century and influenza at the beginning of the twentieth, spread partially due to the extensive overseas exchanges of those times, were just as lethal. They disappeared as the result of collective hygiene, better health education, medical progress, the spread of mass vaccination, and the growth of international cooperation. Recently, many of the great nations of Europe and Asia have recovered from two particularly devastating world wars within a few years. Countless eras were marked by a particular portion of humanity being affected by a series of calamities at the same time, such as war followed by famine, epidemics, poverty, and ruin. Recovery has been due largely to the perfect social integration of populations, with each individual having a particular place so that all coordinate their efforts toward the same goal. This would have been impossible with a mass of solitary individuals, each working for himself according to his own desires.

Socialization has avoided the extinction of man and has made him able to evolve progressively and culturally, a process based not on biological changes as with other animals but on the accumulation, exchange, and perfecting of techniques and cultures. Man's principal value is that of being social. "In spite of his individual intelligence, man would not have become what he is if his powers had not been considerably increased by his social life. Man has changed the face of the earth because the most intelligent beings that have ever appeared in the world are capable of living in societies. These two complementary aspects are characteristic of humanity" (Vandel, 1968).

Thus the success of our species in the face of multiple perils has always been due to its socialization. Any human society that lets its structures be eroded away is threatened in its very existence, and history offers us many examples of civilizations that have disappeared. We are now facing another phase of perils, due to the very rapid acquisition of new technology that has upset our ancestral situation, based on a long historical evolution, and has not taken into account the basic requirements of mankind.

Nowadays the most obvious sign of social disintegration is unemployment and its consequences, such as family instability and delinquency. We too often think of unemployment as a simple material drama, causing economic hardship that can be rectified by giving monetary benefits. This attitude avoids the true nature of the problem, for unemployment is above all a social drama. An unemployed person is excluded from his group. He has no further role to play and has lost his place. He becomes marginal, almost always against his own will, and loses every man's fundamental right to occupy a clearly determined function.

No society that has to bear a high level of unemployment for too long can survive. In 1975, after two years of crisis, the world contained some 300 million unemployed. Since then this figure must have risen by at least 20 million per year. Today there are probably more than 500 million unemployed. If this continues there will be between 1 and 2 billion by the end of the century. During the same period the profits of the large industrial societies have increased regularly.[14] It is unlikely that the civilized world can survive such undermining of its structures. Apart from an atomic war, unemployment constitutes the greatest short-term danger that threatens us. It is the modern form of plague.

The Technological Answer

Many people criticize technology when discussing the rise in unemployment, thinking that robotization and computerization lead to the loss of jobs. This argument is fallacious; otherwise, logically, trucks and trains would have to be replaced by wheelbarrows and backpacks, which would mean work for everybody but would cause an immediate economic collapse and generalized poverty. In fact, what is of prime importance is the creation of wealth, or rather, wealth that is useful to man; for in the end one can only consume what one produces and, as far as production is concerned, modern technology is characterized by extraordinary efficiency. Today machines can carry out the most complex programs and the hardest forms of labor 24 hours a day, 365 days a year. They need no rest and a minimum of human intervention. Factories equipped with such machines have a productivity that could never have been imagined a few decades ago. This development has only just begun. More and more, machines will do the sweating to increase productivity, while man will work less. (This image, which is probably an accurate one of tomorrow's world, was suggested by Jean-Pierre Dumont.) But this poses a new problem, at least in a liberal economy: that of the distribution of wealth. For a rise in productivity should not simply be reflected by increased profits for a few industrial

14. This is true on average, although certain sectors especially affected by crises, such as the automobile industry, have suffered recession. At the beginning of 1982 official figures referred to 10 million unemployed in the European Economic Community.

chiefs, bankers, and shareholders; the whole of the community should benefit from it.

The first way in which this can happen is to reduce working hours, something that should be an inevitable part of technological progress and permit the creation of new jobs. Contrary to what some moralists proclaim, not always in an entirely disinterested way, work for its own sake has no intrinsic value. Its value lies insofar as it allows us to meet our needs and particularly to integrate individuals into their social background. But paid work is not the only form of integration. There are many others that involve free time, and we shall speak of them later.

Retirement

In addition to the unemployed, who are, in a way, "official" marginals, there is another category that is more difficult to acknowledge: the retired people who often also live on the margins of society. The phenomenon of retirement is a recent one. The number of retired people has increased considerably in a short time for two reasons:

1. The rapid increase in life expectancy. For Europe the following mean figures may be quoted: in the seventeenth century a boy's life expectancy was 27 years, a girl's 28. In 1900 these figures were respectively 43 and 47 years; in 1945, 62 and 68 years; and today they are about 70 and 78 years. These increases have thus been regular, while the difference between the sexes has also increased from 1 year in the seventeenth century to 8 years today.

2. The trend toward a reduction in retirement age, which will become even more marked before the end of the century in most industrial nations. This has contributed to reducing work time. Early retirement is related to generalized automation and increased productivity. This tendency will cause a modification of the sociodemographic situation that can already be detected.

In the past men were born, were educated in their families, their villages, and their schools before adopting a trade and usually dying at an age when they were still fully active. Today many workers who reach retirement age still have a long period of life before them. Even if economic measures have been taken to provide decent material conditions for most of them, nothing has been done to ensure their integration into the social group. At present, retirement is often a form of social death, and the retired person is excluded from society.

At the sociological level our model is the one inherited from the beginning of the industrial era, a century and a half ago, which apart from certain material benefits provides nothing for old age; for when the model was developed the problem of a long life after retirement was hardly an important one. No governments have been able to adapt this archaic model to modern conditions.

Some retired people, who have the good fortune to live with their families, survive this last stage of their existence very well. This is particularly true in rural areas, where old people always have their place and their usefulness in the exploitation of a farm. But in urban circles, where the family is reduced to the basic nucleus of parents and young children, the perspectives are bleaker. Retired people remain on the margin of society, isolated and useless. Frequently they degenerate quickly and disappear. More and more find new social structures in sickness, particularly of a psychosomatic nature. They become part of a "health network" that gives them something to live for, a well-defined place that adopts them and protects them. When they enter the hospital they become someone again, for they are once more part of the social pyramid. This solution, or the related one of putting old people into nursing homes where they can watch television while waiting to die, is sociologically absurd, morally indefensible, and economically ruinous.

In fact now is the time to rethink the whole policy of retirement, with the aim of giving people who are no longer working not only a minimum standard of living but also real social integration. For this they should not be deprived of responsibility and of such activities as they may desire. A graded form of retirement, "à la carte," leaving a certain amount of choice to the persons involved, would certainly be the most democratic, humane, and advantageous solution for society. This would mean in some cases delaying retirement age rather than advancing it, thus allowing a worker to remain active as long as he desires. Present day Malthusianism about working hours — which are sometimes thought of like a fortune that should be divided equitably — represents real nonsense. In modern society the most important thing is not to distribute work sparingly, as food rationing coupons were distributed during the war. No, it is to create and offer employment so that no one is excluded from the social order. Even part-time activity by retired persons represents a profit for everyone. Today we live in a society that is turning its back on the best-established principles of anthropology.

Free Time and Quality of Life

Whatever method is adopted, a reduction in working hours, an increase in holidays, and a lowering of the retirement age will pose a new problem — the management of our free time.

The Western work-ethic rat race, nowadays just as prevalent in the

East as in the West, will give way to the duo of work and leisure, the second of which will assert itself more and more over the first. The free time thus created may be devoted to culture, such as art, theater, or reading, to sports, or to social and family activities, which have been neglected in modern times.

We should give more importance to exchanges, journeys, and communication between men, nations, and civilizations — the things lacking most in today's world. People's intellectual level would be improved and their consciousness of belonging to a rich and diversified community would be enhanced.

Part of the new resources created by increased productivity would be consecrated to the creation of indispensible services designed to improve the quality of life and ensure better social integration of those groups that are at the moment neglected. This would involve advising and educating young people looking for jobs, and giving some material help, but also moral assistance, to old, solitary, handicapped, and sick people, who would be kept as long as possible in their normal home surroundings. Social assistance, promotion of culture, and household assistance would create many jobs oriented toward specifically human activities that no machine could ever fulfill. Nowadays we can play games with small home computers and even have "dialogues" with them. They can answer many questions, give us lists of figures, solve many problems, and help us organize ourselves. They allow us to kill time and amuse ourselves, but they in no way compensate for solitude or replace brotherly companionship. One may love one's friends or even one's cat or dog, but it is difficult to love a machine. The human being needs love; he needs to give it as well as to receive it.

Our age has brought security, but it has not brought hope. Thus we must look upon life in quite a new way. This implies both changing our habits and modifying our social relationships.

The International Danger: War

We spoke earlier about the division of modern humanity into four worlds. They are not obliged to remain permanently immobile; each, in its own way and on its own scale, is in turmoil. First of all the superpowers, the Soviet Union and the United States, that have carved out for themselves enormous empires using ideology as a cover. Their empires are economical, political, and military, and other countries have become their satellites, serving them willingly or not. The world is cut in two. This split is also found within the so-called nonaligned countries.

But the two superpowers go even further. They are continually trying to nibble away at each other's zone of influence wherever they can, thus provoking numerous local conflicts of greater or lesser importance. The war between Iran and Iraq that broke out in 1980 was the 127th armed conflict since the so-called peace of 1945. According to the Brookings Institution of Washington, D.C., they have been responsible for thirty-two million victims. This permanent tension makes most nations, even the poorest, pursue ruinous arms races. They invest a growing proportion of their national income in useless expenses under the complacent eyes of the great powers, whose customers and prisoners they remain. This policy supports the strong and enriches the rich while keeping the poor in their material and moral misery. It is striking to note that the only common denominator that unites the four worlds is a frantic search for overarmament. This suicidal pathway that our civilization is treading is due to the fact that no country has yet managed to free itself from social-Darwinist ideologies. All believe that their survival and progress necessarily involve a struggle with others, leading to their ultimate subjugation and even destruction. This behavior, which makes us live on a volcano, is a major obstacle to development. If all nations could accept the lessons of history and admit that nothing can be resolved permanently by armed conflict — that is, if they would abandon a Darwinian view of the world that only recognizes the forces of conflict in nature — the crisis in which we find ourselves plunged would soon be resolved. Military expenditure is today the principal obstacle to progress. Humanity will only be able to progress in its evolution if it accepts generalized disarmament. We are still far from this goal, and it is difficult to see how the double hegemony implied by the Yalta agreements could be questioned, because it is frozen into a balance of terror. In 1981 experts estimated that the atomic powers had stockpiled about a hundred thousand nuclear warheads, each fifty to a hundred times more powerful than the Hiroshima bomb — equivalent, in sum, to four tons of TNT per human being, enough to kill all contemporary mankind several times over and perhaps even all life on earth.[15] For the first time in his history man now has the power of collective suicide, a thing no other species is capable of. To achieve this result ruinous investments were necessary. Even if these bombs are never used, as everyone must hope but of which no one is sure, their manufacture and stockpiling will have frozen a large part of the world's economy and been partly responsible for famine. Such is the price of the balance

15. Today man is probably stockpiling the equivalent of ten or fifteen million Hiroshima-sized bombs. According to experts, by 1990 at least fifty countries, including some of the poorest, will be able to field nuclear weapons. What will happen to the world the day that some irresponsible petty dictator rejects the domination of his protectors and decides to launch a nuclear attack on his neighbors?

between the two great rivals. What was terrifying, now becomes absurd.[16]

The Phenomenon of Protest

The failure of the system is obvious and only gets worse with time. In the West with its capitalist crisis, some people are ruined and others make fortunes, thus provoking an overall imbalance in the world economy. Multinationals, which now represent the real monopolies, have put an end to free competition. "Capitalism was led by the effect of its own logic to break the laws of supply and demand, to reduce free competition to the level of extinction, and thus to reject the principles on which its system was founded. Capitalism can only live by eating itself" (Mitterrand, 1978). We might wonder about the limits of this self-cannibalism. The new industrial nations in the hands of large corporations are beginning to ruin the old industrial countries that were manufacturers (and therefore competitors) but also customers. These former customers, impoverished by the crisis, buy less and less and have difficulties with their payments. The Fourth World cannot be relied upon to replace them as customers, for we have seen their dismal prospects, and their means are constantly declining. Many nations that are enjoying prosperity today may sink into poverty tomorrow.

It would be illusory to think that nations could continue to base the whole of their policy on unlimited overproduction in a world incapable of continuing to overconsume or even to consume normally. We are trapped in a system that can only spell bankruptcy. And what is there to say for Marxism and Leninism, whose failure is of the magnitude of the enormous hope that they kindled?

So it is not surprising that in a world blocked in the East as in the West the youth upon which the future depends should be disillusioned. Our youth are becoming more and more conscious of the impasse in which we are caught, and they refuse to go further. Speeches seem useless to them, for they rarely result in action. Youth has lost confidence in traditional methods and cannot rely on people in the establishment to change direction. They have always made promises but almost never kept them.

This defiance — in the East of the single party and its leaders, in the West of pluralism and political games — has led to protest movements of several different types.

16. Without reaching the level of a widespread nuclear holocaust that would make the superpowers hesitate, *local* nuclear conflicts are not improbable within the next ten years. Let us not forget that medium-sized nations like France or Great Britain could be wiped off the map today by four or five large hydrogen bombs. Only one would be necessary to eliminate nations like Iraq, Syria, or Israel.

Peaceful Protest: Ecology

First of all there is peaceful protest, which in Western democracies has been *ecological.* Vincent Labeyrie (1978) wrote, "Ecological protest is a disordered perception of the global crisis of capitalism."[17] Faced with an absurd situation, the movement is quite sane but not always realistic, and it is unable to resolve its contradictions.

The ecological movement refuses to sacrifice our natural environment for the quick profit of a few people, but we must clearly realize its nature and its limits and not fall into the romantic trap of the "utopia of the oil lamp." Two points should be emphasized:

1. Not all changes are necessarily bad. Since the dawn of time nature has been changing. These changes have permitted evolution. Present-day modifications, however disturbing they may be, are minor compared with the upheavals that the biosphere has already undergone. For instance, in the primitive period of the world the environment was very *reductive* in nature. When clorophyll-based synthesis began it became *oxidative,* so that living matter abandoned its anaerobic form of life for an aerobic one. This transformation had incalculable consequences by allowing the emergence of higher organisms with large energy requirements that could only be provided by aerobic metabolism. Or later, Pangaea emerged and then broke up into multiple, varied continents. Other topographical and ecological changes accompanied the ice ages; and what about the Neolithic revolution, which transformed enormous wild areas into fields?

The dynamism of ecological changes provoked changes in life. Without all the modifications we have just mentioned, the world would still be populated by bacteria.

Mankind represents one of these modifications. By putting nature and society, man and environment in opposition, Western philosophy has long rested on an ambiguity, as was emphasized by Serge Moscovici (1968). In reality man is part of the contemporary natural environment, even if his emergence has greatly modified this environment (although much less than the arrival of the first cell capable of photosynthesis). The "victory" of man over nature, a commonplace of the speeches of Marxists and liberals for a century, has little significance for a biologist. Any new species that comes to occupy a new niche always creates a

---

17. In fact this quotation is correct but rather restrictive. Although the degradation of the environment was first reported in the large capitalist nations as a backhanded result of the lust for profit, it would be wrong to believe that the socialist countries, themselves lured on by the mirage of consumption, have escaped it. And the two superpowers, the United States and the USSR, although technologically the most advanced, still possess large unspoiled regions thanks to their enormous areas and their environmental protection programs.

"victory" for itself compared with what existed previously. Man with his very particular ecological niche is integrated in the movement of evolution just as much as other living organisms. The ecological movement as it is constituted today sometimes seems to ignore evolutionary processes. It appears to be preoccupied with the protection of "traditional" nature, such as we can imagine it on the limited scale of the history of mankind, but which represents only a tiny fraction of time in the history of the world.

2. In fact evolution, either biological or cultural, cannot retrace its steps. Its law of irreversibility has practically no exceptions. It was invoked by Dollo for the organic world but can be applied, although less rigorously, to the world of culture. The industrial era is a direct result of man's technological evolution. Computers were foreshadowed by stone axes; they were intentionally formed for a precise use and needed to be used in a certain way. But one should not confound industrial development with liberal ideology, as is sometimes done. As Maurice Duverger pointed out in his *Lettre ouverte aux socialistes* (1976): "Today, only modern technology is capable of fulfilling the needs of a humanity that has become numerous, exigent, and thirsty for justice and progress." Industrialization has caused sociological and economic harm, but its benefits cannot be denied. In the end industrial development has led to a considerable improvement in the standard and expectancy of life. It has allowed social protection to increase, has favored the development of services and the spread of culture. As we saw earlier, the future of mankind still depends on progress in technology and always will.

As an estimate of the significance of technological progress we may compare the overall situation in industrial nations and that in countries that are not yet industrialized — and may very well never be, in view of the present economic situation. For them ecology in its simplest sense has no meaning. It is a foreign concept found in rich nations. One has often heard in Africa or elsewhere statements like: "We would be happy to accept some pollution if our children no longer starved to death."

Poor populations who envy our situation are astonished to hear that we wish to return to a "biological" form of agriculture. This is the only form that they practice, and it provides them with hardly enough to survive, and sometimes *not* enough, as they struggle desperately to develop an industrialized form of agriculture, the only way of escaping from their poverty, ignorance, and dictatorship. Industrialization for them represents the only hope of attaining humane conditions. Today we should not call to account techniques that will be essential for the survival of the seven or eight billion people living in the world at the end of the century, but rather utilize them with discernment in the service of mankind.

In this domain the ecological movement can play a unique role by reminding our society, which is oriented toward quick profits, of the necessity of protecting the natural environment and the incalculable biological capital represented by the variety of animal and plant life that is our only guarantee of long-term survival. We have described elsewhere the considerable genetic impoverishment brought about by domestication (Ruffié, 1976). For example, in 1939 twenty-one varieties of cattle could be found in France, while in 1972 there were only seven (Vissac, 1972). Since then the number has decreased further. The same phenomenon is found with sheep, goats, pigs, horses, donkeys, chickens, and many others. Each of these varieties was adapted for a particular ecological situation, that of the region in which it had developed. If we take into account the predictable needs of the world's populations in the next few centuries, many species that were formerly raised for food but have now been abandoned must be reintroduced. Unfortunately it is far from certain that the so-called industrial varieties, the only ones that we have preserved because they are well adapted to intensive mechanized agricultural techniques, will be easily adaptable to all microclimates (see Feldman and Sears, 1981). We run the risk of losing a rich potential supply of resources forever.

But human ecology must consider not only the effects of man's destruction of plant and animal varieties but also — and especially — the effects that current forms of cultivation have on societies and individuals (Raffestin, 1976). Ivan Illitch (1973) pointed out this risk when he wrote: "The monopoly enjoyed by the world of industrial production makes man the raw material that tools work upon."

Violent Protest: The Red Brigades

Another type of protest uses violence. We might call it the "phenomenon of the Red Brigades." It would be facile and useless to see in such movements a sort of subversive activity orchestrated by some unknown powers to destabilize the Western world. The true situation is both simpler and more serious. Such "subversive" activities are due to the growing awareness of the younger members of society that they are prisoners of a harmful system, and to their conviction that they can never escape legally. Originally, many of these people were disillusioned protestors of May 1968.

All political structures, whatever ideology they belong to, seem to be accomplices in a situation that guarantees their own survival and that they maintain. Confronted with the impossibility of acting legally, some people fall back on violent action, but in practice the result is often the opposite of what they desired. Terrorism frightens the common people and provides a pretext for repression. In the long run it tends to consolidate the structures it wanted to destroy.

The many forms of the rejection of modern conditions, which offer little hope for the future, show clearly that the economic and political system we have inherited from centuries gone by, and that is trying desperately to perpetuate itself, will no longer be adapted to the conditions of tomorrow's world. We are going through the death throes of one society and the birth of another.

the populational society

THE POPULATIONAL ALTERNATIVE

We have seen how Darwinism was interpreted as giving scientific approval to both the liberalism underlying the development of nineteenth-century industrialization and the Marxism that attacked it. At first sight it may seem paradoxical that two such opposite concepts of the world should derive support from the same source. In fact, the contrast between them is more apparent than real and concerns the way in which they are applied rather than their principles. The two systems have the same fundamental premise: a typological way of thinking that divides mankind into a number of groups based on characteristic types (holotypes), groups that are condemned to eternal competition, with the best and strongest destined to win. This hierarchy may either be inherent in an animal's natural, biological characteristics, or may be based on socioeconomic structures imposed by human historical circumstances. Whatever the situation, Marxists and liberals argue in terms of struggle and elimination, which they consider indispensable for progress. We now know that this holotypic and competitive view does not correspond to reality.

Throughout this work we have emphasized the consistency and extent of genetic polymorphism in natural populations and explained how fallacious the very idea of a holotype is for a biologist. We have seen how, rather than wage internal wars that would result in a threat to their existence, genetically polymorphic groups exploit the possibilities given by polymorphism to progress beyond the frontiers of their niche and explore new, unused resources. Thus new species emerge, from which in turn other species develop. Living beings become diversified not by mutual destruction but rather by new, noncompetitive, ecological conquests. Throughout its history life has destroyed much but created much. This trend toward diversity is the very basis of the mechanics of evolution. With this upheaval in our concepts we must reconsider our idea of human society, its origin, its structures, and its relationships. Everything based on social Darwinism must be reconsidered as fundamentally as was the elementary and naïve way in which biological Darwinism was treated for so long, which led to exaggerated claims that Darwin himself would certainly have denied. We have seen how this brilliant naturalist, one of the most perspicacious minds of his century, was worried and even annoyed by the way in which some over-zealous or incompetent disciples applied his theory, deforming it in a caricatural way. Nowadays philosophers and politicians can no longer use as arguments models inspired by natural sciences as they were understood a hundred years ago, which we now know do not correspond to reality.

This challenge — seen by some people as a sacrilege — is consistent with what Engels stated in 1886: "With every discovery that marks an era in the domain of natural sciences, materialism must inevitably change its form." Today we are entering a historical period of revision. In certain contexts the term "revisionist" has a very bad press. During the quarrel between Stalin and Tito it represented the supreme insult, and was later used by Khrushchev and Mao Tse-tung. This is the wrong way of understanding human cultural evolution. Revision does not attack historical materialism as a way of thought and a scientific approach to history, but it condemns its sclerosed form, which has not changed for a hundred years.[1] And it also attacks "advanced" liberalism, still entangled in the Darwinian model that gave it a degree of scientific credibility at its birth.

In fact Western traditions have great difficulty in freeing themselves of their Manichean heritage, which has shaped mentalities for thousands of years. Whether Marxist or liberal, we have seen how the white nations live with the notion of a permanent struggle between good and bad, the righteous and the evil. All our systems have their ideal holotypes: the saints, the just, the good stewards. Very few admit pluralism. But life is synonymous with variety. We are too often exclusive or sectarian, whereas one of our greatest qualities should be tolerance. We have seen that neither Malthus nor Darwin nor Marx could have existed in a fundamentally agnostic society, even though Marx himself rejected faith.

The time has come to abandon chimeras. Dogma does not change, but life and all the ways it is expressed by individuals and societies constantly change. Nobody who asks themselves about present-day problems — philosophers, economists, and politicians — should fear

---

1. Jacques Monod (1970) was more severe when he wrote: "The only hope for socialism is not in a revision of the ideology that has been dominant for more than a century, but in the total abandonment of this ideology."

this confrontation, but rather should welcome it. For man now has the power to influence his destiny and that of his neighbor. He is no longer subject to evolution, he controls it. The future will not depend on chance or miracles but on our decisions.

ETHICS AND SCIENCE; MIND AND MATTER

Before going further we should consider for a moment a question that is often asked: can we today construct an ethics from science?

Certain naturalists, including such distinguished persons as P. P. Grassé in France, categorically separate the domains of life and morality. They consider that biology has nothing to teach us concerning human conduct, and they put mind and matter in opposition (Grassé, 1980). Their attitude is in part a reaction to the excesses and stupidities of social Darwinism based on an erroneous biological model, including its latest form, sociobiology. Others, like Jacques Monod (1970) and earlier, Albert Vandel (1968), believed that after the rapid development of life sciences and human sciences in the last few years, these two could no longer be put in two separate categories and that biology could provide significant data to elucidate our nature and our place in evolution.

This was expressed by Theodosius Dobzhansky, one of the fathers of the "populational revolution," when he wrote in 1973: "Scientific knowledge should lead to moral truth." No recent discoveries have refuted this point of view — that is, as long as the knowledge involved is at the human level, which is to say the level at which evolution affects sociocultural phenomena. Neither the hepatic cells that "cooperate" to ensure the functioning of our liver nor the bees that participate in the activity of the hive are governed by ethics but simply by genetic programs that trap them and leave them no choice. They form a rigid association, not a voluntary, premeditated group.

Groups of this type are found only among warm-blooded vertebrates, but it is in man that they find their full expression, and in him they imply a certain number of moral rules: "Thou shalt not kill; thou shalt love thy neighbor as thyself," and so on. Any transgression would threaten the system. This moral truth, the origins of which are beginning to be clarified by modern biology, implies no metaphysical options. Science can help man define his behavior better, but can never explain religion. They belong to two different domains. The weakness of Teilhard de Chardin's work — although it is excellent in many respects — is that he wished to include God in evolution at any price and establish a convergence between science and faith. But as Claude Cuénot wrote: "The problem of God is a rigorously extrascientific problem. God is not a scientific hypothesis. Science does not lead to theism nor to atheism" (cited by Vandel, 1968). In fact Teilhard, the eminent paleontologist, could never shake off his Jesuit background. "Human nature" as evoked by P. P. Grassé (1980) as opposed to "animal nature" represents more an article of faith than objective data. Man has no essential distinguishing particularity. Everything he is can already be found in a primitive state in the higher primates. Man has developed and exaggerated preexisting qualities; he has not innovated (Ruffié, 1976).

If a biologist has faith, no laboratory observations will shake his conviction; if he does not have this faith he will never find the proof of God, or that of his nonexistence, in experimental data. Metaphysics is on a different plane. We are familiar with the conversions of famous scientists. They usually develop with advancing age and often imply other motives than rational ones. They may represent a response to the anguish of the passing of time and the gradual narrowing of the circle of familiar faces. They attempt to explain the absurdity of death and the injustice of life, of which some people refuse to admit that there will be no continuation. The faith that is derived from this refusal is more of an adherence than a proof, but this does not lessen its value as long as the faithful respect the belief or disbelief of others. Unfortunately, this was rarely the case in the history of man, where all too often faith and intolerance advanced together.

TECHNOLOGICAL IDEOLOGY: "TECHNOLATRY"

The opposite of "spirituality" is technological ideology, called "technolatry" by Pierre Drouin (1981c). It has been inherited from the scientist ideology that reigned supreme throughout the nineteenth century, following essentially in the footsteps of developments in physics, chemistry, and later, biology. It involves a desire to explain all human phenomena, whether political, social, or moral, by such deterministic and simple relationships as those that emerge from experimental or natural science. The conclusion is that applied technology will resolve all the problems that we face or shall have to face. Confronted with the rapid progress in our knowledge over the last century and the varied ways in which it has been applied, many people have fallen into the trap of an almost religious scientist philosophy. Such blind faith marked Spencer, Marx, and many others. Some of our contemporaries still share their views. We can find it in some biologists, like Wilson, who try to explain complex human social behavior in terms of the infinitely simpler behavior of insect societies. It marks the entire rich and generous work of Jean-Jacques Servan-Schreiber, who in Le Défi mondial (1980) earnestly pleads the cause of the Third World. In his view, the widespread use of computers and distribution of microprocessors could favor a rapid spread of knowledge and technology, finally helping the poorest nations to escape from their underdevelopment. This type of discourse is not new — it appeared throughout the second half of the last century.

On April 6, 1866, the railway from Perpignan to Collioure was inaugurated. When the first train arrived in the station of the little Catalan port, the parish priest, Father Clerc, explained to the enthusiastic crowd that the train would allow men to get to know each other better and that there would be no wars in the future, for war was an absurd phenomenon that was the result of ignorance. Men who knew each other would no longer fight. They would discuss and agree (Ruffié, 1981). This good priest's logic is dazzling. We know what happened to it. The first half of the twentieth century saw the two cruelest conflicts that ever shamed the whole of our blood-stained history — despite our having trains, electricity, and telephones to bring man together technically.

There is a double confusion here:

1. Technology has nothing to do with morality. Technology has no ethical message. In itself it is neither good nor bad. All depends on the use we make of it. With the same knife I can peel potatoes or kill and rob my neighbor. With atomic energy one can cure cancer or destroy cities. Technology has increased our creative and destructive capacity.

2. On the other hand, biology, and particularly human biology, today enlightens us on the origin of man and his place in evolution. It teaches us how the development of mental faculties has led to the progressive replacement of genetically programmed, innate behavior by learned behavior, the fruit of experience, example, and education.

We have already seen that this trend, which reaches a climax in our species, has permitted the organization of cultural societies consisting of individuals united by subtle bonds that ensure them great liberty. Part of man's originality is that he remains free to choose.

In any particular situation an individual can choose among several solutions, which will not all have the same consequences for his own future and the future of the group to which he belongs. He is bound by his decisions, which thus take on a moral value. Man acquires *responsibility*. We have already mentioned the replacement of the innate by the acquired, and for us this is an essential stage in evolution. We should consider it in more detail, for it represents a fundamental division in the animal kingdom.

THE PRIVILEGED MEANING OF EVOLUTION

Evolution is not "neutral." When considered in terms of its major principles — that is, not at the level of the emergence of new species but at that of new types of organization — a particular direction is revealed: the regular progression of intellect. As we have already said, this phenomenon can be found in all lines and is in no way mysterious, because it allows an individual to know his environment and thus exploit it more fully. It is a process that parallels the enlargement of ecological niches and the creation of new niches, fundamental trends in living beings. This "pioneer" movement is in the same direction as that of selection. It implies the development of organs and functions, but particularly behavioral patterns, that are better adapted. It is most marked in warm-blooded vertebrates, the birds and mammals, which can maintain their internal milieu at a constant temperature, irrespective of the external conditions. The appearance of new behavioral patterns and their direct improvement by apprenticeship or indirectly through example and education implies considerable growth of intellectual faculties, which can only be ensured if the brain is sufficiently complex. This organ is made up of a large number of nervous centers linked through a multitude of neuronal interconnections, and in order to function correctly it needs a permanent, large volume of circulating blood at a constant temperature.

Selective pressures must have exerted strong influences in favor of such developments. In spite of its fragility, the acquired represents a considerable progress over the innate. When faced with new constraints — or, in other words, with new selective forces — it allows a rapid voluntary response that can spread quickly through the whole group and be constantly modified and thus improved. It is just the opposite of rigid genetic characteristics. With the development of acquired behavior living beings escape from the tyranny of DNA. This privileged trend is nothing other than the beginnings of liberty in the world (Vandel, 1968).

THE CONSEQUENCES OF ACQUIRED BEHAVIOR FOR SPECIATION

In most groups the organization of a new niche implies the development of new abilities that were either absent or existed in the form of mere precursors. Often this involves a more or less profound modification of the genome—the genetic revolution discussed earlier. Its extent can be such that the pioneer group whose patrimony has been modified can no longer interbreed with its original line; thus new species emerge. But as soon as an individual becomes capable of inventing or learning sufficient new behavioral patterns to respond efficiently to selection, the genetic revolution loses its importance. Adaptation no longer belongs to the organic domain but becomes part of culture. From then on the occupation of a new niche does not necessarily mean the emergence of a new species with particular biological features, but the adoption of new behavioral patterns—in other words, a new culture. This trend can already be seen in many carnivores and even in certain birds, where it constitutes the protoculture, the ensemble of learned behavior that can be used to adapt the animal to its milieu without forcing speciation. It reaches its maximum development in man. This phenomenon has far-reaching consequences. Our species is the only one that has invaded all parts of the earth without breaking up into daughter species with their own individual ecologies. In wild animals, cultural development is not sufficiently developed to allow such an economy of organic adaptation. So those groups that become ubiquitous are forced to become involved very rapidly in adaptive radiations, which, as we saw, involve the breaking up of ancestral lines into several new branches each of which diversifies and becomes established in a particular environmental situation. This "specializing" evolution is marked, in those groups that spring from it, by an impoverishment of the genetic patrimony compared with the common ancestor. This does not apply to man, who has been able to colonize virtually the whole of the landmasses merely by changing his culture. This is a basic feature of man that explains his monospecific structure and his genetic richness.

Let us consider an example that demonstrates the very different extents of biological and of cultural evolution. The Canidae are a common family of hunting mammals. They include wolves and their domestic form, dogs, and also the foxes and jackals found throughout the world from the poles to the tropics. Foxes are particularly widespread. They exist in North America, the whole of Europe, Asia, and Africa and are divided into about ten species. Our common fox, *Vulpes vulpes,* inhabits Europe, Asia, and North America and was for long the terror of chicken farms. Although the red fox is very widely distributed, it is practically restricted to temperate regions. There is an arctic fox, *Alopex lagopus,* characterized by a white coat in winter and a bluish-gray coat in summer. Its physiology is perfectly adapted to cold climates. On the other hand there is the desert fox, the fennec of the Sahara *(Fennecus zerda),* found only in the hot desert climates of Africa and Arabia. During the day it remains buried in its hole in the earth away from the heat. It only emerges at night to hunt insects and small vertebrates for food. It uses its enormous ears to detect the sound they make, while other species use essentially smell.

There are many such examples in all groups, including vertebrates and invertebrates. Species living in cold climates near the poles are not the same as those found in temperate countries or the tropical savanna.[2] Even if they have split off from their common stem quite recently, as is the case for the foxes just described, they have all undergone biological changes to adapt them to their milieu, involving the appearance of new organs or the transformation of old organs, or changes in function and behavior that imply modification of the genome and sexual isolation. They have formed independent species and even genera. Things are very different where man is concerned. Although born in tropical regions and best adapted to a warm, humid climate, *Homo sapiens* was able to colonize temperate lands, where he underwent an enormous development, as well as deserts and polar regions. It was enough for him to adapt his habitat, his clothing, his food, his behavior, and his social structures to the constraints of the environment. In his case specialization was a conscious, voluntary, and reversible process. There was no necessity for organic modifications.

An Eskimo can leave the polar regions with his family and live in more clement climates without needing a genetic revolution. It is enough to change his way of life. He need not even invent a new culture; he can simply adopt that of the natives already living in the land he has moved to. In the same way the Tuareg who decided to move to cold or temperate areas had no major difficulties as long as they adapted to the local culture, best suited to their new environment.

The arctic fox and the fennec belong to two species and even two genera, but the Eskimo from the Arctic and the Tuareg from the desert are both *Homo sapiens*. With very minor exceptions, the physiology of all men is the same (Lambert, 1968). Only populations living at high altitudes, such as the South American Indians of the Andes and the Sherpas of the Himalayas, appear to manifest subtle trends toward real physiological adaptation (Ruffié, 1970). This is probably due to the fact that the low air pressure and reduced concentration of oxygen at high altitude are the only ecological parameters to which man cannot adapt purely culturally. All other factors, such as temperature, humidity, sun, and food, were the objects of technical solutions very early in the history of man's ancestors, thanks to which they were able to take with them almost everywhere the tropical microclimate they needed, a pro-

2. For example, the wolf *(Canis lupus)* lives in cold climates such as Alaska and Canada. In the tropics we find jackals *(Canis aureus, C. adustus, C. mesomelas),* and on the American continent several species of coyote (such as *Canis latrans* in North America). The Himalayan dog *(Ceron lupus)* is a mountain variety living between Kashmir and Nepal.

cess that resulted in the appearance of the first hominids. In this way they defied selection. Had things been different the human race, as it invaded all parts of the earth, would have split initially into autonomous races, then into subspecies, and finally into daughter species, as has happened in most other groups.

Only domestic animals, living in close contact with man and thus benefiting from the constancy of the human environment, have been able to follow their masters throughout the world without dividing into different species, although they may have formed different races as a result of voluntary selection by man. An Atlas sheepdog can perfectly well breed with an Eskimo's husky. All the canine races, wherever they are found throughout the world, are genetically interfertile. The only barriers to interbreeding include, for instance, size differences, but these are only relative problems.

THE BALDWIN EFFECT AND HUMAN RACES

In 1896 J. M. Baldwin suggested that an advantageous behavioral pattern might initially be adopted voluntarily, and then after this period of "approval" become integrated in the genome as its usefulness was confirmed. According to Baldwin, acquired behavior was a sort of "test-bed" for the innate. This phenomenon was later called the *Baldwin effect*, and certain evolutionists consider it an important element in the dynamics of speciation.

We now know that this does not correspond to reality; for if an acquired behavioral pattern is effective, that is to say if it corresponds to the selective pressure that brought it about, the pressure will disappear, and there is no reason for innate behavior to develop. To have it in the genome would be of no advantage; on the contrary, this would fix the individual and the whole group in a permanent ethological context, whereas learned behavior allows it to change at any time. *Homo sapiens* is the champion in learned behavior, thanks to which our species has been able to spread over virtually all dry land without any particular organic specialization.

There must have been a trend toward biological diversification at the beginning of mankind, when our ancestors were dispersed in small groups, each subjected to divergent selective constraints that could only be partially compensated for by cultural development, which was still very elementary at the time. In such circumstances natural selection and chance played a diversifying role. It was at this time that the racial differences we can still observe today appeared. They allow us to distinguish a white man from a black or yellow man, in spite of there being intermediate stages.

The racial characteristics that were emphasized by typological anthropologists represent archaic features, distant witnesses of ancient times. Thanks to the development of culture and the strengthening of migration and exchange, the trend toward homogeneity has for long been stronger than that toward diversification. Modern man is moving towards "counter-raciation."

THE DIFFERENT SPEEDS OF BIOLOGICAL EVOLUTION AND CULTURAL EVOLUTION

Speciation is an ecological necessity and requires a lot of time. In order to respond to a particular constraint, many generations are needed for favorable genetic combinations to emerge and impose themselves on a population, and even more time is necessary for them to spread throughout the species. There is no comparison between cultural evolution with its speed, efficiency, and power of immediate diffusion, and organic evolution. The former has replaced the latter and has not waited for it.

Let us take a recent example concerning the resistance of mammals to certain viruses. The common disease myxomatosis was introduced deliberately in France a quarter of a century ago to combat the proliferation of wild rabbits that were destroying the crops. The disease had and still has devastating effects. As soon as the size of a wild population exceeds a certain threshold that permits contagion, epidemics break out and decimate the population. There will always exist a few resistant individuals that will escape, but a very long time and repeated mass slaughter will be necessary before resistant genotypes will be found throughout the group. As a comparison, at about the same time a vaccine against poliomyelitis was perfected. In one generation, almost the whole of humanity could be protected. Today, poliomyelitis is practically unknown in those countries that have effective health services.

Cultural adaptation to a new milieu in no way compromises the biological future of mankind, for it does not imply genetic upheavals. When he returned to earth after walking on the moon, Neil Armstrong took off his astronaut's suit and became once again the average American whom the cinema and the media have made familiar to us. In the middle of a crowd no anthropologist would be able to point him out as a spaceman. There is little to distinguish him from his contemporaries. By contrast, another species of mammals would have to undergo unimaginable transformations to adapt it to the lunar climate without the help of technology. The differences in the environment are so great that such a performance would certainly be beyond the capacity of organic evolution.

THE END OF THE BIOLOGICAL EVOLUTION OF MANKIND

Anthropologists periodically wonder about the capacity of our species to evolve further and perhaps give rise to a posthominid species. When they do this they are applying the standards of animal evolution to man. These standards led to birds and mammals replacing certain reptiles, which had themselves replaced amphibians derived from fish that had left the water.

This is a false problem. Because of his intelligence, which allowed him to respond culturally to selective constraints, and because of his ability to improve his knowledge over the generations, man has broken away from the limits of organic evolution, which are no longer applicable to him. Nietzsche's *Übermensch,* or superman, represents a romantic vision of the world, not a scientific concept.

Instead of creating species, man has created civilizations. Erickson called this *pseudospeciation;* thanks to it mankind has remained a single species, although genetically polymorphic, thus developing enormous potential.

Even if it were still possible, the biological evolution of the human race would not have time to take place. It would need long periods incommensurable with the never-ending acceleration in cultural evolution. Faced with new constraints, man quickly finds a response, well before there has been any time for organic modification to occur. The beginning of mankind's psychosocial development represented the end of biological evolution. This is the meaning of the occasional comment: "Man has no nature; he only has a history."

THE ACCELERATION OF KNOWLEDGE

Throughout prehistory and history, the technical and cultural patrimony of man has increased regularly, but in an ever-accelerating fashion.

Four or five million years were necessary for marked improvements in the working of stone. The use of fire spread ten times more rapidly; while industries based on metals and agriculture took only a few thousand years to develop. Later, only a few centuries passed between the manufacture of paper — discovered in China in the eighth century — and the development of printing. The first book — the famous forty-two-line Bible published by Gutenberg — appeared in Germany in 1455, and the first weekly newspaper was published in Amsterdam in 1620. In 1690 Denis Papin discovered the motive force of steam. Less than a hundred years later, in 1765, James Watt perfected the steam engine, which was to play an essential role in the industrial revolution. In 1814 George Stephenson introduced the first locomotive. Within scarcely thirty years Europe and America were covered with a network of more than 100,000 kilometers of railroad.

A similar acceleration accompanied the use of the internal combustion engine, made practical by Hugon in 1858 and Étienne Lenoir in 1860. The first automobile using gasoline was produced by Daimler and Benz in 1887, and the first real airplane by the Wright brothers in the United States in 1903. The first Sputnik, weighing 88 kilograms, was launched by the Russians on October 4, 1957; the second, containing the dog Laika and weighing 508 kilograms, was launched on November 3. Less than three years later, on April 12, 1961, Yuri Gagarin left the earth's magnetic field in the satellite *Vostok.* On July 21, 1969, two Americans, Neil Armstrong and Edwin Aldrin, landed on the moon. Thus, scarcely a century elapsed between the invention of the internal combustion engine and space flight.

This modest list could be extended to cover all areas of knowledge. It highlights two phenomena:

1. For a long time, great discoveries were made at the crossroads of different civilizations, where cultural exchanges were the most intensive. For the Western World, the Near East, at the meeting point of Europe, Asia, and Africa, had a privileged position. It is not pure chance that this is where the Neolithic revolution and the development of metallurgy, the wheel, the plough, the sail, writing, and the alphabet all took place. A similar phenomenon can be observed in Southeast Asia, where Chinese, Indian, and Indonesian civilizations come together, and in Central American, where the North and South American subcontinents meet.

2. The volume of our knowledge has increased regularly throughout history, in parallel with developments in communication and exchange. This trend has accelerated over the course of time. Although it took more than a hundred thousand years for the stone axes of Moustier to develop into the fine flint arrows of Mas d'Azil, only fifty years were necessary for the first fabric airplanes of the Wright brothers to evolve into space rockets — in spite of the fact that the technological problems encountered in the two cases are not of the same order of magnitude. Modern man is no more intelligent than his Paleolithic ancestors living in the caves of Mas d'Azil, but he has access to an infinitely larger volume of knowledge and a much better way of dealing with it, using mathematics, computer science, and so on. We are now caught up in an endless mechanism.

The whole of human history consists of a series of feedbacks, in which progress increases the possibilities of exchange, thus favoring the diffusion of technology and leading to new discoveries. This trend accounts for the constant acceleration in our cultural acquisitions. As they become greater they sweep away what is left of our innate behavior, for culture is infinitely more capable of responding to any form of constraint. As technology progresses, cultural response grows more effective still.

At present, man is much more what the milieu he was born into and the education he has received have made of him, than a product of the genomes inherited from his parents. The most important thing is what he knows. Thanks to the possibilities of instantaneous communication among almost all peoples of the earth, mankind is developing into a single sociocultural ensemble of an extremely diverse and rich nature. This type of evolution has consequences for our ecological niche. Because of its ability to respond effectively in almost any situation, our species has broken down many barriers. Today man's envirnoment includes the whole world. He has gone even further because he has walked on the moon. This trend is continuing before our very eyes with no visible sign of organic specialization. Humanity is, and will always be, a single species, rich in its diversity, but efficient in its unity.

THE MEANING OF GENETIC POLYMORPHISM IN THE CULTURAL SOCIETY.

We can now see how a cultural type of socialization essentially based on acquired, transmissible behavior adds to the value of genetic polymorphism. Indeed, because of this polymorphism not all members of a given species have the same tendencies or abilities. We saw in Chapter 1 that this variety was such that in sexually reproducing groups there has never been and will never be two genetically similar individuals, with the exception of identical twins. Thus when confronted with the same problem in a given situation, each individual will look for a solution that corresponds to his ability. He will "invent" what is best for him — that is to say, what he is *able* to invent. But he will not keep this invention for himself alone. Through example — and, in man, through teaching — new discoveries are communicated to the whole of the group. Biological polymorphism attains its full meaning through cultural dissemination, for it thus acquires a collective dimension. A genetically monomorphic society would not offer the same advantages. Since it would be composed of individuals with identical abilities, the spectrum of invented responses would be much less varied. Such a society would be unimaginative and frightfully monotonous. All its members would wish to be active at the same time, live at the same place, choose the same food, and have the same sexual partner. All would have the same desires and would know the same things; none would have anything to learn from their neighbors and nothing to teach. An individual's "capital" would be the same as that of the whole group. There would be a terrible competition within the population, and the ecological niche would be particularly narrow. Although it does not escape completely from its genetic program and its basic attitudes and aspirations, the cultural animal receives an enormous quantity of information from its social milieu from which it chooses the response that seems the most fitting. How can we doubt that this status confers an enormous advantage?

The United States of America is a striking example. Historically the country forms a biological and cultural melting pot, initially based on eighteenth-century Anglo-Saxon Protestant liberal ideas of civilization. It has been enriched by wide-ranging imports from Central Europe, Mediterranean lands, North Africa, the Far East, India, and other areas, not to mention the little that remains of indigenous Indian influence.

An astonishing mixture can be found in most American laboratories. One can often meet, working alongside researchers of English or Dutch origin, Slavs (Russian or Polish), Asiatics (Japanese and Chinese), Latins (Italians or South Americans), blacks, Puerto Ricans, Cubans expelled by Castro, and Chileans who fled Pinochet. Some are Protestants, other Catholics or Orthodox, Jews or Moslems, Hindus, Buddhists, or agnostics. Each brings his own way of feeling, thinking, and acting, which in the end enriches the whole group. This transcultural mixture is certainly not unrelated to the dynamism and creativity of the American people.

Another equally remarkable example, although in a different context, is offered by Japan. During the almost three centuries that its "closed" period lasted, the Empire of the Rising Sun remained totally introverted, refusing all contacts with the outside; but with the emergence of the Meiji in 1868, the archipelago opened up to the world, and even more so after the Second World War. This resulted in the reception of many foreign scientists in Japan, but particularly the sending of a large number of its own researchers to the most advanced nations, particularly the United States and Europe, where they learned, and sometimes copied. Today one finds the Japanese everywhere. Thanks to this extraordinary accumulation of knowledge gathered from all countries, Japan has become one of the most important centers for technical advancement at the end of the twentieth century.

THE POSSIBILITIES OF THE PRESENT TIME

All this gives a glimpse of the extraordinary possibilities offered by contemporary humanity. The knowledge accumulated by a society depends on:

1. The number of individuals in the society capable of communicating their experience and their inventiveness.

2. The number of generations that have transmitted their cumulated experience.

This value can be estimated by calculating the number of *man-years* — that is, the product of the number of years lived by humanity in a given period, multiplied by the number of individuals making up mankind during the same period.

If we take the time between man's appearance on earth and the present day, the number of man-years is of the order of 2,000 billion. Today the earth has almost 4.5 billion inhabitants, and their mean life expectancy can be estimated at about 50 years. Thus our generation alone will account for 225 billion man-years — which is to say more than one-ninth of those in the whole of man's history. This trend will continue to accelerate, until during the twenty-first century our descendants will live, in effect, *half of the whole of human experience* (Jacquard, 1978). This means that they will have at their disposal a volume of knowledge, and therefore a power, infinitely greater than ours. Nevertheless, we should correct these figures a little by taking into account the fact that their knowledge will not be equally divided. At present approximately a quarter of mankind can neither read nor write and thus is cut off from the major channels of communication. These individuals are isolated from cultural progress and live on the margins of the evolutionary movement. They are useless brains, deprived of the benefits of a host of acquired knowledge to which they have no access. Although biologically they belong to *Homo sapiens,* they are not full cultural members of the race. And it is easy to predict that their number will certainly increase, at least in absolute terms.[3] Apart from the intolerable moral aspect of the situation, this marginalization deprives our society of an irreplacable capital. Immediate international action is needed.

This situation is another demonstration of the archaic nature of

3. According to experts at the fifteenth international meeting on illiteracy, organized by UNESCO in Paris in 1981, the *relative level* of illiteracy is decreasing. It was 32.4 percent in 1970, 28.9 percent in 1980, and will be 25.7 percent in 1990 for the whole of mankind. But the *absolute number* of illiterates is increasing; from 814 million in 1980 it should exceed a billion in the year 2000.

political frontiers, a subject we shall discuss in the last chapter. Natural frontiers are indispensable for development and diversification of species; they have triggered and stimulated much of evolution. On the sociocultural level, however, their role is much weaker. We have already seen how man, thanks to his capacity for adaptation, need not heed natural obstacles. But he has created new barriers by establishing political frontiers that are the remains of an archaic stage of evolution. They are now slowing the development of culture, which needs exchange and communication in order to blossom fully. *Knowledge has no homeland.* Unfortunately, many of our contemporaries are still living according to an out-of-date model. They manifest a suspicious form of patriotism that reflects their territorial instinct, an essentially animal feature that is a remnant of our ancestral egoism. This is a serious obstacle in the path of cultural evolution.

By making knowledge a worldwide phenomenon, humanity can hope to continue its progress. All men should attempt to form a single nation with multiple cultures.

EQUAL OR IDENTICAL

It is common to confuse *equal* with *identical.* Just like other animals that reproduce sexually, all men are genetically different, and it is fortunate that this is so, because this difference gives us different abilities, decreases competition, and broadens our niche. We have seen how our cultural diversity comes on top of our biological diversity. Every human life is a unique adventure that has never occurred previously and will never occur again. Every death is an irreparable loss. It has been said that when an old man dies, a library burns. But we might add that this library contains books of which there is only one copy each. We have emphasized how this cultural diversity enriches all members of a group because, thanks to communication, example, and education, individual experience soon becomes part of the collectivity.

Thus everyone has a duty to contribute to the community in his own way. However modest, this contribution is of inestimable value; it is unique. This is why, although men cannot be *biologically identical,* they are nevertheless *morally equal.* Each individual has the same value and the same rights of living, acquiring culture, working, having a family, moving, setting up home, and choosing his reading matter, his friends, his opinions, and his leisure, as well as his mayor, his deputy, or his president.

Diversity is an objective phenomenon; equality is a logical concept with moral overtones. The two are in no way in opposition; on the contrary, diversity is of no value if it cannot assert itself—that is, enjoy liberty and equality. If an individual cannot express himself and add his stone to the common edifice, diversity is useless and society impoverished. No lamp, even the most modest, should remain hidden under a bushel. Diversity justifies equality. But an authoritative form of egalitarianism that would attempt to make everyone identical is just the opposite of a cultural society; it is a barrier to evolutionary progress.

THE NATURAL HISTORY OF OUR LIBERTY

Cultural polymorphism, which valorizes man's genetic polymorphism by giving it a collective dimension, plays its full role only in a society in which individuals are free to think, conceive, achieve, and communicate according to their abilities, their tendencies, and their desires. The progress of humanity is built on the accumulation and transmission of individual initiative. In this way civilizations have been patiently built day after day, century after century.

Liberty is not gratuitous. It cannot be lived fully by an individual alone but only in societies. An individual's liberty ends where that of others begins. In order to be credible, liberty must be shared. All forms of dictatorship impoverish, whether civil or military, lay or religious, right-wing or left-wing. Dictatorship replaces the tyranny of DNA, from which we were liberated when the acquired replaced the innate, by a new program coming from outside: that of ideology. By imposing uniformity on all individuals, by introducing political or religious holotypes, it lowers the level of the collectivity dangerously and amputates its real value: the right to be different, which is the source of its wealth.

Ecumenism is always better than religious wars. The commissar is as much an anachronism as the ayatollah or the preaching friar. Throughout the course of history all dictatorships finished badly, for progress cannot be halted for long. On the human level tolerance is not a luxury, but a necessity.

Escherich's Error

Long before Wilson, Karl Escherich established an almost caricatural confusion between the innate and the acquired. Just like the father of sociobiology, Escherich was a brilliant entomologist, and was rector of the University of Munich in 1933 and 1934. He was a collaborator of the National Socialist party whose theories suited the philosophers of the Third Reich. They can be summed up as follows (Escherich, 1934).

Insect societies, such as those of termites, are much older than human society because they date from about eighty million years ago, whereas ours is scarcely more than one or two million years old.[4] So insect societies were subjected to the forces of selection much longer, allowing them to evolve more than us and progress more toward "perfection." This is why members of insect societies maintain a rigorous order and strict discipline. They serve the collectivity with total abnegation. Within the hive or termite's nest operate systems of regulation involving hierarchies, numbers, work, and so on that no individual would ever question. Escherich saw therein a model of the ideal state, backed up by an ideology shared by all. A termite's nest has no Jews or freemasons, no Communists or homosexuals. In his view it represented the ideal totalitarian nation, the most perfect expression of progress.

By reasoning in this way, the German entomologist committed a glaring error. He ignored the replacement of the innate by the acquired and the fundamental innovations represented by the emergence of learned behavior in a social group. All members of a hive are linked by innate behavior. Their relationships obey genetic programs. Bees, ants, and termites have no need of a registry office; they are not individuals but units forming part of the same organism. A worker bee is no more devoted to its hive than a hepatic cell is devoted to the liver. Even if a termite's nest can represent a perfect society in some ways, it is trapped in this state of "perfection" and cannot progress. Bees' hives and ants' and termites' nests have varied little in tens of millions of years. The insects are condemned to endless repetitions of the same gestures and utilization of the same techniques. In contrast we have already seen how culture and civilization have progressed since the first hominids appeared on earth and how this progression is constantly accelerating. Escherich's mistake, like Wilson's, is to have transposed insect social structures to the human level — an error of interpretation, because in these cases we are dealing with radically different structures. Although both are called societies, they represent two formations that have almost nothing in common.

This entomological view of humanity can lead to the worst excesses. Nowadays certain regimes consider their opponents, or even simple innovators, as mentally ill. History is rich in examples of how the nonconformist has always been suspected by church and state. At different times he has been treated as a criminal or a madman. Sometimes he was burned at the stake or thrown into prison; at other times he entered a monastery or an asylum.

Nowadays it can easily happen that the establishment shelters behind a pseudomedicine that has in fact become an instrument of repression.

---

4. These dates were accepted in Escherich's time, but we now know that the first hominid societies were much older.

The psychiatrist is an assistant of the police. This deflection of mental health for political ends is unfortunately not the prerogative of openly totalitarian regimes. It also flourishes in a baser and more hypocritical way in so-called democratic states.

Such a regime denigrates all that is of value in humanity and plunges into the worst form of social Darwinism: that which eliminates all nonconformism with the help of new pharmacological methods.

Humanity can only progress through tolerance and reason, away from all sectarianism and constraint. We are far from this ideal situation. Almost always, as soon as they attain power, the so-called revolutionaries suppress liberty. But it is liberty that is revolutionary because only liberty allows the individual to disseminate innovation. Socialization based on constraint rejects its true nature. "Socialism is only real if it liberates man" (Mitterrand, 1978). Without this liberation, those who claim to be socialists are simply usurpers. By suppressing liberty totalitarian regimes regress dangerously. They take us a hundred million years back to the era when animals were governed by innate behavior, based on rigid genetic programs. They refuse to recognize all that was acquired by evolution during the long Tertiary era, which saw the innate replaced by the acquired and, ultimately, the emergence of mankind.

Polymorphism and Democracy

Pluralistic democracy is the only conceivable system for human society. It is the translation into political terms of cultural polymorphism. Churchill was right when he said, "Democracy is the worst form of government, except for all those other forms that have been tried from time to time." Democracy, the only way for our species to develop and progress, implies:

1. A minimum of material welfare and social justice. If they are in want, men obey survival reflexes first. Their innate behavior predominates. Someone who is drowning or dying of starvation does not reason. He clings to anything that prevents him from going under. According to traditional good sense, a hungry stomach has no ears. There can be no democracy unless a minimum of legitimate needs are covered. Hitler's coming to power in 1933 by legal means, in a country with an old civilization and a socialist tradition, was provoked by the material and moral crisis that Germany had been suffering since the Treaty of Versailles.

2. Democracy also implies a certain level of culture. Freedom of decision needs enough objective and varied information. To choose, one must be informed.

It is striking to see how people aspire irresistibly to liberty as soon as their economic and cultural level is sufficiently high. After thirty years of indoctrination and subjection, the peoples of Czechoslovakia, Hungary, and Poland still insist on a truly democratic status. These claims are now made by a population most of whom were born after the war and have never known any other regime but the present one, hailed as the ideal solution by officialdom.

But this "ideal" is in opposition to man's real nature. On the death of Franco, Spain, which had been subjugated by forty years of Francoist "peace," switched completely and almost without friction to democracy, as did Portugal. Many other examples spring to mind, for the same phenomenon can be observed all over the world, in many nations of Latin America, Asia, and Africa that have never known anything but absolutism or colonization. Liberty is a fundamental aspiration of man.

Homo sapiens always strives to develop his personality and to benefit from those of others, thanks to the extraordinary variety of the human milieu. But to do this everyone must be able to express himself. A termite's life is not for us. It is incompatible with our nature and our aspirations.

To be effective, democracy must not remain purely formal. The constitutions imposed in Portugal by Salazar and in Russia by Stalin were theoretically representative and liberal, but a great distance separated the spirit from the letter of the law, the texts from the facts. True democracy implies alternation, without which variety of opinion has no meaning. Nations that live thirty years or more under the yoke of the same party or the same man are not truly democratic. All monolithic systems, whatever their nature, lead in the end to sclerosis and abuse. This is a historical law, and even the best societies cannot escape it.

At present, in many regimes that claim to be democratic, those in power remain persuaded that they possess the truth, and they therefore attempt to suppress the opposition, depicted as ambitious and irresponsible. The opposition returns the compliment. No dialogue is possible. We translate our religious fantasies into politics. We judge our representatives not by what they do but by their labels. All members of the same party are put in the same bag. We are back to typological ways of thinking. In practice, although it is legitimate for the majority to exercise power, it is essential that the opposition should be constructive and respected. The government must listen to its critics in order to improve its own activities where necessary. An opposition that can never change governmental policy is quite useless. It might as well not exist, and in such circumstances we effectively slip back into a one-party system.[5]

5. It is useful to remember that under Salazar's regime — one of the cleverest and most cynical covert fascist regimes of this century — the governing powers carefully maintained a small parlia-

A similar absence of dialogue is often found in the world of labor relations, whether between managers and unions in capitalist countries, or within the party, between the leaders and the membership in people's republics.

Recent events in the East as well as in the West demonstrate the catastrophic results of such a lack of dialogue.

The Fragility of Cultural Structures

Protected by their genetic programs, termites in their nest and bees in their hive run few risks. The queen is not threatened by a *coup d'état*. The chromosomes that bear their programs are reproduced indefinitely by replication. There is no risk of losing them. This stability gives the group complete security but prevents any form of progress. Only cultural society is free, to progress but also to collapse, for its structures are acquired and therefore enjoy no organic protection. Democracy is fragile. A society subject to too much tension almost always turns toward totalitarianism to establish a new equilibrium, at least temporarily. During recent history dictatorships have usually followed great crises. In man wisdom is not a biological necessity. Of course, our intellectual capacity has an organic base. Its anatomical and physiological framework is the brain; but one cannot reason in a vacuum. One needs learned data elaborated by apprenticeship and education. Reasoning is a perfect example of an acquired process.

But — contrary to what the romantics, from Jean-Jacques Rousseau to Marx, thought — the triumph of reason is not inevitable. It is not supported by any genetic "obligation." We should compare the perenniality of chromosomes with the precariousness of civilization.

THE SENSE OF RESPONSIBILITY

Liberty represents our nobility but also our servitude. It imposes *responsibility*. The bee cannot choose. However it executes its task, it cannot be held responsible and merits neither reprimand nor congratulations. In the case of man, on the contrary, not all responses have the same value or promise the same results.

---

mentary opposition, composed of a restricted number of deputies who could express a few modest criticisms and now and then even vote against the government — an action that had virtually no effect in view of their small number, and their prudence! But they formed a useful shield. Their mere presence sanctioned the "liberalism" of Salazar in the eyes of his powerful allies who, during the Second World War, frequently used Portugal as an escape route for prisoners and as an intelligence center against the Axis. Like his neighbor, Franco, but more subtly, Salazar knew how to play his cards.

Our choice is not "neutral"; it does not only affect the individual who makes the choice but has consequences for the whole of the society, both in the present and in the future. All our actions influence the society in which we live and those that will follow. This "longitudinal" historical conscience is not always found in statesmen. In the past one could invoke ignorance and hide behind mysticism. Now, no excuse is possible. Our knowledge is sufficient to understand the meaning of evolutionary processes and our place within them. The behavior of nations, and particularly those that possess power, knowledge, and fortune, will determine the future of the world. Unfortunately, very few people seem to be conscious of these fundamental principles.

Technocracy and Computers

Computers permit rapid access to an almost unlimited volume of data by allowing us to analyze them almost immediately and through multiple pathways. They represent one of mankind's most important acquisitions since the invention of writing and printing.

The consequences of the spread of computer techniques to countries subjected to the problems of underdevelopment were emphasized in 1980 by Jean-Jacques Servan-Schreiber in a work already mentioned. One can also consult the report of Simon Nora and Alain Mink on the computerization of society (1978). This technology can certainly help underdeveloped nations, but, as we emphasized earlier, it would be dangerous and useless to rely entirely on technology to resolve all their present difficulties. Technology has no inherent moral value; everything depends on how it is used. Technology can be the servant of man or his master. As techniques become more powerful, the danger becomes greater that they will corrupt.

Nowadays our daily life is becoming heavily computerized. The fact that all administrations, and not only the police, can have almost immediate access to information concerning every one of us represents a serious threat, for this information is permanently recorded.[6] A cultural society is superior to an organic society because any individual can at any time modify his behavior and thus adapt to new constraints. We can continually correct our errors and set out on new pathways. Compared with a genetically programmed animal, man's great advantage is

6. According to the commission on Information and Liberties, at the end of 1981 the gendarmerie possessed between 200 and 250 million records concerning French citizens and residents, including certain confidential information (particularly about their private lives) that had nothing to do with administrative details. For instance, some information about Jewish families is said to have been amassed during the German occupation ("Les fichiers illégaux de la gendarmerie," *Le Monde*, December 17, 1981).

his ability to change and to forget. Chateaubriand recognized this in *Atala:* "We are not even capable of being unhappy for long."

In 1842, exiled from Spain by the Carlist war, Peire de Boisia fled from Barcelona and crossed the frontier secretly across the Alberes to reach Perpignan. De Boisia told the old wolf-hunter who sheltered him one night in the mountains about the events that had brought fire and bloodshed to his country, which had to be forgotten if Spain was to live again one day. "The faculty of forgetting is one of the great forces of man. It frees us of our own life. Without forgetting we would spend our whole existence pursued by our past. Forgetting allows us to look to the future and to engage in it. This is even truer for nations than for individuals. Children learn history in school, but fortunately adults forget it. If people's memories were too good, life would be impossible. Since Louis XIII France would be at odds with the House of Austria, since Louis XIV with the whole of Europe, and since Napoleon with the whole world" (Ruffié, 1981).

If the whole of society was recorded on magnetic tape, man would become the prisoner of his past. He would be trapped in a permanent framework from which he could no longer free himself. When he was assassinated at Tamanrasset in 1916, Charles de Foucault was a hermit conquered by the spirituality of the desert. What traces did he still bear of the agnostic playboy officer appreciated by his comrades and even more so by the women of the towns where he was garrisoned?

From 1936 to 1938 Josef Stalin undertook enormous purges within the regime, resulting in a million deaths by firing squad and nine million deportations. Ten years later, at the beginning of the Cold War, all the uncertain elements of his satellite democracies were eliminated with the same efficiency (Fontaine, 1966). Could one recognize in this last Red Tsar the timid, pious seminarist, admitted to the Orthodox school of Tiflis in 1894 because of his good character? Would computerization have preserved Foucault's reputation as a playboy and Stalin's as a minister of the church?

In truth, our systems of relations and identification must be sufficiently flexible to allow everyone to evolve. At the cultural level progress is only possible if we forget what we were. Our past is the worst of prisons. Our magnetic identity card would condemn us to immobility and make us into termites.

Another danger of the computer is the risk that data would rapidly become a monopoly. The cultural individual benefits from the information he receives, utilizes, and exchanges. His efficiency depends in large part on the quality of this information, its accuracy, objectivity, and variety. For the true value of human polymorphism to be expressed, sources of information must be very diverse. Nowadays in pluralistic democracies we have a choice of books and newspapers of many different tendencies. We can read through the same text several times and reexamine our ideas. We have time to reflect and criticize. This is less true of audiovisual media—in particular television, which brings us our information suddenly in "real time," with no possibility of going back. Thus it is indispensable that television companies, whose power of diffusion is enormous, provide access to all forms of thought. In this context it is important that all tendencies—whether spiritual, political, or philosophical—should be allowed to express themselves. Nothing would be compulsory, and everyone could choose what he wanted to achieve his own synthetic view. This diversity would only be possible if the information came from varied sources. Before long even the most modest family will have a terminal at home, connected to a data bank, so that through a simple keyboard we shall be able to ask questions and receive an almost immediate reply on a television screen. Imagine the danger if all the data banks were in the hands of a totalitarian party or other powerful interest. First of all, such information would hardly be interpreted critically but according to criteria drawn up by a class or an ideology. There would also be the possibility of limiting access to these data for certain categories of citizens. This form of communication, with its prefabricated concepts, would imply a relatively limited number of data and would prohibit all discussion. Society would rapidly become impoverished. All political tendencies fear a reductionism that would threaten what, at the level of the individual, is commonly called "general culture."

Olivier Giscard d'Estaing wrote: "Why should we bother to memorize quotations from famous men if it is enough to touch the keyboard of a computer to produce a list of all thoughts pertaining to a given subject on an individual screen?" Georges Lambert (1979) summed up the risks of the computerized society at the cultural level. "Loss of individual memory, manipulation of knowledge through the discriminatory storing of data, isolation of the individual, reduction to a type of univocal conversation, creation of arbitrary software, sterilization of the psychosocial personality. What is more, nothing proves that the form of dialogue that the stupidity of the machine's language requires (the only criteria being quantitative: speed and capacity to store data) is compatible in the long run with the limitless wealth and variety of logic and imagination."

Another danger, according to Lambert, is neuro- or psychophysiological. Normally our perception of and exchanges with the outside world utilize all our sensory systems,—hearing, smell, touch, and so on. This diversity of signals ensures considerable redundancy, thus diminishing errors and losses, and allows all these signals to be integrated at different levels of the nervous system, including a permanent comparison of this information with previous experiences thanks to memory. All this allows us to choose an action freely and after reflection. However, computers involve a continuous utilization of the visual sense alone. We might wonder what will be the overall psychological diminution of an individual whose other sensory functions are little or badly utilized.

What is more, this new source of information can permanently modify social structures by threatening traditional hierarchies, most particularly the family, which is still the basis of our society. For thousands of years we have respected old people because of their knowledge, just like the medieval scholar or the modern schoolteacher. What will happen to these solid social values in the face of this new technology, which brings us many things but will tend to isolate men rather than unite them? We should not overlook the danger that in a few years — or, at most, in a few generations — stereotyped individuals may emerge from these data banks who will no longer be familiar with our greatest treasure, cultural polymorphism.

The danger is no less on an international scale. Instead of helping poor nations to develop and become free, as Jean-Jacques Servan-Schreiber so desires, the powerful nations that possess the most perfected machines and the most complex programs will make their satellite nations even more dependent. One should recall the exclusive use of English required for access to American data banks, which at present monopolize more than 70 percent of the world's information capital. Its power and speed make this a new form of cultural imperialism, of which we have never seen the like even at the most agitated moments of our history.

By taking over information, education, culture, knowledge, and interhuman relationships, computers — particularly television-based systems — constitute one of the most important phenomena of modern times. We are faced with the eternal dilemma of François Jacob's knife. Will we use it to peel a potato or to kill our neighbor?

RESEARCH AND TEACHING

The two conditions necessary for evolutionary progress of our society, both of which are closely related, are a regular increase in knowledge and its dissemination: that is to say, in research and teaching, two specifically human activities.

"Research," wrote Albert Vandel in 1968, "is the way in which evolution is manifested in a human society that has reached a certain psychosocial development." The cultural capital that we possess dates from long ago. Our everyday life is based on agricultural techniques that have not changed fundamentally for thousands of years, even if they have been improved in detail. They are the fruit of an inestimable number of individual initiatives and inventions that have been integrated into civilization. Thanks to them our Neolithic ancestors remain with us. The condition of humanity has constantly improved thanks to research. *Invention constructs the future.*

The Scriptures tell us, "Remember that you are dust and you will return to dust." But this applies only to our biological selves, fleeting and mortal, not to our sociocultural selves, the only aspect of us that will have any influence on the future of humanity because it lives after us. Each of us only exists by what we give to the community. This gift is our way of surviving beyond the limits of our physical life. However modest it may be, it represents immortality, for death represents the end of what we were but not of what we did. Our body may become dust again but not our creations, however insignificant. We survive by our research and its communication.

In the last few decades the industrial nations have tried to make research into a profession and the researcher a sort of specialist. Such an idea might be justified in a few domains of high technology, but it would be a dangerous abuse if it became generalized. In practice it is impossible to separate research and teaching. For a century or more most great inventions have come from universities, and the role of government research organizations should be to help universities by establishing collaborative contracts, as is becoming more and more the custom, rather than by creating their own laboratories, which always tend sooner or later to become isolated or sclerosed.

Some modern nations, led on by a desire for short-term profit, have mainly, or even exclusively, encouraged *applied* research leading to immediate results. This attitude shows a lack of understanding of the true meaning of evolutionary trends. Man should seek knowledge for knowledge's sake. There will obviously always be applications of any discovery, but it is impossible *a priori* to predict their importance. If we limit ourselves to applied research, we are amputating a large part of the originality of human activity. When birds look for the best site for their nest or foxes organize their hole or their hunting, they are undertaking applied research in their own way. Very often the desired result is achieved. In this way the future of the group is ensured, but little progress is made. They do not innovate but remain stationary. Fundamental research, the only true way of expressing the victories of cultural evolution, will always be beyond their grasp.

Because discoveries have no value unless they are publicized, research and teaching must always go hand in hand. "Teaching" should be understood in the widest possible sense. In the same way that all men are researchers, they are also teachers and students. For the characteristic of a cultural society is a constant spread of information. We learn and teach throughout our whole life. When we reach the threshold of old age we are very different from the adolescent of forty years earlier. We are constantly building ourselves from birth to death.

All individuals, even the most modest, have something to invent and to communicate to others. The egoist, the solitary, he who only works for himself, is socially dead. By isolating himself he remains outside evolution. When he dies his work dies with him, as if he had never lived. Although zoologically part of the species *Homo sapiens,* he excludes himself from humanity and does not merit the name of man.

In the last few centuries our knowledge has developed so much that the education of youth is entrusted to a specialized body of teachers concentrated in establishments that may either belong to the state or be private. Formerly, when libraries were few and far between and books were rare and expensive, the role of education was mainly to give a maximum amount of knowledge. Today, information is easily accessible through books and data banks, and education should aim more at explaining ways of thinking, methods of analyzing and synthesizing, and at the definition of concepts. However, we still tend to behave as if we were living when books were available only to the privileged elite.

At present, the industrialized nations educate their administrative chiefs in specialized schools where the education provided is essentially technological, in the widest sense of the word. Very little attention is paid to the humanities. It is not surprising that when these young businessmen enter public life, they are very rapidly swallowed by technological bureaucracy, because they were never prepared for anything else.

These young people will not only be called upon to manage businesses, factories, or banks; they will, above all, be responsible for the administration of their nation and thus will have to make decisions that will bind society for the next thousand years.

Unfortunately, the teaching of the humanities has never occupied a sufficiently important place in colleges and universities. But education must be rich, varied, and objective, for the dominating ideology will always be tempted to take over education for its own needs. By imposing the same political or philosophical profile on everyone, the whole wealth of human polymorphism is lost. Education should lead to a pluralistic and ever-changing view of the world that will allow everyone to develop his own individuality and integrate himself into society.

This is the only way that the individual and the group he belongs to will become richer. Unfortunately even the most liberal countries have often fallen into the trap of Darwinian elitism in their school systems. Almost always the model is the same — that of the best scholar. This elitism can be found in careers, for which recruitment is by competition that defines and consecrates a holotype and eliminates those who deviate therefrom by labeling them dreamers or fools. Such people do not correspond to the sociocultural, and even political, profile the ruling class desires and needs to maintain its power. What is more, industrial nations tend to favor more and more specialized technological education in very narrow fields, to the detriment of culture and a type of education that will provide openings to a wide range of subjects, thus permitting a better comprehension of social life. Education is oriented strictly toward the needs of production and has become ever narrower. It is a mere professional training, sometimes with highly political implications. Education and culture are being separated, and the individual is trapped by a series of preestablished techniques and concepts. Professional mobility is hindered and reorientation made very difficult. This is the basis of the dramatic situation now experienced by members of the managerial classes who lose their jobs and have difficulties in readapting.

An objective form of education should be open to the extreme variety of different cultures and therefore must renounce ethnocentric inclinations. Nobody should consider himself the center of the universe, whatever the color of his skin, the nature of his political system, or the age of his frontiers. All peoples have a right to the same respect. Everyone must try to understand what comes from elsewhere. One should not be afraid of the strange or the stranger but observe them and listen to them. They always have something to offer. Our view of humanity would be incomplete if we did not add a temporal dimension to this spatial notion of variety: thus we should not hesitate to leave our time in order to evaluate better what we owe to our predecessors, whose knowledge, transmitted and improved throughout the ages, has permitted the construction of the present-day human environment. In this way we would see that we are not the sole inheritors of a race and its crowning glory, but simply links in a chain that will continue after us. We must therefore think of those who will follow, and prepare for them a hospitable and fraternal world. The idea of the solidarity of generations allows us to understand the meaning of birth and of death.

This is why teaching and research are indivisible and should figure among the absolute priorities of our modern world. This is the only hope for evolutionary progress by mankind.[7] Those who devote themselves to the task will never see the end, for there will never be an end. The mid-nineteenth century saw staggering discoveries in the life sciences. In less than thirty years Charles Darwin, Louis Pasteur, Claude Bernard, Gregor Mendel — to mention only some of the most

---

7. In any country seeking progress, research and technological organizations and universities should be the responsibility of the same government department, to avoid redundancy and waste.

prestigious names — opened up the way to modern biology. They were followed by the wave of scientism we have already described, with people believing that science would soon know everything and resolve all problems. At that time there were *scientists;* now there are only *researchers,* and it is fortunate that this should be so, for if we *knew* everything we would no longer *search.* The prospects for our future knowledge are unlimited. The only certain thing about science is its precarious nature, but this does not detract from its value.

"The mind can only advance in its comprehension of facts if it avoids making its explanations acts of faith. Science is a way of perfecting the human mind, not a certain way of knowing the world. This is what distinguishes it from sorcery" (Soula, 1953).[8]

Research, linked to the development of the cerebral cortex, our so-called gray matter, has no real value unless it is published, exchanged, and spread widely: first of all between individuals, but also between nations. This should be easy, given modern communication techniques and the overall increase in the cultural level of most of humanity. And yet we are surrounded by aborted projects, disappointed hopes.

Have we forgotten the almost negative results of the agreements on security and collaboration in Europe, formulated in Helsinki in 1975, that recommended, among other things, the free passage of men and ideas between East and West? What progress have we made ten years later, compared with the progress that humanity would have made if these decisions had been put into practice?

At present a nation's contribution to civilization cannot be judged by the strength of its army, the size of its gross national product, the stability of its currency, or its raw materials, but by the number and quality of its teachers and researchers and the ways their knowledge is distributed both within the country and outside its frontiers: which is to say, in the cultural level of the society and its openness toward the world.

---

8. Throughout history, men have often confused hypotheses with postulates.

future possibilities

PROSPECTS

An animal lives in the present, remembering the past, but hardly making projects for the future. Thanks to the development of his intelligence, man is capable of planning ahead. Possessing imagination, he can conceive future situations and prepare for them. Einstein said, "Imagination is more important than knowledge." Throughout his history man has devoted himself to prospection, considering his past experience and imagining future possibilities. The hunters and harvesters of the Paleolithic era slowly developed their tools and their arms over thousands of years. Later, the Neolithic farmer plowed and sowed his fields, the shepherd began crossbreeding his flocks in order to create new races; in so doing they were planning ahead. In the old days projects were conceived within the family or the tribe; now they are made at the level of the nation, for all modern countries plan ahead. Some of these plans are on a worldwide scale, such as the projects of the United Nations or the programs of its agencies — WHO, UNESCO, and the others.

La Fontaine, the brilliant teller of fables and great psychologist, breathed humanity into his animals in order to teach his contemporaries without offending anyone. His grasshopper learned the cost of not planning ahead, while his ant, who stocked its provisions for the bad season, was behaving like a human.[1]

Among the questions that have always troubled our race are what we are, where we are, and particularly where we are going. An animal

1. Many animals save up for winter, like the common squirrel, adopted as the symbol of the savings bank in France. In fact, this hoarding is not rational and voluntary as in man, but an automatic pattern depending on a genetic program preserved by natural selection, since it is an advantageous feature for the survival of the species. Such hoarding behavior is commonest in groups living where the seasons are most marked, for the animal must survive during the cold season when nature offers little or nothing to cover its essential needs. Without the hoard the whole species would disappear.

is not the master of its own future. It follows the path of selection or chance passively. Biological evolution is not voluntary. Cultural evolution, on the other hand, leaves a wide margin for free choice. Almost always one has to choose between several possible alternatives. Today we are confronted with a grave crisis. We can and must find effective solutions, whose nature we can already foresee; but before discussing this problem let us imagine what the future of mankind could be.

THE END OF HUMANITY

Many of our contemporaries maintain, although often unconsciously, a fixist view of the world. They behave as if things have always been what they are and will never change. Nothing is further from the truth. All history has a beginning, a middle, and an end. Natural history is no exception. Life began, continues, and will end; like many others, our species will disappear one day.

We can already imagine the conditions under which mankind may die out. In the long run all life will disappear from earth. The sun will finally be extinguished. Its temperature, estimated at about thirty-six million degrees centigrade, is maintained by a series of nuclear reactions that slowly but surely are using up its mass. It has been calculated that in seven billion years our star will be entirely depleted.[2]

The source of the energy that allowed the emergence, development, and extraordinary diversification of living groups on earth will have disappeared. Our planet will be plunged into absolute cold, under a sky studded with stars even brighter than today but that no dawn will ever fade again. It will belong to a dead system like the myriads that already exist in the universe. It will be an enormous cemetery, perhaps preserving a few vestiges of disappeared civilizations, but there will be no more archeologists to study them. Like the legendary *Flying Dutchman,* manned by a crew of wraiths and condemned to voyage eternally at the mercy of the waves, the earth will continue its silent orbit to the end of time unless we manage to invade other planets, which is unlikely. Man's adventure will be over.

It will have been a mere vicissitude in the history of life, which itself is a minor incident in the history of the cosmos. Mankind was the consequence, but of what use was it? Has it a meaning? Depending on our individual opinions, to answer affirmatively may appear pretentious; to answer negatively, absurd. It is not the task of a researcher to take sides in such an argument; that is beyond his competence. Science

---

2. It is estimated that the sun loses four million tons of matter per second. But its mass is such that it will take seven billion years to disappear completely!

may explain the universe; it cannot justify it. "The world is what it is. We possess the means of scrutinizing nature and understanding its organization, but to ask why the world exists is a false problem that has no answer" (Vandel, 1968).

In fact, the reply lies in another domain, that of metaphysics — which, as we saw earlier, is independent of knowledge. Everyone has his own opinion about the questions we have just raised, but whatever an individual's view is it concerns him alone, and in any case remains impossible to verify because it is inaccessible to experimentation.

In the shorter term, our species could disappear by a sort of biological "wearing out," which brings about the aging of most widespread, little-differentiated groups. Earlier we saw that relict species that have survived for very long periods without change — the "living fossils" — are rare, specialized, and almost always hidden in isolated niches away from all competition. Biologically man has no feature that may predispose him to become a living fossil. We know little about the causes of the aging of species. It may be connected with an upset of the genetic machinery — that is, the genetic program of each species — as a consequence of errors accumulated with time. Thus the more complex a group, the more fragile it may be. Only a few bacteria, with their single, ringlike chromosome, have been able to survive for more than a billion years without too much difficulty. But it is unlikely that man, and even most mammals, will be able to do likewise.

Before being concerned by aging, our own species must survive many other obstacles. We have already discussed the ice ages that have periodically caused the cooling of the surface of the earth, only to be followed after varying periods of time, by a new warming. We do not understand the origin of these variations in temperature. We have only barely emerged from the last Ice Age, the Wisconsin or Würm glaciation, which ended 12,000 years ago. During each cold period an enormous mass of water is trapped in the form of ice around the poles or in mountain glaciers, which lengthen considerably. The level of the oceans falls, uncovering landmasses that were formerly below sea level. Continents that were once separate became joined together, while submerged mountain chains form archipelagos. Even a quite moderate drop in the mean annual temperature of five or six degrees Centigrade, if maintained long enough, would cause the sea level to fall by one or two hundred meters or even more. Figure 31 shows what the surface of our planet must have been like at the time of the Wisconsin glaciation.

This was the era when man was able to invade America, Australia, Tasmania, and many other hitherto inaccessible regions. When the earth warmed up again, the sea level rose, flooding continents and transforming mountain chains into islands. Today we are living in a

Figure 31. Map of the world during the Wisconsin or Würm glaciation. Land that was then above sea level but is now flooded is cross-hatched.

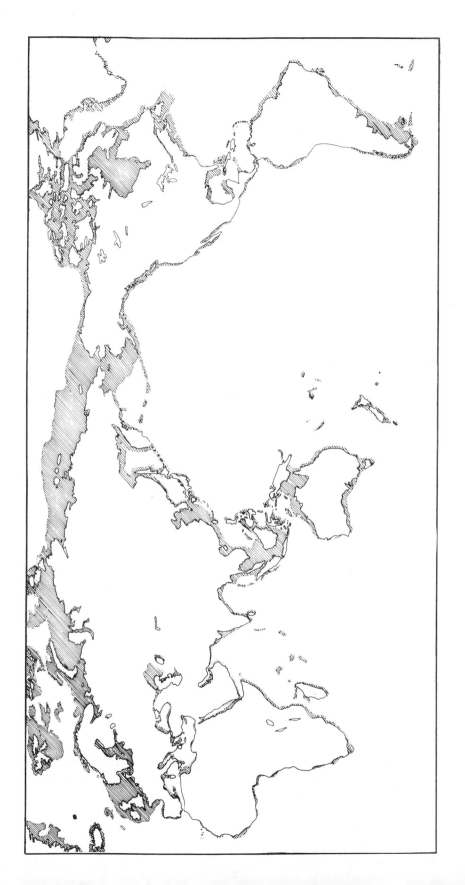

period of warming that is, however, irregular, like all climatic phenomena. This warm period was interrupted between 1550 and 1850 by a slightly colder phase—one or two degrees Centigrade lower, on the average—that is sometimes referred to as a "little Ice Age." Between 1900 and 1950 there was a new increase in temperature, but now the trend is reversing again. It has been forecast that this relatively cold period will probably last for twenty years or so, with temperatures increasing again around the year 2000. But to what extent? If in the centuries to come the increase in temperature is sufficient to cause the oceans to rise about a hundred meters—which is not at all unlikely—many of the capitals of the modern world will disappear beneath the sea; Paris, London, Brussels, the Hague, Lisbon, and Rome in Europe; Montreal, New York, Washington, Los Angeles, San Francisco, and many others in North America; in Asia, Tokyo, Osaka, Shanghai, and Canton; and most cities situated along the African coasts. Seen on a geological time scale, the map of the distribution of our species could be profoundly modified.

The topographical distribution of mankind as we see it today is quite recent, dating from the warming that followed the last Ice Age and is still with us. But conditions were different during prehistory. Our remote ancestors had to face great climatic variations, with subsequent modifications of the regions they could inhabit. In the last 500,000 years, three major ice ages caused enormous upheavals in man's occupation of the earth. During these ages men were predators, relatively few in number, and widely dispersed. They lived as hunters and harvesters and were of necessity nomadic or seminomadic with no fixed abode, although a certain number of particularly favorable areas might have been permanently occupied or at least inhabited regularly, although only periodically. This would have depended on the richness of the animal and plant resources of the environment, the existence of natural shelters, and the ease with which the area could be defended against invaders. This is why many Paleolithic civilizations occupied the foothills of mountains or the coastlines of oceans, particularly around river estuaries, which for long were the easiest means of penetrating into the interior, particularly where there were forests, as was the case over much of the earth's surface. Thus progressive climatic modifications were spread over a long period of time and did not influence the topographic distribution of man over periods of a few generations. They caused only minor inconveniences for these men, who were relatively few in number and quite mobile. They moved around as they wished. But all this began to change a few thousand years ago at the time of the Neolithic revolution, when man began to settle in those zones that were most favorable for agriculture and the rearing of animals, particularly with respect to the terrain and the climate. From then on, mankind became generally sedentary, in the geographical and ecological context of the end of the Würm period, whose conditions were very similar to those of today.

First villages grew up, then towns. The first walled cities were founded as recently as seven or eight thousand years ago. New agricultural techniques increased resources, favored the storing of produce, and permitted a considerable demographic expansion. Security and the standard of living increased. Soon the first agricultural empires of the eastern Mediterranean emerged. Trade grew up, not only in agricultural but also in industrial products such as cloth, pottery, and, later, bronze and iron objects. The level of culture increased with the development of writing and the alphabet. Over a period of a few thousand years, mankind began to live in its present context. Apart from a few more or less legendary traditions — like the Great Flood — climatic changes after this period were relatively modest and have hardly left any trace in the collective memories of nations. All this indicates the recent and precarious nature of the way in which modern humanity is distributed — since it represents the result of a particular moment in the world's climatic evolution. If the great civilizations had emerged in a different era — for instance, during a warm period, in a damp, tropical world with greater rainfall and a higher sea level — the topography of the land, the sites of cities, the frontiers of empires, would have been quite different. It is difficult to know how mankind will react to future ecological changes. Such modifications will extend over thousands or tens of thousands of years, thus giving our species, which is already very advanced on the pathway of sociocultural evolution, plenty of time to adapt to the very slow variations that will certainly take place.

In fact, the most dangerous threat to our civilization, because we are already living it — one that can threaten the very future of humanity within quite a short time, of the order of a few decades or at most a few centuries — is the situation of crisis already examined, in which we seem inexorably bogged down. Social-Darwinist ideology, as still pursued by many people, can only offer us rivalry, conflict, and elimination, which could conceivably end in an atomic apocalypse.

Nevertheless, the level of conscience attained by modern humanity and its enormous capabilities should point to a much less tragic outcome.

THE POPULATIONAL CHANGE

Accepting a populational model, and at the same time abandoning the typological approach, has important consequences. First of all we must abandon the idea of the holotype. No one can impose a model that does not exist. All groups are made up of widely varying individuals. At the level of man these variations are both biological and cultural. We must accept them, study them, and understand them because they are our greatest form of wealth and represent our hope for the future.

We must also reject the idea that rivalry, confrontation, and elimination are necessary for progress. On the contrary we must promote mutual aid, exchange, and fraternity within nations and between nations, for no one must be left by the wayside. We must construct a worldwide humanity from the mosaic of classes and nations that man has created artificially. This represents the only valid approach at the sociocultural level reached by modern man. Thus action is necessary at two levels: within nations and between nations.

INTRANATIONAL MODIFICATIONS

At present the existence of rival classes — created entirely by an illegitimate system — is an unnatural phenomenon. All social groups must have their hierarchies, but they must remain flexible at the cultural level in order for polymorphism to play its full role. The existence of rigid holotypic categories puts us at the level of the termite's nest and takes away the most original aspect of human nature, its flexibility and variations. Tyranny did not die with Louis XVI, and it must be replaced by cooperation. Everyone must play his part according to his abilities, desires, merits, and needs. All must share the responsibility in a decentralized and self-governing system — which is to say, a system worthy of the human condition.

The abandonment of the class system implies the abandonment of profits, the basis on which we have lived for so long and which has been amplified since the industrial era. Henceforth we must look not for monetary interests but for human interests.

RELATIONSHIPS BETWEEN NATIONS

To be fully effective, integration must be achieved at an international level. An isolated country, however powerful, that decides to be an example will soon fail when faced with the aggressive competition of its neighbors.

The first and most urgent measure is to bridge the gap that separates the four groups of nations defined in Chapter 13: those that have both raw materials and technology, those that possess technology alone or raw material alone, and the other quarter of the world that possesses nothing.

Soon after the end of the second World War, Léon Blum stated in New York on April 12, 1946, on the occasion of the first anniversary of Roosevelt's death, "Prosperity has become indivisible, like justice and peace. Solidarity between nations has become a physical necessity as much as a moral obligation. No nation can maintain lasting prosperity in a world pervaded by a spirit of hate and conquest, just as no nation can preserve lasting freedom in a world subject to tyranny and oppression" (cited by Lacouture, 1977).

Geopolitics: The Rights of the First Occupant

For the naturalist, national sovereignty is just as relative as are property rights. The geological evolution of the earth is the fruit of chance. This is what Lyell had imagined; he was the first nineteenth-century geologist to be strongly influenced by the evolutionary theories of Darwin. Man was in no way responsible for the way in which Pangaea broke up during the Jurassic era to produce the outlines of the modern continents. At that time mammals did not exist. They represented a mere potential in certain reptiles. We can take just as little credit for the geological structure of the continents. When human populations began to establish themselves much later as a result of the events of prehistory and history, they knew nothing about what was under the surface of the land on which they were living. When they arrived the Carboniferous deposits that gave rise to coal, oil, and gas had already been there for ages. These human populations chose their sites according to their surface features, such as the presence of natural shelter, drinkable water, and other resources permitting harvesting and hunting, and later, husbandry and agriculture. In other cases they sought refuge in isolated or inhospitable areas with difficult access, in order to protect themselves from aggressive neighbors. Much more recently, at the time that Europe and America entered the industrial era, men became interested in the riches of the soil on a grand scale. This occurred first in their homelands, and involved particularly coal, which for more than a century was the essential source of industrial energy. Later, they explored overseas, using techniques and capital brought in from the outside.

Numerous studies have been devoted to the origin of property rights, either individual or collective. An occupant may have legal rights, but for the anthropologist these are as arbitrary as national frontiers. Those who control the riches of the earth by accident do not have the right to starve others, for these riches belong to everyone and must be at the service of the whole of humanity. The same is true of science and technology. Certain regions have been favored by the progress of knowledge following a series of chance events. The Neolithic revolution rapidly invaded Europe, whose climate was propitious for the plants and animals that had been domesticated in the Near East. It also extended to the coasts of North Africa, but spread only much later to the rest of the African continent, for the plants imported by the first settlers were not suited to the damp, tropical climate.

The development of Neolithic culture in the southern Sahara was due either to endemic foci or to domestic species imported from Southeast Asia through Madagascar. Thus Africa became involved in the Neolithic era more slowly than Europe for purely climatic and geographic reasons.

A few thousand years later Renaissance Europe, urged on by its traders and justified by religious proselytism, embarked on its colonial adventures. This was the time of the great discoveries, which saw the growth of the first overseas empires that drained a large part of the world's wealth toward the white homeland. All this favored the entry of Europe and later North America into the industrial era. We have seen how the hierarchies that grew up between nations were legitimized in the nineteenth century by the Darwinian view of humanity, based on a difference in value between races, leading to inevitable competition.

The World Torn Apart

Although it may appear less tense, the international situation has hardly improved in the last century. It has even worsened, if we consider the deep gulf that separates the poorest nations from the most powerful — a gulf that is constantly widening.

The morals of international politics are particularly disturbing. The governments of civilized countries quite rightly forbid piracy, but every day we see cases of legal and accepted piracy in the relationships between nations. Frequently a head of government can be responsible for initiatives that, if perpetrated as a private citizen, would send him to prison for the rest of his days. At present there exists an individual morality that we try to follow and apply as best we can, but there is no international morality. The law punishes theft, bodily harm, and murder, but all of these are permitted on a grand scale at the level of states. If our neighbor is being murdered, we normally try to help him or at least alert the police; but when faced with the systematic massacre of a third of the population of Cambodia, we make speeches in the name of a sacrosanct national sovereignty that in any case no one respects. For a long time there have been limits to private property, but there are no limits to the maneuvers adopted by a government in its relations with other countries. The only limit is the fear of reprisals. This is the morality of dissuasion. But woe betide those who cannot fight back. This mad politics incites even the poorest countries to procure ruinous armaments. They waste the resources they should devote to development and to raising the living standards of their inhabitants. But this comforts a few local potentates, the servile lackeys of the great powers, who would be swept away if the living standards and the education of the people were improved. The relations between developed nations are hardly more encouraging. The Eastern and Western blocs pursue an underhanded war, each trying to nibble away at the empire of the other without daring to openly admit the split that resulted from the Yalta agreements.

In his annual message on the occasion of the day of peace, January 1, 1982, Pope John Paul II denounced "the unbridled nationalism that attempts to achieve hegemony, in the context of which relationships with other nations seem faced with an impossible choice, either satellization and dependence, or competition and hostility." Emphasizing the value of communication, he added: "They work for peace who, by the diffusion of information, remove the screen of distance, so that we feel really concerned by the fate of those men and women who, far from us, are victims of war and injustice." He concluded: "War is the most barbarous and inefficient means of resolving conflicts." Indeed, wars have never resolved anything.

THE PLANETARY CONSCIENCE

The road of evolution, whether biological or cultural, has never been easy. It is full of pitfalls. History has always been borne painfully; thus, in spite of the anguish of our time, we must neither abandon hope nor be discouraged. An army can stop an invasion; it can never stop ideas. As we said earlier, to be effective a change in society must involve all nations, failing which, those that might be tempted to follow a new pathway risk being crushed by the more powerful ones.

Example is as contagious as the plague. The most horrible persecutions did not stop Christianity from invading the Roman Empire. Later, liberal ideas originating in France penetrated throughout Europe. No *cordon sanitaire* could stop them. A century later socialism spread in the same way.

With the informational means mankind now possesses, no government, even the most severe police state, can keep the truth from the people for long. The speed of communication that Pope John Paul II mentioned in his speech is perhaps one of the most important new factors of this century. For thousands of years every human society accepted its fate. Peasant risings were rare and almost always very localized. Even the the hardest situations seemed bearable, for it was difficult to imagine that things could be different. What is more, although poverty was more widespread than today, absolute poverty was less common, and differences within a group, the only ones people knew about, were accepted as part of natural events. Apart from a more or less informed elite, most people knew nothing of what was going on elsewhere, and everyone tended to consider his fate as normal or inevitable, with religion confirming this idea.

This era has gone. At present, information travels through the world in all directions and at every moment. It knows no frontiers: neither geographical, for radio waves overcome all barriers, nor cultural, for information is not destined, as before, for an elite that knows how to read, but is broadcast by the spoken word in all languages. Later we shall judge the importance for the mid-twentieth century of the "transistor revolution," which allowed the spread of news from the whole world to all peoples, even the most isolated. The nomads of the Sahara, the peasants of the High Andes, the inhabitants of the African forests, all of whom lived in a situation thought to be unchangable, learned what was going on beyond their own horizons. Thereafter injustice seemed intolerable, and the exploited rose up against the exploiters, often violently. Today everyone desires a minimum of comfort, security, health, education, and liberty. The situation in many Third World countries is not freely chosen by the people but is almost always imposed by pitiless dictatorships, backed by some great power that reaps the benefits. Certain countries in Central America, Africa, and Southeast Asia are sad examples of these endless, violent situations. Nowadays guerilla warfare is spreading rapidly. This trend will continue as more men achieve a higher cultural level and at the same time develop a political conscience.[3] We can foresee quite surely that these trouble

3. At the end of the last chapter, we spoke of the miserable results of the Helsinki agreements. This indicates the necessity of establishing some form of morality that would be respected and followed. As individuals, the representatives of the states participating at the conference were doubtless honest citizens in their everyday life. So why do they not keep their word when nations are involved rather than individuals? This unresolved question is asked periodically. The answer lies in the fact that it is still rare to find men who feel themselves part of the same global ensemble as the whole of the rest of humanity. We stick to the notion of rival factions, grouped into competitive nations or empires. Most of our contemporaries are still in the social-Darwinist phase.

Nevertheless, the notion of the common heritage of mankind is slowly making itself felt, at least in territories that have not yet been appropriated. Thus the General Assembly of the United Nations adopted a resolution on December 18, 1970, that the seabed outside territorial waters would not belong to the first nation to set foot there. It belongs to our common heritage and can only be placed under the authority of an international organization. Any profits from the exploitation of the seabed are to be distributed among all states, with priority to the least developed. In the same spirit agreements signed between 1969 and 1979 decided on the nonappropriation of the moon and other heavenly bodies, of the orbits for geostationary satellites, and of radio frequency bands. Thus, legally, even the poorest nations cannot be denied the utilization of these new milieux and any advantages they might possess. But they must be technologically capable of exploiting them—otherwise these resolutions will remain pious wishes, of no practical value.

spots will increase in number, particularly in the tropics, where the majority of underdeveloped peoples live.

DIVERSIFICATION OR FINAL STRUGGLE

We have reached the hour of choice. Our species is unique and so is our ecological niche, and we are all in the same boat. Two attitudes are possible.

The first is based on social Darwinism. It involves dispute, military if necessary, for raw materials and industrial power so that the strongest can preserve their advantage or take over that of others. This competition may involve social groups. Then it is class struggle as defined by nineteenth-century Marxists — about which, however, our late-twentieth-century Marxists are more discreet, having learned by experience. The competition can also involve nations. This is class struggle at a world level, as has been widely practiced by industrial nations for a century. This politics has been largely unsuccessful. The danger to the world should be obvious.

The other attitude is populational, replacing struggle by cooperation, creating a new international order, and choosing the enlargement and diversification of our ecological niche instead of beating ourselves to death over what is left of it. This attitude is the only one that corresponds to the ideals of evolution.

Enlargement of the Human Niche

The enlargement of our niche in new directions has become a necessity. Since time immemorial we have behaved as if natural resources were inexhaustible. This is an idea inherited from prehistory that, in spite of what we now know, still marks the minds of our contemporaries very strongly. In this realm we have proved improvident, and we are now gathering the bitter fruits.

Let us take a topical example: oil. Oil has played a great role in our ecological niche for the last few decades. Now it occupies the forefront of our economy, having almost entirely replaced the coal that helped launch the nineteenth-century industrial revolution. Oil has not only become our major source of energy, it also intervenes in the manufacture of many utilitarian products. Today the petrochemical industry represents an essential part of our activities, either directly or indirectly.

As oil reserves become scarcer, there looms the danger of a battle for the wells. Many people believe this, and the armies of the major nations must have plans drawn up along these lines, although for the moment the superpowers confront each other through the intermediary of small countries in the Near East or Central America, for example. The danger is so much the greater in that oil is not only an indispensable source of energy for the industrialized nations but also a prime strategic material. Today any power that controlled the whole of the Arabian Gulf area could impose its wish throughout the world. Another policy, better representative of the laws of progressive evolution, would be to extend beyond this overcrowded niche and prepare others. The first alternative that occurs to many people is to replace oil by atomic energy. In fact this option also has its inconveniences, although they are more in the domain of sociology and politics than in that of technology.

Atomic Energy and Democracy

Almost everything must have been said about the real or supposed dangers of atomic power plants. It would be impossible to summarize in a few pages the tons of documents published on this subject, which is often discussed with more passion than reason. The construction of numerous nuclear power stations will create a dangerous situation because whatever we do they will represent a possible means for a dangerous madman to obtain massive destructive power.

Unfortunately, our present situation (and even more so that of tomorrow) does not allow us to reject atomic energy, at least for the time being. If we really were to suffer a lack of energy in ten or twenty years, the resulting crisis would be incomparably greater than the one we are subjected to today, and the poorest members of the human race would be in grave danger of plunging into a situation whose tragedy would exceed anything we can imagine just now. But once again the supporters of atomic energy must not have too many illusions. Nuclear power can help us cover our present needs and those of the immediate future. It cannot resolve the long-term problems. Uranium is no more inexhaustible than oil. The widespread use of atomic power will lead to a new energy crisis in a century or two, resulting from the exhaustion of the fuel, which will have simply changed its state. Our successors will know a situation worse than our own, for their energy needs will be much higher.[4]

The danger of atomic energy to modern society is not really that of the risk of accidents. That risk is fairly remote, certainly much less than the daily aggressions that the traditional power stations represent for

---

4. In the meantime the great powers are stocking enormous potential nuclear energy in the form of bombs. If the rich hypermilitarized nations abandoned their armaments, they could make available to the people a considerable mass of energy that would help the poorest sector of mankind emerge from underdevelopment.

our organisms. Whether they use gas, oil, or, worst of all, coal, they are continually discharging highly carcinogenic smoke into the atmosphere. The danger of atomic energy is that its sources are so concentrated. It is more political than technical. Let us imagine what would happen in an average country like France if nine-tenths of its energy were furnished by five or six supergenerators controlled by a few dozen specialists. The desires of a handful of men or the action of a small commando force could paralyze all of the nation's activity in a few seconds. There would also be the danger of blackmail that a moderately well-organized terrorist group could exercise. Whether we like it or not, this type of industry would lead inevitably to the development of a police state in the name of national security and the protection of the citizen.

Technological Decentralization

A genuine widening of the human ecological niche means a diversification of its wealth, either energetic or otherwise.

As we have seen in the preceding chapters; such a trend has a highly positive selective value because it reduces competition, which is always dangerous within a given group, while increasing its wealth. This should encourage us to diversify our energy sources very rapidly, first of all by turning toward those that are renewable indefinitely, such as watercourses and the sun. Most industrial nations have already developed hydroelectricity, at least in mountainous areas. On the other hand the potential energy of rivers and tides is still very little exploited.

Solar energy, which in the present state of our technical knowledge can hardly be utilized industrially, can, however, already cover a part of our domestic needs for hot water and heating. The advantages are numerous. Solar energy does not pollute and is inexhaustible, at least for a few billion years, which will certainly account for much more than the rest of the history of our species. Only the decentralization of our energy supplies can guarantee our individual liberty.

The same argument is valid for all sorts of activities, even if they concern less essential sectors than the production of electricity. Small and medium-sized enterprises scattered widely throughout a country, thus dispersing production, play an irreplaceable role in ensuring a more homogeneous sociological structure that is more regular and more flexible than a network of a few large enterprises and supermarkets (Drouin, 1981a). They also make a country less vulnerable. Unfortunately, except during preelection periods, little is said about small and medium-sized enterprises, which often die out slowly because they cannot face the competition of multinationals.

Ecological Limits and Technological Progress

The energy crisis, which has today taken the form of increases in the price of oil and tomorrow will see a rise in that of uranium, is only one facet of the problem humanity is facing. It is a single tree in a large forest, for the true problem is much wider and touches on the whole of our environment.

The production of large quantities of low-priced energy, using new techniques such as the fusion of hydrogen, is quite likely in the next few decades. This could mean that all nations, even the poorest, could attain the standard of living of the Americans. But our environment might not be able to tolerate such a situation, there would be the danger of an ecological "disintegration," due in particular to a change in the world's climate resulting from the liberation of a large amount of heat. This would have catastrophic consequences on the balance of nature, and penury would certainly follow.

Whatever we do, our environmental patrimony is limited. We must begin to manage it like a "good father" and finally get rid of the competition that leads to waste and enmity among men. Does this therefore mean that we must resign ourselves to permanent immobility in a world whose resources have reached a ceiling and thus refuse all growth, however gradual? This fixist view of the phenomenon of man is contrary to the cultural evolution that has manifested itself for thousand of years by new technical developments. There will be many others that will permit us to derive more benefits from the same elements, whether living matter or minerals. We will do this by improving their utilization, increasing productivity, inventing new forms of production —particularly in the domain of genetic engineering—and developing methods of recycling to recover as much raw material as possible and reduce to the minimum irreplaceable losses of waste products. This is why, more than ever, we must encourage scientific research that can give us the means of better utilizing what our earth has to offer us and ensuring a slow but constant growth while using and reusing the same basic elements. At present, in spite of a skepticism that pervades certain sectors, casting doubt on the interest and utility of scientific progress, research is no longer a luxury but a necessity.

THE NEW WORLD ORDER

The development of new techniques of exploitation and recycling will be of little use if they must be available principally, and almost exclusively, to those who already have technological knowledge, as has been the case until now. Humanity cannot survive if it continues its egoistic path, both within nations and internationally. After a century and a half of social Darwinism we must find another model for civilization.

The stupefying rivalry on which our predecessors built their wealth is now turning against us. Inundated by unsalable overproduction, the industrialized nations will in the end destroy each other in this merciless struggle, while the countries of the Third World will be ruined and starve to death. The process has already begun.

Today a change is necessary, but in which direction? We might try here to reply to a fundamental question. In this industrial age, is man's aim to continue to accumulate new wealth, new profits, that will benefit only a chosen few, who represent the "best" specimens in the sense of Galton or Vacher de Lapouge, or should he increase his knowledge and disseminate it to as many people as possible in order to widen the human niche to new horizons? Throughout the generations the nature of the cultural revolution has hardly changed, even though, in the euphoria of technological success, the industrial revolution has sometimes hidden the true meaning of life. Modern conditions demand a more realistic approach. The civilization that has been built up over a century on competition and waste is declining. We must now invent the civilization of the third millennium. It will involve upheavals in relations between people for which recent history has hardly prepared us. But the facts are there: just like truth, that "daughter of the times." We are living a paradox in a world where some producers, flooded by excess goods that they cannot sell, and being ruined, live side by side with great potential consumers who can neither produce nor buy, for they lack the means. This second group is constantly increasing. More and more countries are heading toward a state of absolute poverty, toward famine. We must end this immoral and absurd situation, which is holding back all progress by humanity and can lead to its disappearance. It is essential to give all nations the economic, cultural, technical, social, and moral means of becoming associated with the ascent of humanity. No one must be excluded. We must reshuffle the cards internationally and prepare a worldwide "new deal," as Claude Cheysson described it in 1981. What we must try to do is open the enormous potential markets represented by the Third World to the agricultural and industrial products and the services offered by the industrialized nations. At the same time we must banish famine and underdevelopment on the one hand, and inflation and unemployment on the other. This resurgence of the world economy is within our grasp, but it needs an international effort that no one can escape. This is not just a fanciful dream. To be persuaded of this, one need only think of the history of Europe and America over the last fifty years. The exceptional growth that Western nations underwent in the 1950s and 1960s resulted from technological progress in the postwar years, but also from the opening up of the markets of culture and leisure to formerly underprivileged classes. In France this rise in living standard was in large part the fruit of the social measures taken by Léon Blum and the Popular Front of 1936. Twenty years later, in spite of the Second World War and the occupation and sacking of France by Germany from 1940 to 1944, they spelled prosperity. This happened despite the fact that at the time of the socialist government, financiers and conservatives foresaw an impending apocalypse. A little earlier, Franklin D. Roosevelt had saved America from the most serious crisis in its history by preparing *his* New Deal, which, as its name suggests, involved a new distribution of wealth. At the end of the Second World War, America put forward the Marshall Plan, which represented a gift of $45 billion to the European nations to help them rebuild their ruins and start up their economies again. This generous action also opened a vast commercial market to America by replacing exhausted allies with solvent partners. Today, the same operation is needed but on a worldwide scale. It will require much more money — probably more than $100 billion per year for several years, according to Maurice Lauré (1981). This effort is not beyond us if we consider that in 1980, military expenses throughout the world must have been on the order of $500 billion. So a quarter or a fifth of what we devote to industries of death is sufficient to save mankind from misery and injustice and put it back again on the path of progress.[5]

If all the heads of the great nations were suddenly inspired with wisdom and agreed to demilitarize the world and use the enormous resources that would thus become available for education and development, the world crisis would soon be resolved. Such an operation would involve rigorous planning. Too often much of the aid given to poor nations, particularly when it is bilateral, goes to the elite that holds the power and ends up in Swiss or American banks, while the population only receives the crumbs.

It would be necessary to know the specific needs that must be satisfied as a priority for each country. These would include saving the population from starvation and then evaluating their productive capacity to allow them to integrate themselves efficiently into the international economic scene.

In such an enterprise education would play an essential role and must be considered as an absolute priority. It would valorize and perpetuate the results of aid, which without education would be lost like water in the desert. An old Chinese proverb says, "If you give a poor man a fish,

5. As a comparison, while the world armament budget in 1980 was more than $500 *billion,* that of UNESCO, the United Nations agency responsible for education, science, and culture, had a ceiling of $100 *million!*

he will not be hungry for a day. If you teach him to fish, he will not be hungry for the rest of his life."

Some people raise the objection that the industrialization of poor nations will increase competition for our own industries. We must reply clearly that massive planned aid must be accompanied by social controls. In particular there must be a statute that offers sufficient guarantees to workers. A properly organized international aid program must not play one country off against another.

If all these conditions could be fulfilled, the development of the Third and in particular of the Fourth World, far from creating a danger of competition, would represent not only a great moral victory but also a beneficial situation as far as the Old World was concerned, which in the midst of its crisis would see the opening of enormous long-term markets: 80 percent of the world's population at the end of the century will be more or less underdeveloped and will need enormous quantities of consumer goods and equipment.

THE REALISTIC UTOPIA

The plan of development that we have just described is not utopian. We could achieve true cultural socialization involving all people and make a world where everyone would have his share of liberty, exchange, and dignity without distinction of color, creed, or language. Some liberal minds of the nineteenth century had already counted on this worldwide republic and the abolition of frontiers and classes. We know what happened to their desires, and the disasters brought about by short-sighted egoism in recent history.

The failure is obvious, but it does not abolish hope. Whether claiming to be humanistic or "scientific" — that is, Marxist — the last century's utopia was an impossibility. Born in the West, in nations undergoing rapid industrialization and made up of opposing classes vowed to bitter rivalry, this white utopia hardly bothered about what was happening elsewhere, unless "elsewhere" meant regions that could be useful to it. Our colonial adventures had no other justification. Our epoch has been based on a social-Darwinist model of which we have never been able to rid ourselves. This explains the failure of all attempts to create a new order.

In 1960 John Kennedy launched a "decade of development." In 1961 the General Assembly of the United Nations proposed to accelerate development in the most backward regions. In 1974 there was another declaration for a new world economic order. In 1980 the United Nations suggested worldwide negotiations for a better economy. On September 1, 1981, UNESCO held in Paris a United Nations conference on the least-developed nations. The North–South Conference held in Cancun in Mexico in October 1981 was hardly more successful, in spite of the general declarations that resulted. This will continue to be the case as long as mankind remains prisoner of a system that appeals to his ancestral instincts for territory, dominance, defense, or attack, which are no longer tempered in man by innate behavioral patterns formed through natural selection that would have the effect of protecting the species from self-destruction.

Our intellectual development brought us liberty but makes us conscious of and responsible for our actions. This is why the construction of a new world order that would translate the sociopolitical principles of internal democracy into an "external" democracy, or true international freedom, is not at all impossible.

The day must come when we can look forward to this universal multicultural republic, uniting men through mutual aid, exchange, and progress, instead of separating them through rivalry, isolation, and struggle. The major obstacle to its birth will still remain, for a long time, the rigid frontiers — the former scars of history that today have become refuges of national selfishness and tomorrow will become instruments of multinational interests. Many politicians are conscious of the necessity of such evolution at the end of the present century. In 1976, Willy Brandt declared: "I remain convinced that all peoples have a great need to mobilize all their reserves of knowledge, creativity, and intellectual responsibility. This must be done together — that is, beyond national frontiers, states, and political blocs. More than ever we need close communications between minds. The problems of our time cannot be resolved in a narrow framework or in isolation."

At present all forms of nationalism, and particularly the hypernationalism that we see here and there, constitute an aggression toward our species, for it prevents our attaining worldwide supranationalism and cooperation. Einstein said that "nationalism was a childhood disease of humanity," but *this* childhood disease can be fatal.

The difficulties in arriving at a worldwide entente are aggravated by the division of the world since Yalta into two blocs, each trying to impose its hegemony according to a different ideology.[6] But no wall, no curtain, whether of iron or of bamboo, can stop the spread of information all over the world. Consciences are waking and proclaiming truth, sometimes at the risk of lives. Let there be no doubt. At present mankind has the technical and material means to escape from its crisis but does not yet possess the political or moral means.

Now is the time to acquire them and to take our destiny in hand. This attitude alone corresponds to the principle of cultural evolution, which

6. This situation has often been analyzed by historians (see particularly Conte, 1964).

will permit us to enter a new phase of history, when yesterday's utopia will become tomorrow's reality.

BIOLOGICAL IMPLICATIONS: STEPS IN INTEGRATION

As we near the end of this work, we might ask ourselves whether the constitution of a universal republic integrating all peoples on earth into a single group corresponds to an idealistic but unreal vision, or whether the emergence of this type of society is practical and corresponds to the laws of life. In other words, is it a philosopher's dream or a biological probability?

Before replying to this question we must go back and look at what nature has achieved since the dawn of time. When we look at this parade, extending over hundreds of millions and even billions of years, we see the most grandiose and moving spectacle that the human mind can conceive. It is the visible image of evolution.

At this level, evolution would seem to represent two processes with very different implications.

1. Speciation, the creation of new species; and typogenesis, the creation of new types of organization, which can be considered as a chain of accumulated speciations, within which many links remain unknown.

2. Megagenesis, a series of small numbers of steps toward integration spread throughout time. Its consequences are infinitely greater than those of the emergence of new species or even of new types (Ruffié, 1976).

At each level we find the same elements as at the previous level, but in more specialized form and strongly integrated into a whole, making up a sort of "superindividual" with new qualities. Thus the semantic content of each level becomes richer without the basic elements that make it up changing in nature. This enrichment is simply due to better and more complex integration. Let us reconsider the example of the alphabet we spoke of earlier. A series of letters taken individually, such as *e, f, o, r, t,* have little semantic value; their significance is purely phonetic. The same letters arranged into words, such as *fort, effort,* and *free,* have a much greater value. They have even more value when the words form sentences: "It was an effort to free the fort." Using sentences made up of similar letters I can write even more significant texts, as Karl Marx did in writing his *Kapital,* Darwin in the *Origin of Species,* and St. John in his Gospel. Each has a much higher semantic value because it describes a philosophy, a theory, or an ethical principle.

The letters may be said to correspond to structural genes, or rather the peptide chains that they produce. Thanks to these chains, living matter is built up through a series of steps, each of which introduces new properties without the need for new basic information in the system after a very early stage.

The major integrative steps that can be recognized begin with the *molecular* stage, the first to appear in evolution. The macromolecules that formed in Haldane's "primeval soup" have a very modest semantic value. They have enzymatic properties or reproduce by self-duplication. Thus the possibilities of these macromolecules are limited and soon reach their ceiling. As soon as they attain a certain complexity they become fragile because of their size and, as a result, break up. Nature cannot progress unless it takes another step. Thus these complex macromolecules group themselves and specialize into a functional ensemble, the *cell,* a sensorimotor unit capable of reacting in an adaptive fashion. The most complex cells, the ciliated Protista, even show signs of an elementary "memory." If the macromolecule can be compared to a letter, the cell is a word. But an isolated cell that must face all the constraints of life alone soon reaches its own limits. The only way of progressing is to specialize further and form closely integrated groups. Thus *animals* are born, and with them intelligence.[7] Pursuing our comparision with the alphabet, animals represent sentences. Just like the isolated cell, the solitary animal obliged to face all the problems of existence soon exhausts its possibilities. It can only go further by forming *societies* made up of specialized individuals but held together by integrative systems that are sufficient to prevent the group splitting up. A society implies a certain number of rules, which are innate and rigid in primitive forms such as insects, but acquired and flexible in the cultural forms that culminate in man, whose conscious, calculated rules constitute *ethics.* Compared with letters, the alphabetic units, societies represent the texts of St. John, Darwin, or Marx.

This model of evolution by stages that periodically change in nature is not surprising. Our world is limited and could not tolerate an indefinite quantitative evolution. In a closed system a uniformly accelerating trend soon reaches its limits. The only way to progress is to modify *qualitatively* and introduce a new organization of the elements making up the original system, rather than relying on a simple numerical increase. Thus new stages can be reached without the whole being threatened.

There can be little doubt that present-day humanity, marked by overconsumption on the one hand and overpopulation and famine on the other, split up into unnatural classes and nations, is now entering

---

7. We will not consider the plant kingdom here. As we have seen, its development was essential for the evolution of animals, but in the present context it forms an evolutionary dead end.

a phase of limitation that neither growth nor conflict can resolve permanently. Our contemporaries are beginning to realize the extent of the crisis, which is both material and moral. This appears as a form of discouragement in adults and skepticism in the young, a generation that "could not care less." All wish to advance beyond the level of the political games that seem so mediocre and powerless. The social structures with which we are struggling are out of date. Whatever the good will and talent of those who are responsible for the present system, we are incapable of getting out of our rut. All our attempts have failed because they have not been able to respond to the constraints of life. We are at a dead end that could finally mean the end of civilization and even the disappearance of our species.

Superman or Superhumanity

Nothing condemns us irrevocably to early destruction — or rather, nothing but our ignorance. Nothing actually indicates that evolution, which has reached a high sociocultural level over several million years, is nearing its end. We may rather consider that humanity has exhausted its historical possibilities and must enter a new phase. This will not be achieved by a more robust and intelligent superman invading our niche and replacing our species — as the Darwinian model would have it — who would be replaced in turn by another superior individual. A superman would resolve none of the problems but would simply accelerate the final catastrophe, because he would be even more efficient than we in plundering the environment and indulging in murderous battles. The racists who desire the victory of "superior" beings are living in the wrong age. Man's future depends not on biological evolution but on sociocultural evolution. The future belongs not to a *superman* — the eternal, legendary hero — but to a *superhumanity* that we must now construct. This is impossible in the social-Darwinist context in which too many of our contemporaries still live. Evolution teaches us that to reach a new stage of integration there must be modifications in the relationships between preexisting units. At the level we have reached, this means a fundamental change in the relationships between societies and nations, which must abandon their rivalry for cooperation, their mistrust for esteem, their fear for confidence, their hate for love. This concept was proposed by Jean Hiernaux in 1964 and independently by Albert Vandel in 1965 and Jacques Ruffié in 1966. These ideas were hardly listened to, for they were developed in the decade from 1960 to 1970 when the Western world was enjoying the euphoria of an unprecedented growth that many people considered "normal" and permanent. Our civilization was living in opulence that seemed permanent. So why talk of change? And yet certain premonitions already indicated that we were reaching the end of an epoch.

The first half of this century was particularly cruel. It was marked by two world wars, the October Revolution, the Spanish Civil War, revolution in China, the establishment of totalitarian regimes among ancient civilizations, and the Great Depression in America. Everywhere one had the impression that democracy was tired. Since Yalta the world has been arbitrarily cut up into two powerful, rival blocs. It soon became clear that we would have to choose another pathway rapidly or risk seeing the end of the human adventure. The law of megagenesis shows that change is possible. It involves the effective integration of men and of peoples, renouncing eternal confrontation and choosing cooperation, thus placing our species on a new level. For the first time since life began such a trend would not be the fruit of natural selection but the result of a deliberate conscious action.

When macromolecules united to form cells, and then cells formed the first metazoans, and later animals organized societies, none of them made a deliberate "choice." They simply adopted the most favorable alternative in relation to selective forces. Their progression to a higher level of integration was the best solution for them, and probably the only one that enabled evolution to continue. They obeyed factors that they did not understand.

For us things are different. Nothing is forcing us to progress to a new stage. We are free to remain in the past or to enter the future. Our destiny is in our own hands.

What form might this superhumanity take over the next thousand years, created by integrating all the world's peoples in a richly diverse, but harmonious, ensemble? Our present level of consciousness cannot even guess, because at every integrative stage we discover new qualities, hitherto unsuspected, that permit us to overcome barriers that were insurmountable for previous generations. In this way, evolution can continue. There is no *objective* reason to think that this will not happen again. For centuries man has been conscious of his imperfection and has aspired to an advancement that has often been confused with betterment.

Throughout history philosophers have been dissatisfied, but they were rarely understood. Man has often stoned his prophets. Our century has accumulated horror and multiplied its martyrs, sometimes on the scale of whole populations. But it is marked by hope. Perhaps Hiroshima and Dachau, Biafra and Cambodia, announce the end of the world, but they may represent the last convulsions of the old world and the birth of a new. If they allow us to become conscious of our responsibility, no sacrifice will have been in vain. Suffering no longer seems so absurd.

ANGUISH AND JOY

We are too conscious to be happy. Our ancestors already lived in anguish resulting from what they did not know and did not understand. They feared the future, which they alone among living beings were able to imagine. Anguish was the motive force for cultural evolution. It encouraged research and innovation. This continuous quest permitted man to improve his techniques and thus affirm his existence. Little by little he learned to fashion stones and use arms, to make fire and master it. Later he domesticated plants and animals. We are well aware of the enormous progress in all domains of science over the last century. Our dissatisfaction helps us lead the world (see Vandel, 1948). Humanity will probably never be happy again. Progress is the fruit of worry; but no discovery has provided answers to our fundamental questions: the unknown, the future, death, and, to sum up everything, the meaning of life. The modern era raises other questions: the signficance of evolution and of the very existence of the universe.

Religions are born when knowledge reaches its limits. They conjure up superior forces in inaccessible realms. We transpose our feelings to the gods that we form in our image. Through our allegiance to the celestial powers or their representatives on earth, the priests and the kings, we looked for favor and protection. Often metaphysical anguish was turned to profit to ensure power and reap its benefits. God lived with Caesar. But the role of religions has not been negative only. They all have their sages and their saints, who see the blemishes of our condition and preach altruism, justice, and love in the language of their time.

This theme is a true, cultural "invariant" corresponding to a fundamental requirement of progress. No religion has avoided being swallowed up by temporal powers, and none has kept its promises. The philosophical movements that claimed to replace them have done no better. In many ways we are prisoners of our organic structures, torn between our material state and our aspirations. We are still attached to our animal nature by too many bonds. We allow ourselves to be guided by egoism, whereas the future lies in a universal task. We remain too often slaves of passion, and our reason gives way to our instincts. We wish to accumulate material wealth instead of devoting ourselves to exchange and creation. Like animals, we desperately seek to prolong our life, whereas immortality resides in what each of us contributes, however modestly, to the collective work that is civilization.

Man has not attained the plentitude of his "cultural humanization," which can never be the fruit of a biological modification but only be derived from new social integration. It implies abandoning a number of concepts that are still the basis of our life, as well as changing national and international structures. If we are prepared to pay this price we may hope to emerge from our present impasse and discover new horizons. The promised land that we have been awaiting for so long will cease to be a myth and become a reality. There is nothing utopian about such a vision. It is founded on the facts of megagenesis. But such an upheaval is not part of a genetic program; it is not imposed upon us by any selective force; it does not depend on chance or miracles but on our consciousness and our will. We are alone in the face of destiny.

Overcoming the uncertainty and the misery of our time, it is up to us to prepare this superhumanity of the next millennia. When he at last shakes off his animal instincts, man might finally enter a world built in his own image.

bibliography

Allison, A. C. (1954). Protection afforded by sickle-cell trait against subtertian malarial infections. *Brit. Med. J. 1*, 250–294.

Amouroux, H. (1977). *La grande histoire des Français sous l'occupation,* vol. 2. Paris: Robert Laffont.

Anxolabéhère, D. (1976a). Heterosis, overdominance and frequency-dependent selection in *Drosophila melanogaster* at the sepia locus. *Evolution 30,* 523–534.

Anxolabéhère, D. (1976b). Rôle des valeurs sélectives variables dans le maintien du polymorphisme chez *Drosophila melanogaster. C.R. Acad. Sci. Paris,* Series D, *283,* 1449.

Anxolabéhère, D., & Périquet, G. (1972). Variation de la valeur sélective de l'hétérozygote en fonction des fréquences alléliques. *C.R. Acad. Sci. Paris,* Series D, *275,* 2755.

Aron, J. P. (1969). *Essais d'épistémologie biologique.* Paris: Christian Bourgois.

Aron, R. (1964). *La lutte des classes.* Paris: Gallimard.

Ayala, F. J. (1971a). Competition between species: Frequency dependence. *Science 171,* 820–824.

Ayala, F. J. (1971b). Environmental fluctuations and population size. *Nature 231,* 112–113.

Ayala, F. J. (1972). Frequency dependent advantage in *Drosophila. Behav. Gen. 3,* 85–91.

Ayala, F. J. (1977). Variation génétique et évolution. *La Recherche 8,* 736–744.

Ayala, F. J. (1978). The mechanisms of evolution. *Sci. Amer. 239,* 48–61.

Ayala, F. J. (1982). *Biologie moléculaire et évolution.* Paris: Masson.

Baker, R. J., & Bicham, J. W. (1980). Karyotypic evolution in bats: Evidence of extensive and conservative chromosomal evolution in closely related taxa. *Syst. Zool. 29,* 239.

Baldwin, J. M. (1896). A new factor in evolution. *Amer. Nat. 30,* 441–446.

Barthelemy-Madaule, M. (1979). *Lamarck, ou le mythe du précurseur.* Paris: Seuil.

Bettelheim, C. (1974–83). *Les luttes des classes en URSS.* Paris Seuil (translation: *Class Struggles in the USSR.* New York: Monthly Review).

Blandin, P. (1977). Le problème de l'espèce chez les araignées. In *Les problèmes de l'espèce dans le règne animal. Mém. Soc. Zool. Franc 39,* 13–56.

Blondel, J., & Bourlière, F. (1979). La niche écologique: Mythe ou réalité? *Terre et Vie. Rév. Ecol. 33,* 345.

Bocquet, C. (1972). La spéciation des *Jaera albifrons:* Etat présent des connaissances et des problèmes. In *Fifth Europ. Marine Biol. Symp.,* Padua, 131–137.

Bodmer, W. F., & Bodmer, J. G. (1978). Evolution and function of the HLA system. *Brit. Med. Bull. 34,* 309.

Boesiger, B., Benoît, R., & Boesiger, E. (1975). Comparison des moyennes de la variabilité de poids de douze muscles bilatéraux de la caille japonaise *(Coturnix c. japonica)* au cours de quatre générations de croisements consanguins. *Acta Anat. 91,* 612–630.

Boesiger, E. (1969). Homéostase du développement et homéstase génétique. *Ann. Biol. 8,* 581.

Boesiger, E. (1974a). Le maintien du polymorphisme et de la polygénotypie par l'avantage sélectif des hétérozygotes. *Mém. Soc. Zool. France 37,* 363–416.

Boesiger, E. (1974b). Polymorphisme chez les drosophiles. *Mém. Soc. Zool. France 37,* 167.

Boesiger, E. (1975). Relation entre le degré d'hétérozygotie et l'homéostase chez les embryons de caille incubés à basse température. *Rev. Suisse Zool. 82,* 27.

Boesiger, E., & Ruffié, J. (1971). La préadaptation génétique. In *Préadaptation et adaptation génétique. INSERM Colloquium,* 19–34.

Boyer, S. (1972). Primate hemoglobins: Probes into the events of evolution. *XVII Cong. Int. Ass. Zool.,* 1–32.

Brandt, W. (1976). Solidarité mondiale, culture et politique. *Cultures 3,* 44.

Britten, R. J., & Davidson, E. H. (1969). Gene regulation for higher cells: A theory. *Science 165,* 349–357.

Britton, J., & Thaler, L. (1978). Evidence for the presence of two sympatric species of mice (genus *Mus* L.) in southern France based on biochemical genetics. *Bioch. Gen. 16,* 213–225.

Britton-Davidian, J., Ruiz Bustos, A., Thaler, L., & Topal, M. (1978). Lactate dehydrogenase polymorphism in *Mus musculus* L. and *Mus spretus* Lataste. *Experientia 34,* 1144–1145.

Buettner-Janusch, J. (1973). *Physical anthropology: A persepctive.* New York: Wiley.

Buican, D. (1981). Lucien Cuénot et la redécouverte de la génétique. *Pour la Science 45,* 21.

Cain, A. J., & Sheppard, P. M. (1950). Selection in the polymorphic land snail *Cepaea nemoralis. Heredity 4,* 275–294.

Camin, J. H., & Ehrlich, P. R. (1958). Natural selection in water snakes *(Natrix sipedon* L) on islands in Lake Erie. *Evolution 12,* 504–511.

Cartier, R. (1966). *La seconde guerre mondiale,* vol. 2. Paris: Larousse.

Cassagnau, P. (1977). Quelques problèmes relatifs aux critères spécifiques et à la notion d'espèce chez les collemboles. In *Les problèmes de l'espèce dans le règne animal. Mém. Soc. Zool. France 39,* 57–93.

Chaline, J., & Thaler, L. (1977). Le problème de l'espèce chez les rongeurs:

Approche biologique et approche paléntologique. In *Les problèmes de l'espèce dans le règne animal. Mém. Soc. Zool. France 39,* 359–381.

Chernyshev, A. (1966). Acclimation of tropical species of monkeys in the suburbs of the Moscow area. *Int. Zool. Yearbook 6,* 217–251.

Chesnais, J. C. (1980). Vers la stabilisation de la population mondiale? *J. Soc. Stat. 121,* 3.

Chetverikov, S. S. (1926). On certain aspects of the evolutionary process from the standpoint of modern genetics. *J. Exp. Biol. A2,* 3–54 (translation: *Proc. Amer. Phil. Soc. 105,* [1961]: 167–195).

Cheysson, C. (1981). Un "new deal" planétaire. *Le Monde,* April 30.

Chiarelli, B. (1967). Caryological and hybridological data for the taxonomy and phylogeny of the Old World primates. In *Taxonomy and phylogeny of the Old World primates with reference to the origin of man* (Chiarelli, B., ed.). Turin: Rosenberg & Sellier.

Chiarelli, B., & Capanna, E. (1973). *Cytotaxonomy and vertebrate evolution.* London and New York: Academic Press.

Christen, Y. (1976). A la recherche des bases moléculaires de l'hétérosis. *La Recherche 7,* 970–972.

Clarke, B. (1962). Balanced polymorphism and the diversity of sympatric species. In *Taxonomy and geography* (Nichols, D., ed.). Oxford: Systematics Assoc.

Club of Rome (1972). *The limits of growth.* London: Earth Island Limited.

Conry, Y. (1974). *L'introduction du darwinisme en France au XIXe siècle.* Paris: Vrin.

Conte, A. (1964). *Yalta, ou le partage du monde.* Paris: Robert Laffont.

Coon, C. S., Garn, S. M., & Birdsell, J. B. (1950). *Races.* Springfield, Ill.: Thomas.

Crozier, M. (1979). *On ne change pas la société par décret.* Paris: Grasset.

Cuénot, L. (1936). *L'espèce.* Paris: Doin.

Dadzie, K. K. S. (1980). Economic development. *Sci. Amer. 243,* 54–61.

Daget, J., & Bauchot, M. L. (1976). Les problèmes de l'espèce chez les téléostéens. In *Les problèmes de l'espèce dans le règne animal. Mém. Soc. Zool. France 38,* 67–127.

Darlington, C. D. (1939). *The evolution of genetic systems.* London: Cambridge Univ. Press.

Darwin, C. (1839). *The Voyage of the Beagle.* London.

Darwin, C. (1859). *On the origin of species by means of natural selection, or the preservation of favoured cases in the struggle for life.* London: John Murray.

Darwin, C. (1871). *The descent of man and selection in relation to sex.* London: John Murray.

Darwin, C. (1872). *The expression of the emotions in man and animals.* London: John Murray.

Dausset, J. (1981). The major histocompatibility complex in man: Past, present and future concepts. *Science 213,* 1469–1474.

De Grouchy, J. (1978). De la naissance des espèces aux aberrations de la vie. Paris: Robert Laffont.

De Grouchy, J. (1980). Les facteurs génétiques de l'évolution humaine. In *Les processus de l'hominisation.* Paris: Coll. Int. CNRS.

Delamare-Deboutteville, C., & Botossaneau, L. (1970). Formes primitives vivantes. Paris.

De Lesse, H. (1970). Le nombre de chromosomes à l'appui d'une systémique du groupe *Lysandra coridon* (Lycaenidae). *Alexanor 6,* 203–224.

De Rosnay, S., & de Rosnay, J. (1979). *Le mal bouffe.* Paris: Olivier Orban.

De Vries, H. (1901). *Die Mutationstheorie.* Leipzig: Veit.

Diamond, J. M. (1966). Zoological classification system of a primitive people. *Science 151,* 1102–1104.

Dobzhansky, T. (1950). Origin of heterosis through natural selection in populations of *Drosophila pseudoobscura. Genetics 35,* 288–302.

Dobzhansky, T. (1951). *Genetics and the origin of species.* 3d ed. New York: Columbia Univ. Press.

Dobzhansky, T. (1973). *Genetic diversity and human equality.* New York: Basic.

Dobzhansky, T. (1977). *Colloquium on ethological mechanisms of evolution* (Mechons, J., & Boesiger, E., coord.). Paris: Masson.

Dobzhansky, T., & Boesiger, E. (1968). *Essais sur l'évolution.* Paris: Masson.

Dobzhansky, T., & Epling, C. (1944). *Contributions to the genetics, taxonomy, and ecology of Drosophila pseudoobscura and its relatives.* Washington, D.C.: Carnegie Inst. Publ. No. 554.

Dombrowski, H. (1963). Organismes vivants du paléozoïque: Problèmes géologiques, biologiques et méthodologiques. *La Presse Méd. 71,* 1147.

Doolittle, R. F. (1980). The evolution of vertebrate fibrinogen. In *Protides of biological fluids* (Peeters, H., ed.). Elmsford, N.Y.: Pergamon, pp. 41–46.

Dorozynski, A. (1981). La co-évolution ou le gène du béret basque. *Sci. et Vie 771,* 28.

Dorst, J. (1965). *Avant que nature meure.* Neuchâtel: Delachaux & Niestlé.

Dorst, J. (1971). *Les oiseaux dans leur milieu.* Paris: Bordas.

Dorst, J. (1979). *La force du vivant.* Paris: Flammarion.

Dreux, P. (1977). L'espèce chez les orthoptères. In *Les problèmes de l'espèce dans le règne animal. Mém. Soc. Zool. France 39,* 95–136.

Drouin, P. (1981a). La statue du commandeur. *Le Monde,* April 2.

Drouin, P. (1981b). Des bouches inutiles. *Le Monde,* August 27.

Drouin, P. (1981c). Gare à la technolatrie. *Le Monde,* October 20.

Dubinin, N. P., & Tiniakov, G. G. (1946). Structural chromosome variability in urban and rural populations of *Drosophila funebris. Amer. Nat. 80,* 393–396.

Dubois, A. (1977). Le problème de l'espèce chez les amphibiens anoures. In *Le problème de l'espèce dans le règne animal. Mém. Soc. Zool. France 39,* 161–284.

Dubois, A. (1980). Populations, polymorphisme et adaptation: Quelques exemples chez les amphibiens anoures. In *Recherches d'écologie théorique: Les stratégies adaptives.* Paris: Maloine.

Dubos, R. (1974). *Choisir d'être humain.* Paris: Denoël.

Duby, G. (1981). *Le chevalier, la femme et le prêtre.* Paris: Hachette (translation: *The knight, the lady, and the priest.* New York: Pantheon, 1983).

Dufour, L. (1844). Anatomie générale des diptères. *Ann. Sci. Nat. 1,* 244–264.

Dutrillaux, B. (1979). Chromosomal evolution in primates: Tentative phylogeny from *Microcebus murinus (prosimian)* to man. *Hum. Gen. 48,* 314.

Dutrillaux, B., Couturier, J., & Chauvier, G. (1980). Chromosomal evolution of several species and sub-species of cercopithecinae. *Ann. Gen. 23,* 133.

Dutrillaux, B., Couturier, J., & Viegas-Pequignot, E. (1981). Chromosomal evolution in primates. *Chromosomes Today 7,* 1976.

Duverger, M. (1976). *Lettre ouverte aux socialistes.* Paris: Albin Michel.

East E. M., & Jones, D. F. (1919). *Inbreeding and outbreeding.* Philadelphia: Lippincott.

Ehrlich, P. R., White, R. R., Singer, M. C., McKechnie, S. W., & Gilbert, L. E. (1975). Checkerspot butterflies: A historical perspective. *Science 188,* 221–228.

Engels, F. (1845). *Die Lage der arbeitenden Klassen in England* (translation: *The condition of the working class in England.* Stanford: Stanford Univ. Press, 1958).

Engels, F. (1877). *Herrn Eugen Dührings Umwälzung der Wissenschaft.* Leipzig.

Engels, F. (1886). *Ludwig Feurbach und der Ausgang der klassischen deutschen Philosophie* (translation: *Ludwig Feuerbach and the outcome of classical German philosophy.* New York: Intl. Pub., 1941).

Erlanger, P. (1979). *Clemenceau.* Paris: Perrin.

Escherich, K. (1934). *Termitenwahn.* Munich.

Falconer, D. S. (1960). *Introduction to quantitative population genetics.* Edinburgh: Oliver and Boyd.

Faure, J. P. (1978). *Le cas Lamarck.* Paris: Blanchard.

Feldman, M., & Sears, E. R. (1981). The wild gene resources of wheat. *Sci. Amer. 244,* 98–109.

Fischer, R. A. (1930). *The genetical theory of natural selection.* Oxford: Clarendon Press.

Fontaine, A. (1966). *Histoire de la guerre froide,* vol. 1. Paris: Fayard.

Fontaine, A. (1978). *La France au bois dormant.* Paris: Fayard.

Fourastie, J., & Bazil, B. (1980). *Le jardin du voisin.* Paris: Coll. Pluriel, L.G.F.

Ford, E. B. (1945). Polymorphism. *Biol. Rev. 20.* 73–88.

Franc, A. (1977). L'espèce chez les gastéropodes. In *Les problèmes de l'espèce dans le règne animal. Mém Soc. Zool. France 39,* 137–159.

Frelinger, J. A. (1972). The maintenance of transferrin polymorphism in pigeons. *Proc. Nat. Acad. Sci. 69,* 326.

Gallien, C. L. (1970). Recherches sur la greffe nucléaire interspécifique dans le genre *Pleurodeles (Amphibiens-Urodèles). Ann. Embryol. Morphol. 3,* 145–192.

Galton, F. (1865). Hereditary talent and character. *McMillan's Magazine 12,* 319–320.

Galton, F. (1869). *Hereditary genius.* London.

Galton, F. (1883). *Inquiries into human faculty and its development.* London.

Galton, F. (1908). *Memories of my life.* London: Methuen.

Galton, F. (1909). *Essays in eugenics.* London: Eugenics Educ. Soc.

Gasser, F. (1977). La notion d'espèce chez les amphibiens urodèles. In *Les problèmes de l'espèce dans le règne animal. Mém. Soc. Zool. France 39,* 285–333.

Ghent, W. J. (1902). *Our benevolent feudalism.* New York: Macmillan.

Girardet, R. (1966). *Le nationalisme français (1871–1914).* Paris: Seuil.

Giscard d'Estaing, O. *Macastra.* Paris: Plon.

Goldschmidt, R. (1940). *The material basis of evolution.* New Haven: Yale Univ. Press.

Gomila, J. (1976). Définir la population. In *L'étude des isolats.* Paris: Assoc. Anthrop. Int. Langue Franc.

Gorer, P. A. (1959). Some new information on H-2 antigens in mice. In *Biological problems of grafting* (Albert, F., & Medawar, P. B., eds.). Oxford: Blackwell.

Grapin, J. (1981). Des "multinationales rouges." *Le Monde,* December 8.

Grassé, P. P. (1973). *L'évolution du vivant.* Paris: Albin Michel (translation: *Evolution of living organisms.* New York: Academic Press, 1977).

Grassé, P. P. (1978). *Biologie moléculaire, mutagénèse et évolution.* Paris: Masson.

Grassé, P. P. (1980). *L'homme en accusation.* Paris: Albin Michel.

Grjebine, A., et al. (1976). La notion d'espèce chez les moustiques: Étude de quatre complexes. In *Les problèmes de l'espèce dans le règne animal. Mém. Soc. Zool. France 38,* 249–306.

Guillaumin, M., & Descimon, H. (1976). La notion de l'espèce chez les lépidoptères. In *Les problèmes de l'espèce dans le règne animal. Mém. Soc. Zool. France 38,* 129–201.

Haeckel, E. (1868). *Natürlich Schöpfungsgeschichte.*

Haeckel, E. (1874). *Anthropogenie oder Entwicklungsgeschichte des Menschen.* Leipzig: Engelmann.

Haeckel, E. (1876). *The history of creation.* New York: Appleton.

Haldane, J. B. S. (1932). *The causes of evolution.* London: Harper.

Hamilton, T. H. (1961). The adaptive significance of intraspecific trends of variation in wing length and body size among bird species. *Evolution 15,* 180–195.

Harant, H., & Ruffié, J. (1956). Evolution et parasitisme. *Biol. Med. 45,* 382.

Harris, H. (1966). Enzyme polymorphisms in man. *Proc. Roy. Soc., Lond.* Series B, *164,* 298–316.

Harrison, G. A. (1959). Environmental determination of the phenotype. In *Function and Taxonomic Importance* (Cain, A. J., ed.). System. Assoc. Publ. No. 3, Hampton, Middlesex: Classey, pp. 81–86.

Henschen, A., Lottspeich, F., Töpfer-Petersen, E., Kohl, M., & Timpl, R., (1980). Fibrinogen evolution: Intra- and inter-specific comparisons. In *Protides of biological fluids* (Peeters, H., ed.). New York: Pergamon Press, pp. 47–50.

Hiernaux, J. (1964). *L'avenir biologique de l'homme.* Brussels: Cerc. Educ. Pop.

Hill, J. J. (1922). *My memories of eighty years.* New York.

Hillel, M. (1975). *Au nom de la race.* Paris: Fayard.

Hökfelt, T., Johansson, O., Ljungdahl, A., Lundberg, J. M., & Schultzberg, M. (1980). Peptidergic neurons. *Nature 284,* 515–521.

Hood, E. P. (1850). *The age and its architects.* London.

Hunt, W. G., & Selander, R. K. (1973). Biochemical genetics of hybridization in European house mice. *Heredity 31,* 11–33.

Hutchinson, G. E. (1958). Concluding remarks. *Cold Spring Harbor Symp. Quant. Biol. 22,* 415–427.

Illitch, I. (1973). *La convivialité.* Paris: Seuil.

Jackson, I. N. D. (1981). Evolutionary significance of the phylogenetic distinction of the mammalian hypothalamic releasing hormones. *Fed. Proc. 40,* 2545.

Jacob, F. (1977). Evolution et bricolage. *Le Monde,* September 6–8.

Jacquard, A. (1978). New demographic regime and its consequences. In *Mutations, biology and society.* Paris: Masson.

Jacquard, A. (1980). L'homme au péril de la science? La sociobiologie. *Le temps de la réflection 5,* 877.

Jacquard, A. (1982). *Au péril de la science? Interrogations d'un généticien.* Paris: Seuil.

Johnson, F. M., Kanapi, C. G., Richardson, R. H., Wheeler, M. R., & Stone, W. S. (1966). An analysis of polymorphism among isozyme loci in dark and light *Drosophila ananassae* strains from America and Western Samoa. *Proc. Nat. Acad. Sci. 56,* 119–125.

Johnson, G. B. (1976). *Molecular evolution.*

Johnson, G. B. (1977). Characterization of electrophoretically cryptic variation in the alpine butterfly *Colias meadii. Bioch. Gen. 15,* 665–693.

Jones, J. S. (1980). How much genetic variation? *Nature 288,* 10–11.

Kawamura, T., & Sawada, S. (1959). On the sexual isolation among different species and local races of Japanese newts. *J. Sci. Hiroshima Univ. 18,* 17–30.

Kimura, M. (1968). Evolutionary rate at the molecular level. *Nature 217,* 624–626.

Kimura, M. (1968). Genetic variability maintained in a finite population due to mutational production of neutral and nearly neutral isoalleles. *Genet. Res. 11,* 247.

Kimura, M. (1973). Gene pool of higher organisms as a product of evolution. *Cold Spring Harbor Symp. Quant. Biol. 38,* 515.

Kimura, M. (1979). The neutral theory of molecular evolution. *Sci. Amer. 241,* 94–104.

Kimura, M. (1980). A simple method of estimating evolutionary rates of base substitutions through comparative studies of nucleotide sequences. *J. Mol. Evol. 16,* 111.

Kimura, M. (1981). Estimation of evolutionary distances between homologous nucleotide sequences. *Proc. Nat. Acad. Sci. 78.*

Kimura, M. (1983). *The neutral theory of molecular evolution.* London & New York: Cambridge Univ. Press.

Kimura, M., ed. (1982). *Molecular evolution, protein polymorphism and the neutral theory.* Tokyo: Japan Scientific Soc. Press.

Kimura, M., & Ohta, T. (1973). Mutation and evolution at the molecular level. *Genetics 73* (Suppl.), 19–35.

King-Hele, D. (1977). *Doctor of revolution: The life and genius of Erasmus Darwin.* London: Faber and Faber.

Kojima, K., & Tobari, Y. (1969). Selective modes associated with karyotypes in *Drosophila ananassae* II: Heterosis and frequency-dependent selection. *Genetics 63,* 639–651.

Labeyrie, V. (1978). Crise écologique, crise de société et démocratie. *La Pensée,* 198.

Lacouture, J. (1977). *Léon Blum.* Paris: Seuil.

Lamarck, J. B. de (1809). *Philosophie zoologique.* Paris.

Lambert, G. (1968). *L'adaptation physiologique et psychologique de l'homme aux conditions de vie désertiques.* Paris: Hermann.

Lambert, G. (1979). L'informatisation de la société: Technologie libératrice ou technocratie réductrice? *Bull. Inf. Prog. Vis. Tran. IRACT,* Sept., 85.

Lamotte, M. (1959). Polymorphism of natural populations of *Cepaea nemoralis. Cold Spring Harbor Symp. Quant. Biol. 24,* 65–86.

Lamotte, M. (1966). Les facteurs de la diversité du polymorphisme dans les populations naturelles de *Cepaea nemoralis* (L). *Lav. della Soc. Malacol. Ital. 2,* 33–73.

Langaney, A. (1977). La résurrection de l'anthropologie: La quadrature des races. *Sci. et Vie 4.*

Larmat, J. (1979). *La génétique de l'intélligence.* Paris: P.U.F.

Lauré, M. (1981). Plan Marshall ou grande crise? *Le Monde,* March 4.

Le Bras, H., & Todd, E. (1981). *L'invention de la France.* Paris: Hachette.

Lefebvre, H. (1966). *Pour connaître la pensée de Marx.* Paris: Bordas.

Lejeune, J. (1968). Adam et Eve ou le monogénisme. *Nouv. Rev. Théol. 90,* 191.

Lejeune, J. (1975). Sur le mécanisme de la spéciation. *C.R. Soc. Biol. 169,* 828.

Lejeune, J., Gauthier, M., & Turpin, R. (1959a). Étude des chromosomes somatiques de neuf enfants mongoliens. *C.R. Acad. Sci. 248,* 1721.

Lejeune, J., Turpin, R., & Gauthier, M. (1959b). Le mongolisme: Premier exemple d'aberration autosomique humaine. *Ann. Gén. 1,* 41.

Leontieff, W. (1977). *1999.* Paris: Dunod. (See also *Sci. Amer.,* Sept. 1980, p. 166.)

Leridon, H. (1982). Vers une baisse de la fécondité dans le monde. *La Recherche 13,* 521–523.

Lerner, I. M. (1954). *Genetic homeostasis.* Edinburgh: Oliver and Boyd.

LeRoth, D., Shiloach, J., Roth, J., & Lesniak, M. (1980). Evolutionary origins of vertebrate hormones: Substances similar to mammalian insulins are native to unicellular eukaryotes. *Proc. Nat. Acad. Sci. 77,* 6184–6188.

Leroy, Y. (1978). Les signaux de communication. *Sci. et Vie 24.*

Leroy, Y. (1979). *L'univers sonore animal: Rôles et évolution de la communication acoustique.* Paris: Gauthiers-Villars.

Lesourne, J. (1981). Les chemins de la nécessité. *Pour la Science 37,* 8.

Le Taccoen (1979). *La guerre de l'énergie est commencée.* Paris: Flammarion.

Lewontin, R. C. (1976). The fallacy of biological determinism. *The Sciences,* March–April.

Lewontin, R. C., & Hubby, J. L. (1966). A molecular approach to the study of genetic heterozygosity in natural populations of *Drosophila pseudoobscura. Genetics 54,* 595–609.

Lucotte, G. (1977). Le polymorphisme biochimique et les facteurs de son maintien. Paris: Masson.

Lucotte, G. (1978). Variabilité génétique et évolution. *Pour la Science 13*, 83–95.

Lucotte, G., & Dubouch, P. (1980). Étude électrophorétique de l'hybride expérimental entre *Papio anubis* et *P. cynocephalus. Bioch. Syst. Ecol. 8*, 323–327.

Lucotte, G., & Guillon, R. (1979). Génétique des populations, spéciation et taxonomie chez les babouins. *Bioch. Syst. Ecol. 7*, 231–244, 245–251.

Lucotte, G., & Jouventin, P. (1980). Distance électrophorétique entre le mandrill et le drill. *Ann. Genet. 23*, 46–48.

Lucotte, G., & Kaminski, M. (1978). Increased stability of a hybrid molecule. *J. Hered. 69*, 354.

Lucotte, G., Kaminski, M., & Perramon, A. (1978). Heterosis in the codominance model: Electrophoretic studies on proteins in the chicken–quail hybrid. *Comp. Bioch. Physiol. 60B*, 169–171.

Lucotte, G., & Lefebvre, J. (1980). Distances électrophorétiques entre les cinq espèces de babouins du genre *Papio* basées sur les mobilités des protéines et enzymes sériques. *Biol. Syst. Ecol. 8*, 317–322.

Lucotte, G., Perramon, A., & Kaminski, M. (1977). Molecular basis for heterosis in the chicken–quail hybrid. *Comp. Bioch. Physiol. 56B*, 119–123.

Lyell, C. (1830). *Principles of geology.* London.

Lyell, C., (1863). *The antiquity of man.* London.

Marx, K. (1962). *Manuscrit de 1844.* Paris: Edit. Soc.

Marx, K., & Engels, F. (1848). *Manifest der Kommunistische.* London.

Marx, K., & Engels, F. (1973). *Lettres sur les sciences de la nature.* Paris: Edit. Soc.

May, R. (1978). The evolution of ecological systems. *Sci. Amer. 239*, 118–133.

Mayr, E. (1942). *Systematics and the origin of species.* New York: Columbia Univ. Press.

Mayr, E. (1963). *Animal species and evolution.* Cambridge: Harvard Univ. Press.

Mayr, E. (1969). *Principles of systematic zoology.* New York: McGraw-Hill.

Mayr, E. (1976). *Evolution and the diversity of life.* Cambridge: Harvard Univ. Press.

Mayr, E. (1981). *La biologie de l'évolution.* Paris: Hermann.

McMillan (1979). *Direct foreign investment in the Comecon countries.* Ottawa: Carleton Univ.

Miklos, G. L. G., & Nankivell, R. N. (1976). Telomeric satellite DNA functions in regulating recombination. *Chromosoma 56*, 143–167.

Mitterrand, F. (1978). *L'abeille et l'architecte.* Paris: Flammarion.

Monod, J. (1970). *Le hasard et la nécessité.* Paris: Seuil (translation: *Chance and necessity.* New York: Random, 1972.)

Monod-Broca, T. (1980). Paul Broca (1824–1880): Le chirurgien, l'homme. *Bull. Acad. Nat. Med. 164*, 536.

Moscovici, S. (1968). *Essai sur l'histoire humaine de la nature.* Paris: Flammarion.

Muller, H. J. (1950). Our load of mutations. *Amer. J. Hum. Genet. 2*, 111–176.

Naccache, B. (1980). *Marx, critique de Darwin*. Paris: Vrin.

Nevo, E. (1978). Genetic variation in natural populations: Patterns and theory. *Theor. Pop. Biol. 13*, 121–177.

Nevo, E., & Cleve, H. (1978). Genetic differentiation during speciation. *Nature 275*, 125–126.

Nora, S., & Mink, A. (1978). *L'informatisation de la société française*. Paris: Seuil.

Ohno, S. (1980). L'évolution des gènes. *La Recherche 11*, 4–14.

Pasteur, G. (1974). Génétique biochimique et populations; ou: Pourquoi sommes-nous multipolymorphes? *Mém. Soc. Zool. France 34*, 473.

Pasteur, G. (1977a). Endocyclic selection in reptiles. Letter. *Amer. Nat. 111*, 1030.

Pasteur, G. (1977b). Quelques commentaires à propos de l'espèce chez les lygodactyles. In *Les problèmes de l'espèce dans le règne animal. Mém. Soc. Zool. France 39*, 335–358.

Pasteur, G., & Pasteur, N. (1980). Les critères biochimiques et l'espèce animal. In *Les problèmes de l'espèce dans le règne animal. Mém. Soc. Zool. France 40*, 99.

Pasteur, G., Pasteur, N., & Orsini, J. P. G. (1978). On genetic variability in a population of the widespread gecko *Hemidactylus brooki. Experientia 34*, 1557.

Pearson, K. (1905). *National life from the standpoint of sciences*. Cambridge.

Pelt, M. (1977). *L'homme re-naturé*. Paris: Seuil.

Petit, C. (1951). Le rôle de l'isolement sexuel dans l'évolution des populations de *Drosophila melanogaster. Bull. Biol. Fr. Belg. 85*, 392–418.

Petit, C. (1958). Le déterminisme génétique et psychophysiologique de la compétition sexuelle chez *Drosophila melanogaster. Bull. Biol. Fr. Belg. 92*, 248–329.

Petit, C. (1966). La concurrence larvaire et le maintien du polymorphisme. *C.R. Acad. Sci. Paris 263*, 1262–1265.

Petit, C. (1971). Frequency-dependency and the maintenance of polymorphism in insect pool. *Acta Phyto-path. 6*, 137.

Petit, C. (1973). L'avantage du type rare, facteur du maintien du polymorphisme. *Mém. Soc. Zool. France 37*, 417–441.

Petit, C., & Nouaud, D. (1975). Ecological competition and the advantage of the rare type in *Drosophila melanogaster. Evolution 29*, 763–776.

Petit, C., & Zuckerkandl, E. (1976). *Evolution*. Paris: Hermann.

Poliakov, L. (1968). *Histoire de l'antisémitisme*, vol. 3. Paris: Calmann-Lévy.

Powell, J. R. (1971). Genetic polymorphisms in a varied environment. *Science 174*, 1035–1036.

Prenant, M. (1938). *Darwin*. Paris: Edit. Soc. Int.

Raffestin, C. (1976). Le concept d'écologie humaine. *Carrefours d'écologie humaine du Lac d'Annecy*, March.

Rensch, B., (1960). *Evolution above the species level*. New York: Columbia Univ. Press.

Roberts, D. F. (1973). Climate and human variability. *Anthropol.* 34.

Robertson, W. R. B. (1916). Chromosome studies. *J. Morphol. 27*, 179–331.

Romer, A. S. (1970). *L'évolution animale*. Paris: Bordas. (Translation)

Rouzé, M. (1980). Us et abus de la biologie. A.F.I.S., Paris 93.

Ruffié, J. (1966). *Hémotypologie et évolution du groupe humain.* Paris: Hermann.

Ruffié, J. (1970). Studies of population genetics and adaptation in the Andes and other countries. *Soc. Study Hum. Biol.* (London), May.

Ruffié, J. (1973). Les frontières chromosomiques de l'hominisation. *C.R. Acad. Sci.,* Series D, *276,* 1709.

Ruffié, J. (1976). *De la biologie à la culture.* 5th ed. Paris: Flammarion.

Ruffié, J. (1978). Spéciation populationnelle ou spéciation chromosomique. *Ann. Gén. 21,* 69.

Ruffié, J. (1979). Sociobiologie et génétique. *Le Monde,* Sept. 11, 12.

Ruffié, J. (1981). *L'histoire de la louve.* Paris: Flammarion.

Sahlins, M. D. (1976). *The use and abuse of biology.* Ann Arbor: Univ. of Michigan Press.

Salmon, P. (1980). *Le racisme devant l'histoire.* 2nd ed. Paris: Fernand Nathan.

Sanguinetti, A. (1980). *Histoire du soldat, de la violence et des pouvoirs.* Paris: Ramsay.

Sawada, S. (1963). Studies on the local races of the Japanese newt *Triturus pyrrhogaster. J. Sci. Hiroshima Univ. 21,* 167–180.

Schindewolf, O. H. (1936). *Paläontologie, Entwicklungslehre und Genetik.* Berlin: Borntraeger.

Scudder, S. H., & Burges, E. (1870). On asymmetry in the appendages of hexapod insects, especially as illustrated in the lepidopterous genus *nisionades. Proc. Boston Soc. Nat. His. 13,* 282.

Senghor, L. S. (1971). Problématique de la négritude. *Présence Africaine 78,* 3.

Servan-Schreiber, J. J. (1980). *Le défi mondial.* Paris: Fayard.

Shirer, W. L. (1969). *Collapse of the third republic.* New York: Simon & Schuster.

Simpson, G. G. (1944). *Tempo and mode in evolution.* New York: Columbia Univ. Press.

Simpson, G. G. (1952). Probabilities of dispersal in geologic time. *Bull. Amer. Mus. Nat. Hist. 99,* 163–176.

Simpson, G. G. (1961). *Principles of animal taxonomy.* New York: Columbia Univ. Press.

Soula, C. (1953). *Précis de physiologie.* Paris: Masson.

Spencer, W. P. (1947). Mutations in wild populations of *Drosophila. Advances in Genetics 1,* 359–402.

Stebbins, G. L. (1950). *Variation and evolution in plants.* New York: Columbia Univ. Press.

Steegman & Platner (1968). Experimental cold modification of cranio-facial morphology. *Am. J. Phys. Anthrop. 28,* 17–30.

Sumner, F. B. (1909). Some effects of external conditions upon the white mouse. *J. Exp. Zool. 7,* 97–155.

Tanaka, G. (1924). Maternal inheritance in *Bombyx mori. Genetics 9,* 479–486.

Thuillier, M. (1977). Un anarchiste positiviste: Georges Vacher de Lapouge.

In *L'idée de race dans la pensée politique française contemporaine.* Paris: Ed. CNRS.

Thuillier, P. (1981). Bible et science: Darwin en procès. *La Recherche 12,* 710–719.

Timofeeff-Ressovsky, N. W. (1940). Zur Analyse des Polymorphismus bei *Adalia bipunctata. Biol. Zentralbl. 60,* 130–137.

Tinbergen, J. (1976). *Reshaping the international order.* Amsterdam: Sevier.

Tinbergen, N. (1954). The origin and evolution of courtship and threat display. In *Evolution as a process.* London: Allen and Unwin.

Tsacas, L., & Bocquet, C. (1976). L'espèce chez les Drosophilidae. In *Les problèmes de l'espèce dans le règne animal. Mém. Soc. Zool. France 38,* 203–247.

Vacher de Lapouge, G. (1896). *Les sélections sociales.* Paris.

Vacher de Lapouge, G. (1899). *L'aryen: son rôle social.* Paris.

Vacher de Lapouge, G. (1909). *Race et milieu social: Essai d'anthropologie.* Paris.

Vaïman, M. (1980). Le complexe majeur d'histo-compatibilité chez les animaux. In *Le complexe majeur d'histo-compatibilité de l'homme (HLA):* Cours supérieur d'histo-compatibilité. (Dausset, J., et al., eds.)

Vandel, A. (1958). *L'homme et l'évolution.* Paris: Masson.

Vandel, A. (1963). Evolution et autorégulation. *Ann. Biol. 2,* 179.

Vandel, A. (1965). Une prospective de l'évolution. *Ann. Biol. 4,* 367.

Vandel, A. (1968). *La genèse du vivant.* Paris: Masson.

Vissac, B. (1972). La seconde révolution de l'élevage. *Sci. et Avenir 309,* 902.

Von Bertalanffy, L. (1961). *Les problèmes de la vie.* Paris: Gallimard.

Vuilleumier, F. (1976). La notion d'espèce en ornithologie. In *Les problèmes de l'espèce dans le règne animal. Mém. Soc. Zool. France 38,* 29–65.

Waddington, C. H. (1957). *The strategy of the genes.* London: Allen and Unwin.

Wallace, B. (1968). *Topics in population genetics.* New York: Norton.

Wilson, E. O. (1975). *Sociobiology: The new synthesis.* Cambridge: Harvard Univ. Press.

Wilson, E. O., & Lumsden, C. (1981). *Genes, mind and culture: The co-evolutionary process.* Cambridge: Harvard Univ. Press.

index

and power of recognition, 5–6
recycling and, 272–73
Ford, E. B., 97
forests, disappearance of, 265
founder effect, 55
foxes, 293–94
Franc, A., 163
France:
  anti-Semitism and racism in, 229–32, 235–36
  class struggle in, 244–45
  living standard in, 332
  proletarianization in, 223–24
Franco, Francisco, 306
Frelinger, J. A., 119
French Revolution, 244
frogs, 70–71, 100, 183
fruit flies, xiv, 18–19, 22, 28, 30, 39, 77–78, 88, 109–10, 113, 115–16, 129, 143, 144, 146, 162, 168–69, 176, 183, 191
functional complementarity, 127

Gagarin, Yuri, 298
Gallien, C. L., 170
Galton, Francis, xix, 224n, 225–28, 233, 244, 247, 256
gametes, 85, 89
Gase, Jean-Pierre, 247n
Gasser, F., 71
Gaxotte, Pierre, 230
geckos, 36, 163–64
genes:
  conditional, 79
  Darwinist view of, xvii, 13
  dominance and recessivity in, 77–78
  exchange of, between populations, 31, 52, 56–58, 62, 64–66
  exchange of, between species, 62, 67–73, 74, 75
  fixed, 77
  good mixer, 138–39
  heterostasis and, 78–80
  lethal, 28
  penetrance and expressivity of, 81, 82
  recombination of, and natural selection, 85–87
  super-, 96–101, 102
  unfavorable, maintenance of, 76–82
  variations in selective value of, 126–31
  wild, 13, 14, 64

Genes, Mind and Culture (Wilson and Lumsden), 256–57
genetic clines, 47–48, 176–77
genetic drift, 55, 58
genetic homeostasis, 117–21, 143–45, 157
genetic load, 87, 125–26
genetic monomorphism, 13–14, 35, 38
  evolution precluded by, 52
  intraspecies competition fostered by, xvi, 123
genetic polymorphism, xv, 62–82
  adaptation and, 35–36
  advantages of, at individual level, 115–17
  advantages of, at population level, 121–23
  comparison of, in related species, 37–38
  as component of all living organisms, xvi, 23, 29, 30, 36–37, 42–43, 132
  crossing-over and, 96, 101–3
  cultural polymorphism and, 212–13, 303
  distribution of, over long distances, 46–48
  distribution of, over small distances, 50
  distribution of, within species, 38–39
  evolution and, 131–33
  interpopulation gene flow as source of, 62, 64–66
  intrapopulation competition minimized by, xvi, 123
  introgression as source of, 62, 67–73, 75–76
  in man, 36, 37, 48, 50, 66
  measuring size and extent of, 30–34
  mechanisms limiting degree of, 123–25
  mutations as source of, 62, 63–64, 87–88
  as paradox to Darwinism, 105–11
  phenotypic monomorphism and, 144
  and position of species on phylogenetic scale, 33–34
  qualitative, 24, 27–30
  quantitative measurement of, 24–27
  reinforced by cultural polymorphism, 212–13
  selective advantage of, 114–23
  selective pressure and, 108–11
  sexual cycle and, 88–89
  supergenes and, 96–101
  variations in, and population location, 57–58

ABOUT THE AUTHOR

A doctor, biologist, and professor at France's prestigious Collège de France and New York University, Jacques Ruffié is also director of studies at l'École Pratique des Hautes Études and the author of three major scientific works, including *From Biology to Culture*. He lives in France.